Shakespeare's Representation of Weather, Climate and Environment

[. . .] never till tonight, never till now,
Did I go through a tempest dropping fire.

(*Julius Caesar*, 1.3.9–10)

Shakespeare's Representation of Weather, Climate and Environment

The Early Modern 'Fated Sky'

Sophie Chiari

EDINBURGH
University Press

Edinburgh University Press is one of the leading university presses in the UK. We publish academic books and journals in our selected subject areas across the humanities and social sciences, combining cutting-edge scholarship with high editorial and production values to produce academic works of lasting importance. For more information visit our website: edinburghuniversitypress.com

First published in hardback by Edinburgh University Press 2019

Edinburgh University Press Ltd
The Tun – Holyrood Road
12(2f) Jackson's Entry
Edinburgh EH8 8PJ

Typeset in 11/13 Adobe Sabon by
IDSUK (DataConnection) Ltd, and
printed and bound by CPI Group (UK) Ltd, Croydon, CR0 4YY

A CIP record for this book is available from the British Library

ISBN 978 1 4744 4252 7 (hardback)
ISBN 978 1 4744 4253 4 (paperback)
ISBN 978 1 4744 4254 1 (webready PDF)
ISBN 978 1 4744 4255 8 (epub)

Published with the support of Institut d'Histoire des Représentations et des Idées dans les Modernités (IHRIM, UMR 5317), under the aegis of the Centre National de la Recherche Scientifique (CNRS), ENS de Lyon, Université Lumière-Lyon 2, Université Jean-Moulin-Lyon 3, Université Jean-Monnet-Saint-Etienne, and Université Clermont Auvergne.

Contents

Illustrations vi
Acknowledgements viii
Textual Note x

Introduction 1

1. 'We see / The seasons alter': Climate Change in *A Midsummer Night's Dream* 31

2. '[T]he fire is grown too hot!': *Romeo and Juliet* and the Dog Days 57

3. 'Winter and rough weather': Arden's Sterile Climate 80

4. *Othello*: Shakespeare's *À bout de souffle* 111

5. 'The pelting of [a] pitiless storm': Thunder and Lightning in *King Lear* 150

6. Clime and Slime in *Anthony and Cleopatra* 176

7. The I/Eye of the Storm: Prospero's Tempest 217

Conclusion: 'Under heaven's eye' 256

Bibliography 267
Index 298

Illustrations

1. William Cuningham, *The cosmographical glasse*, London, 1559, The Second Booke, fol. 64. By courtesy of the Huntington Library, Art Collections, and Botanical Gardens. 3
2. Thomas Blundeville, *M. Blundevile his exercises containing sixe treatises*, 1594, chap. 18, fol. 194v. By courtesy of the Huntington Library, Art Collections, and Botanical Gardens. 9
3. Luna in *The Kalender of sheephards*, 1585, sig. LIIIIv. By courtesy of the Bodleian Library. 34
4. Caius Julius Hyginus, 'Canis Major', in *Fabularum liber*, 1535, p. 100. By courtesy of the Bibliothèque Nationale de France. 72
5. Noah peeping out from his ark. Illustration taken from John Harding's copy of the Bible translated into German by Martin Luther. *Biblia: das ist, die gantze Heilige Schrifft*, Wittemberg, Gedruckt durch Hans Kraffts Erben, 1584. By courtesy of the President and Fellows of Magdalen College, Oxford, Magd.b.7.6. 94
6. 'Soufflacul', misericord carved by André Sulpice, collegiate church Notre-Dame, Villefranche-de-Rouergue, France, 1473–87. By courtesy of Ciel Bleu. 100

7–8. Sun and moon eclipses in John Blagrave, *The mathematical ievvel*, 1585. By courtesy of the British Library Board, C.60.o.7, pp. 56–7. 116

9. Pieter Bruegel the Elder, *The Peasant Dance*, 1568. By courtesy of the Kunsthistorisches Museum, Vienna. 131
10. Engraved frontispiece of *1607. A true report of certaine wonderfull ouerflowings of waters*, 1607. By courtesy of the British Library Board, 1103.e.58, front cover. 163

11a–11b. Andrea Mantegna, *St Sebastian*, 1457–9; detail of the rider in the cloud. By courtesy of the Kunsthistorisches Museum, Vienna. 195

12. The 1540 map of Gemma Frisius, 'Charta cosmographica, cum ventorum propria natura et operatione', in Petrus Apianus's *Cosmographia*, Paris, 1551. By courtesy of Rare Books, Special Collections, J. Willard Marriott Library, The University of Utah, GT3200 1540 A65. 221
13. A ship caught in a tempest and struck by St Elmo's fire in Thomas Hill's *Contemplation of mysteries*, 1574. By courtesy of the British Library Board, C.56.b.15, fol. 22r. 225
14. A witch causing a storm. Woodcut in Olaus Magnus's *Historia de gentibus septentrionalibus*, Rome: Johannes Maria de Viottis, 1555, p. 117. By courtesy of Cornell University Library, Division of Rare and Manuscript Collections. 238

Acknowledgements

The writing of this book would not have been possible without the research conditions that I have enjoyed over the past three years. My research team, IHRIM ('Institut d'Histoire des Représentations et des Idées dans les Modernités', UMR 5317, Centre National de la Recherche Scientifique), has been a source of intellectual stimulation and has allowed me to travel both in France and abroad for my research. Olivier Bara and Marina Mestre Zaragozá have generously given me the freedom I needed for my work and they have supported me all along. My colleagues at Université Clermont Auvergne (UCA, France) have also provided me with a friendly and welcoming environment which has helped to make things much easier for me. In addition, I have often tested my ideas in my Master and Agrégation classes and I am grateful to my students for their patient attention and their responsiveness.

While I am clearly indebted to ecocriticism at large and to the numerous studies that aim at reconciling early modern literature with early modern science, my collaboration with learned and friendly colleagues such as Mickaël Popelard and John Mucciolo has been particularly fruitful and important to me. Colleagues and friends read portions of the book along the way. Edward Paleit and Jean-Marie Maguin kindly agreed to look at a first draft of this text while Sophie Lemercier-Goddard and Tiffany Jo Werth each commented on separate chapters, and I am deeply indebted to the four of them for their thoughtful remarks. I also warmly thank Jonathan Pollock for his suggestions and clarifications.

I would also like to express my personal gratitude to Roger Chartier and Stephen Greenblatt for their encouragement and for the time they were kind enough to give me in spite of their frantic schedules. My thanks also go to Dympna Callaghan, Richard Dutton, Frank Lestringant and Carla Mazzio for their unfailing support and their immediate adherence to the overall project.

On top of that, I have been fortunate, over the past five years, to organise and participate in lectures, conferences and seminars directly or indirectly related to the subject of this study. Stimulating exchanges with several participants have helped me to go ahead with my project in many more ways than I can say. The various papers I had the opportunity to give in conferences or seminars on weather and climate in early modern England were the starting points of some of the chapters of this book. In particular, the ninth Congress of the Shakespeare Society of Southern Africa in Grahamstown, which offered me the opportunity to present a 'climatic' reading of *A Midsummer Night's Dream*, *Romeo and Juliet* and *King Lear*, proved a decisive moment that gave me the necessary impetus and confidence to turn a budding project into a full-length monograph.

I am also grateful to the curators and librarians for the help and assistance they have provided me with at Bibliothèque Nationale de France, the British Library, the Huntington Library, the Bodleian Library and Cornell University Library. My thanks more particularly go to Anne Chesher, Deputy Librarian at Magdalen College, who helped me locate an engraving in John Harding's copy of the Bible, and to Luise Poulton, Managing Curator at J. Willard Marriott Library, who drew my attention to the 1551 Paris edition of Peter Apian's *Cosmographia*.

The two anonymous readers of the first draft of this book are finally to be warmly thanked since their comments and advice have certainly helped me improve my work and flesh it out with additional details. Last but not least, I warmly thank François Laroque for his kind support and understanding while I was writing this book.

Textual Note

Original spelling has been retained whenever I quote from early modern editions but I have systematically replaced long s's by short modern ones and I have expanded contractions such as tildes in order to make these texts more accessible to the readers.

Unless otherwise stated, the following editions of Shakespeare's plays have been used for the seven chapters of this book:

A Midsummer Night's Dream, ed. Peter Holland, Oxford: Oxford University Press, 1995.

Romeo and Juliet, ed. Jill L. Levenson, Oxford: Oxford University Press, 2000.

As You Like It, ed. Michael Hattaway, Cambridge: Cambridge University Press, The New Cambridge Shakespeare, (2000) 2009.

Othello, ed. Michael Neill, Oxford: Oxford University Press, 2006.

King Lear, ed. René Weis, A Parallel Text Edition, 2nd edn, London: Routledge, (1993) 2010.

Anthony and Cleopatra, ed. Michael Neill, Oxford: Oxford University Press, 1994.

The Tempest, ed. Stephen Orgel, Oxford: Oxford University Press, 1994.

References to Shakespeare's other works are to *The Complete Works*, ed. Stanley Wells, Gary Taylor, John Jowett and William Montgomery, 2nd edn, Oxford: Clarendon Press, 2005.

Introduction

> And as it was in the dayes of Noe [. . .]
> [. . .] the flood came, and destroyed them all.
> Like wise also, as it was in the dayes of Lot [. . .]
> [. . .] it rained fyre and brimstone from heauen, and destroyed them.[1]

Climate generates fears and superstitions generally assuaged by religious rituals and representations. In England, the Reformation tried to eradicate the beliefs of the old religion but, as far as climate and weather are concerned, no really satisfactory results were achieved. The sky remained ominous and often threatening while astrological attempts to read one's destiny in the stars (well before William Herschel's observations[2]) or in the clouds (when Luke Howard's classification was not even conceivable[3]) long continued to flourish. At any rate, in the Geneva Bible, Shakespeare's contemporaries could find plenty of stories telling of the erratic weather caused by God's wrath and were used to reading the changes in the English firmament in the light of the Scriptures. Most of them still thought that, because of men's ingrained tendency to sin, the rain poured down from the heavens, threatening to engulf the peasants' yearly harvests, if not humanity at large, and flashes of lightning cracked in angry skies, punishing human hubris.

All the same, important as this biblical background may have been, early modern scholars and writers began to question the extent and even the reality of the sky's influence[4] upon their daily lives. While apocalyptic discourses[5] and sermons attributing natural disasters to divine fury weighed heavily upon them, men of letters increasingly challenged such assumptions as they reassessed men and women's abilities to act upon celestial forces rather than being acted upon by them.

Plays of the period illustrate this new trend of thought and Shakespeare's *All's Well That Ends Well* (c. 1604) is certainly one of them. After engaging in a series of witty exchanges with Paroles, Helen, the daughter of the famous physician Gérard de Narbonne, concludes the opening scene by stating her determination to try and cure the old French king's fistula and by making clear that this is her own free decision rather than luck or fortune coming from above:

> Our remedies oft in ourselves do lie
> Which we ascribe to heaven. The fated sky
> Gives us free scope, only doth backward pull
> Our slow designs when we ourselves are dull. (1.1.212–15)[6]

If these lines primarily betray the young girl's single-mindedness (she is in love with her social superior, Bertram, and hopes that, if she succeeds in curing the king, she may then win the young man as her husband), they also epitomise some important aspects of the early seventeenth-century vision of the sky and of the relationship between human beings and their celestial environment. As Helen suggests, stellar influences (the sky *is* fated) undoubtedly guide our lives but they nonetheless give us the possibility to make choices and to eschew or to bend predetermined patterns. By combining a deterministic logic with the possibility of free will, Shakespeare's heroine seems to stand halfway between the medieval and the early modern vision of the skies, and she points to the somewhat paradoxical Renaissance perception of the heavens.

Geographical and cultural context

The fact is that, torn as they were between trying to control their own destinies and letting God shape their actions, the Elizabethan and Jacobean subjects continued to look for answers in the skies while they were also anxious to fashion their own lives in more coherent or rational ways. England's geographical position of course played a crucial role in the vision that Shakespeare and his contemporaries had of the heavens, at a time when the earth was still divided into 'frozen', 'temperate' and 'burning' (or torrid) zones.[7]

English writers logically situated their own territory in the temperate area, neither too hot nor too cold, but the climate they lived in kept nonetheless being considered as a rather chilly one by their European neighbours, especially the Mediterranean peoples.[8] This

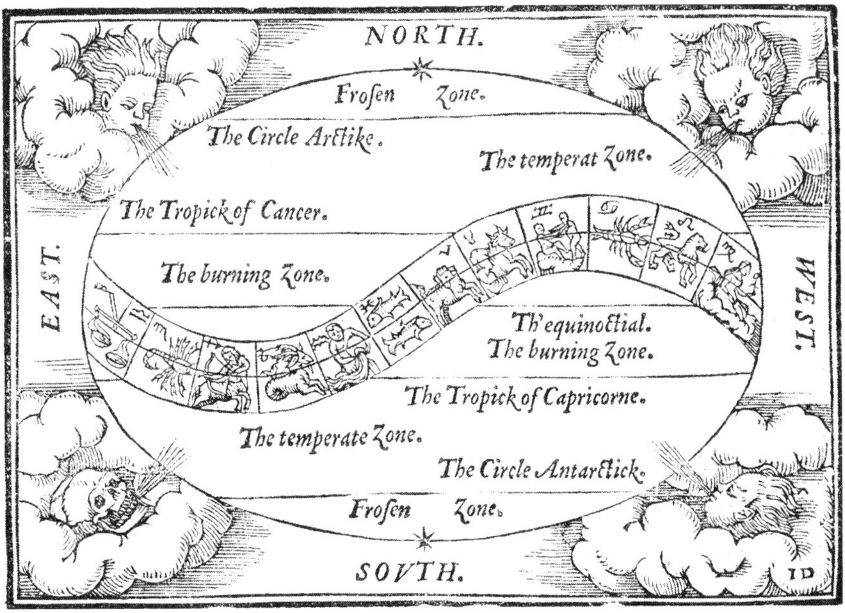

Fig. 1 William Cuningham, *The cosmographical glasse*, London, 1559, The Second Booke, fol. 64. By courtesy of the Huntington Library, Art Collections, and Botanical Gardens.

scheme entailed the logic of what Mary Floyd-Wilson has termed 'geo-humoralism' – that is, the interrelation of geographical and weather issues and their impact on the human conduct and temperament – and it implied that 'humoral temperance [. . .] was held to be attainable only in a temperate clime'.[9] As a northern country, England was not impervious to climatic excesses: Shakespeare's 'sceptred isle' (*Richard II*, 2.1.40) was indeed often exposed to the wind, the rain and the freezing air and, in *Henry V*, the Constable of France famously describes the English climate as 'foggy, raw and dull' (3.5.16). Such characteristics were believed to have a lasting impact on the 'mettle' (3.5.15), mores and habits of the English nation, which, from the French point of view, was as fierce as its weather.

Independently of England's geographical situation, the reason why Shakespeare's contemporaries were so obsessed with the weather was also due to their having to struggle against the adverse weather conditions characterised by what is now referred to as the 'Little Ice Age', a phrase coined in 1939 by the American glaciologist François Matthes.[10] It designates the long 1300 to 1900 period which was not uniformly cold but was marked by several atmospheric shifts. The second half of the sixteenth century and the first half of the seventeenth century proved especially harsh periods for the Arctic explorers seeking new commercial routes for the kingdom. The 1607 journal left by the navigator Henry Hudson and his crew, for instance, testifies to the frigid temperatures that affected the northern hemisphere by repeatedly underlining the presence of 'ice laying very thick all along the shore'.[11] More generally, according to climate historians, 'the years from 1560 to 1600 were cooler and stormier, with later wine harvests and considerably stronger winds than those of the twentieth century'.[12] Mike Hulme acknowledges in particular that, in Shakespeare's lifetime, the climate was much colder than ours and that especially harsh winters had to be endured over the years 1564–5 and 1601–3, while very poor summers plagued crops and inhabitants alike from 1594 to 1597.[13] This must have been all the more difficult to understand and accept for the playwright's contemporaries as, during the first half of the sixteenth century, weather conditions had been relatively temperate.[14] As a result, Philip Armstrong suggests that '[t]he sharpness of this contrast between the climate of Elizabeth's reign and that of her father's means that Shakespeare would have grown to maturity surrounded by a generational sense that a previously fecund, temperate, and reliable natural environment had been replaced by freezing temperatures, blighted harvests, and sudden, wild storms'.[15] Those natural disasters went hand in hand with a general lack of sunlight, a deficiency cogently pointed out in Thomas Nashe's *Summer's Last Will and Testament*, a comedy probably written in 1592 to be performed at Archbishop Whitgift's Palace in Croydon. The play presents a dying Summer who, having to make his will, designates Winter as Autumn's overseer. His efforts are somewhat complicated by the behaviour of one of Winter's sons, the acerbic Backwinter, who gives a cataclysmic and outrageously exaggerated illustration of the adverse weather conditions that plagued the 'Little Ice Age':

> Would I could bark the sun out of the sky,
> Turn moon and stars to frozen meteors,
> And make the ocean a dry land of ice;
> With tempest of my breath turn up high trees,
> On mountains heap up second mounts of snow
> Which melted into water might fall down
> As fell the deluge on the former world.[16]

While early modern spectators naturally perceived the tongue-in-cheek nature of such depictions, the notion of the 'Little Ice Age' would not have meant much, if anything, to them since they did not and could not then acknowledge the existence of distinct climatic cycles. To most scholars of the period, the skies remained a spiritual entity and were, as such, an inexhaustible treasure trove for metaphors and natural imagery. According to James I's court physician, Helkiah Crooke, 'the Soule was infused into us from Heaven'.[17] While being reputedly at the origin of humankind, the heavens were also clouded in mystery and altogether inaccessible to men. Writers, astrologers and natural philosophers thus described and studied meteorological phenomena in order to try and master the sky, one of men and women's eternal dreams. The Neoplatonists knew that the observation of the heavenly vault could not yield complete understanding but were nonetheless particularly anxious to fathom these heavenly mysteries. As he discusses the patterns of the heavenly spheres, and invoking the marvel of Archimedes' famous 'heaven of brass', the Florentine Marsilio Ficino declares for instance in his *Platonic Theology* (1482):

> Since man has observed the order of the heavens, when they move, whither they proceed and with what measures, and what they produce, who could deny that man possesses as it were almost the sole genius as the Author of the heavens? And who could deny that man could also make the heavens, could he only obtain the instruments and the heavenly material, since even now he makes them, though of a different material, but still with a very similar order?[18]

For Ficino, controlling the sky amounts to ruling the universe and to taking God's place. Admittedly, this logic owes much to intuitive faith. But in the sixteenth century, as both Mary Thomas Crane and Carla Mazzio observe, new forms of '*reasoning* within astronomy, physics, meteorology and mathematics were in many ways moving further and further away from an understanding of nature accessible to the senses, the body, the human or intuitive experience of the world'.[19] What Crane more specifically suggests is that, relying as much on classical knowledge as on new discoveries, cosmographers (who studied the heavens and the earth) and, by and large, early modern men of science began to give precedence to a counter-intuitive approach over the traditional intuitive one in their efforts to read the book of nature.[20] This new stance turning science into an abstract and abstruse discipline barely accessible to the common man was, however, countervailed by the literature and the drama of the period which, up to a certain point, offered various means of reconciling science with the senses.

While Crane's statement certainly needs to be qualified – physical observation was gaining rather than losing ground by the turn of the seventeenth century – it places renewed emphasis on the early modern approach to science (one not entirely stripped of religious worship, but taking other standards of truth into account) and it exemplifies its complex relationships with the world of drama and entertainment. So, by the end of the sixteenth century, while Thomas Digges and Thomas Harriot, along with other scholars, embarked on the study of the astronomical and weather patterns in a more systematic way and over a longer stretch of time than their predecessors, Shakespeare and his fellow playwrights transformed the public playhouses and their overstage sky-like canopies into contact zones between the higher spheres of knowledge and the familiar environment of the Elizabethan subjects.

'Environment', 'weather' and 'climate': Some definitions

The colour of the heavens often set the tone of a performance. In *1 Henry VI*, probably staged at the Rose theatre, the opening line of the play, 'Hung be the heavens with black' (1.1.1), instantly refers to a black-hung stage and reminds us that, in the early modern era, the public playhouse's superstructure, then known as the 'heavens' and traditionally decorated with astrological signs,[21] could be adorned with black for tragedies. As actor and playwright, Shakespeare was aware that the sky was as much a theatrical as a natural element, and as provincial man turned Londoner, he must have realised how much the vagaries of the weather fashioned his native environment, with, for instance, the massive Clopton bridge being built in Stratford-upon-Avon under the reign of Henry VII in order to replace the former 'inadequate wooden bridge, perilous at times of flood'.[22] The playwright must have also been conversant with popular weather lore as is shown by his depiction of the intrusion of celestial influences in the daily lives of kings and peasants alike and by his keenness to represent the vibrant natural world of his early years. His characters, as a result, are regularly seen as 'poor player[s] / That stru[t] and fre[t] [their] hour upon the stage' (*Macbeth*, 5.5.23–4), that is, as frail creatures confronted by a harsh environment under the continual pressure of the elements.

This being said, tackling the question of the representation of climate, weather and the early modern environment first calls for a definition of 'environment' and for a clear-cut distinction between 'weather' and 'climate'. The notion of 'environment' already existed in the early seventeenth century but referred to '[t]he action of circumnavigating, encompassing, or surrounding something' (*OED*, 1).[23] The meaning we now ascribe to the word (i.e. '[t]he physical surroundings or conditions in which a person or other organism lives, develops, etc.', *OED*, 2.b) is fairly recent as it dates back to the mid-nineteenth century.[24] Implied in this definition is the fact that nature interacts with human beings – nature, in other words, shapes us as much as we shape nature – and that it can be thought of as a frail and problematic entity. In the following development, I assume that the awareness of this interaction actually emerged in the early modern period, when nature was no longer exclusively perceived as God's creation and when Europeans engaged with it in fresh ways so as to study, mathematise and transform it.[25] In order to examine the links between Shakespeare's contemporaries and their natural surroundings, the elemental forces of earth, air and especially wind, fire and water[26] will be given pride of place as these principles, already present in ancient Greek thought, continued to inform the early modern understanding of landscape and cosmology.

The weather has always fashioned men and women's representations of their environment, and this must have been particularly blatant in the preindustrial era, when daily expectations focused on the colour of the sky, which, in people's minds, was likely to determine agricultural yields[27] and the outcome of battles, especially at sea. 'Weather', as I take it, designates 'instantaneous meteorological conditions'[28] generating immediate emotional and cultural responses. John Florio, in *A Worlde of Wordes* (1598), significantly defined the Italian *tempo* as 'time, season, leisure, opportunitie or occasion, the state of time, commodity or necessitie *of the time present*. Also weather foule or faire' (my emphasis).[29] By contrast, the word 'climate' entails a broader timescale. Beyond its operational scientific definition provided by the World Meteorological Organisation (for which it is, in a narrow sense, 'an average weather condition in an area over a long period of time'), it also works as a cultural concept, 'an ordered container' through which the 'arbitrariness' of the weather could be 'interpreted', according to Mike Hulme.[30] 'Climate', in this perspective, is a concept producing a sense of stability into what would otherwise remain chaotic. It is a rationalising idea which 'organize[s] the [. . .] visceral experiences of daily weather into seasonal rhythms and organize[s] the geography of the world into recognizable zones of difference and regularity'.[31] Shakespeare resorted to it so as to give his plays an overall coherence, primarily based on a pragmatic, even materialistic, approach to the vagaries of the English weather.

Etymologically, the word 'weather' derives from a root word (*vĕtrŭ* in Old Church Slavonic) meaning 'air' or 'wind', while 'climate' comes from the Greek *klineio* or 'slope' (i.e. latitude).[32] Lexicographers and early modern men of letters first used the word 'clime' or 'climate' in this specific sense. Florio regarded *clima* or 'climate' as 'the dividing of heaven and earth, a clime',[33] a definition echoed in Randle Cotgrave's *Dictionarie of the French and English Tongues* where climate is defined as 'a division in the skie, or portion of the world, beweene South and North'.[34] The author and translator Thomas Blundeville, whose numerous fields of interest notably included astronomy and mathematics, provides a more detailed explanation in a substantial treatise published in 1594, and reissued in 1597 and in 1606. 'A clime', he writes, 'is a space of the earth comprehended betwixt two parallels, in which space the longest day doth vary by halfe an houre.'[35]

Spread by the cosmographical treatises of the period, this vision is logically present in a number of early modern plays. Among them, Christopher Marlowe's *Doctor Faustus* (1604) reveals how climatic views were sometimes intertwined with colonial aims.[36] In act 4, scene 2 of the play (i.e. scene 6 in the 1616 B-text), when he manages to bring out-of-season grapes to satisfy the longings of the pregnant Duchess, Faustus provides a simple geographic explanation for what could then pass as a sheer miracle:

> the year is divided into two circles over the whole world, that when it is here winter with us, the contrary circle it is summer with them, as in India, Saba, and farther countries in the East; and by means of a swift spirit that I have, I had them brought hither, as ye see. (4.2.22–7)[37]

What strikes us today is that he incorrectly mentions the east/west hemispheric division instead of the north/south one as he focuses on hemispheric seasonal variation. Yet, for all its erroneousness, Faustus's explanation illustrates the early modern understanding of climate while making it clear that cosmographical knowledge was becoming part and parcel of a barely disguised imperialist rhetoric.

From the mid-sixteenth century onward, another meaning was aggregated to the etymology of the word 'climate', as the term gradually came to designate 'the prevalent pattern of weather in a region throughout the year' (*OED*, 2.a). In actual fact, as Mary C. Fuller

The fecond Booke

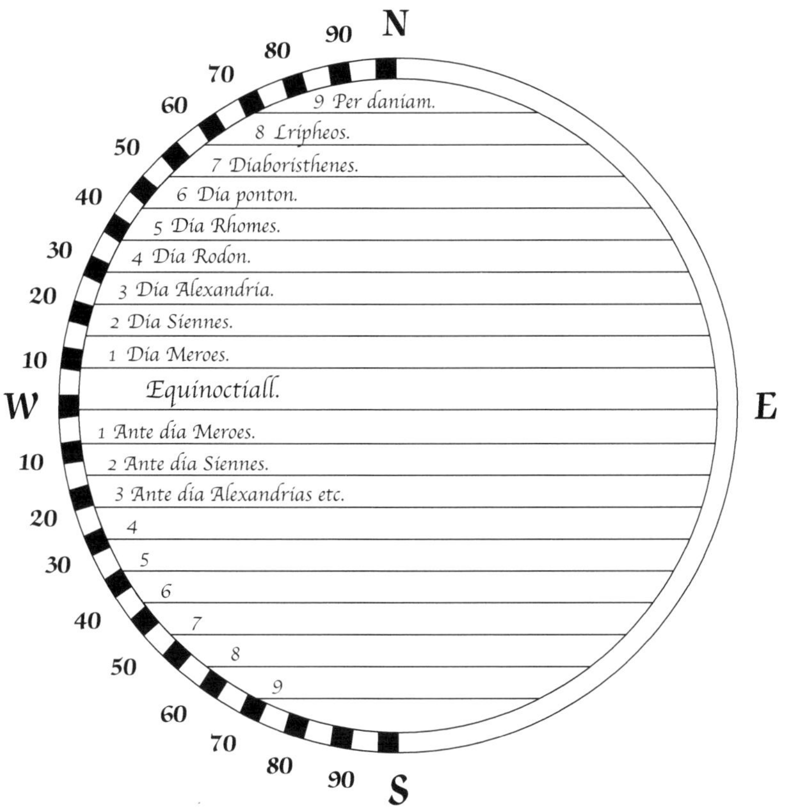

	The beginuing.			The mid ft.			The code.	
Degrees and Minutes of houres.	D	M		D	M		D	M
The firft Clime.	12	45		16	35		20	30
The fecond Clime.	20	30		24	15		27	30
The third Clime.	27	30		30	45		33	40
The fourth Clime.	33	40		36	24		39	0
The fift Clime.	39	0		41	20		43	30
The fixt Clime.	43	30		45	24		47	15
The feuenth Clime.	47	15		48	40		50	30
The eight Clime.	50	30		51	50		53	10
The ninth Clime.	53	10		55	30		56	30

Fig. 2 Thomas Blundeville, *M. Blundevile his exercises containing sixe treatises*, 1594, chap. 18, fol. 194v. By courtesy of the Huntington Library, Art Collections, and Botanical Gardens.

notes, early modern 'perceptions of climate were subjective, value-laden and unstable'[38] and, as a result, the meanings of 'climate' and 'weather' overlapped by the end of the century, especially in popular literature. While not totally interchangeable, they could both be aligned with such notions as geographical location, the cycle of the seasons, the position of the stars, or divine intervention.

Short-term and long-term perspectives

If the beliefs inherited from the Middle Ages were being progressively challenged, they still partly shaped the early modern vision of climate and the apprehension of the weather. By and large, the medieval period had been dominated by a theological vision in which weather and climate reflected God's direct action upon nature. In the first half of the sixteenth century, Luther still thought that the various manifestations of the weather testified to God's presence. In the daily sermons delivered to the parishioners, bad weather continued to be directly correlated with men's sinful behaviour and with the imminent advent of doomsday. Even the seasons, causing 'changes infinite',[39] were regarded as God's punishment for man's sin, and Edmund Spenser's *Faerie Queene* (1590) testifies to the persistence of this view at the turn of the century.[40]

So, spells of dry, wet, cold or stormy weather were still thought of as small-scale, practical theophanies by a number of Elizabethan subjects. In the Middle Ages, God controlled the crops, causing drenching rains, famines and political crises when He was angered, or clement temperatures, good crops and political stability when He was satisfied. Anxious as they were to contain peasants' rebellions, the Church authorities had tried hard to channel people's anxieties by taking into their hands the control of various means to soothe the anger of the skies. Hence the cultivation in monasteries of herbs like mugwort (*Artemisia vulgaris*) as a protection against storms and thunder, the processions to attract God's mercy, and the diverse rites and ceremonies meant to establish a dialogue of sorts between the local community and the threatening natural phenomena that were causing repeated losses as well as putting them in danger. Keith Thomas explains that '[i]n 1289 the Bishop of Chichester ruled that it was the duty of every priest to order processions and prayers when he saw a storm was imminent, without waiting for orders from above'.[41] Consecrated church bells were also used to dispel thunder and lightning, as in 'the great thundering' of Sandwich in 1502 and 1514.[42]

In rural England, after the Reformation, the mysteries of the sky were inseparable from the small local events of daily life as well as from the cycle of the seasons. As Wolfgang Behringer explains, '[f]or people living at the time, short-term changes [in climate] had greater importance than medium to long-term ones'; it is no mere chance that most documented accounts of early modern Europe contain 'observations about the weather [rather] than about the climate'.[43] Most of Shakespeare's contemporaries were essentially concerned by the immediate present because their lives depended upon it, leading them to assuage their fears by seeking to predict the weather and play down its possibly negative effects, thus following a long tradition dating back to antiquity – an attitude which Luther had firmly condemned (like all the traditional beliefs in calendar-divination, prophecies and supernatural protections against bad weather) but which persisted well into the seventeenth century.[44] It is in this context that, in 1591, John Florio translated the *Perpetuall and naturall prognostications of the change of weather*, a small pamphlet which contained various tokens of foul, fair and calm weather and which was successful enough to be reprinted in 1598. The possibility of knowing what would happen the day after was important for practical purposes as it could mean considerable improvements in one's everyday existence. Consequently, prophecies and auguries soon became all the rage and cheap publications like almanacs, which attracted readers from both high and low orders, began to flood the book market.

Basically, two types of predictions prevailed: those in which the weather was an end per se, and those, even more numerous, in which, alongside other celestial events, the weather was just a means to touch on other provinces of foreknowledge. So, while husbandmen, yeomen, artisans and travellers were eager to peruse compilations of weather prognostications[45] sometimes inspired by biblical writings, sometimes based on animal activities such as the flights or the nesting of birds[46] or the swarming of bees,[47] nobles and merchants were equally intrigued by writings which presented the weather as a vector of knowledge enabling them to make sense of their physical environment in connection either with their routine tasks or with the political situation of the country. In such cases, natural events tended to overlap with or to foreshadow supernatural ones, as in *3 Henry VI* when, after three uncanny suns have appeared in the sky, Richard tells Edward that '[i]n this the heaven figures some event' (2.1.32). From the mid-sixteenth century onwards, a number of writers, translators, cosmographers and almanac-makers, capitalising on this fascination for omens and predictions, attempted to simplify and popularise the classical theories linked to meteorology, thereby making them available to an educated, if not specialised, readership.

Robert Laneham's 1575 letter to Humphrey Martin is often quoted in connection with middle-class literary taste. In this letter describing the literary and scientific tastes of a so-called Captain Cox, a Coventry mason who owned a large collection of popular books and ballads, Laneham remarks that, besides the medieval legends of 'King Arthur, Bevis and Sir Guye', astronomical treatises and almanacs occupy a prominent place in his friend's library.[48] As part of the quadrivium, astronomy dealt with 'the movements of the heavenly bodies',[49] a topic that attracted indeed an important readership – one that kept increasing after the advent of the perspective lenses. Astrology, 'the study of the effects of these movements', was no less in vogue, however.[50] First and foremost put forward by educated men but far from being a coterie system of thought,[51] astrological doctrines essentially meant to order confused human affairs into some sort of rational pattern. Nativities, a branch of judicial astrology consisting in 'maps of the sky at the moment of a person's birth, either made on the spot at the request of the infant's parents, or reconstructed for individuals of mature years who could supply the details of their time of birth',[52] were then especially popular. Given this general enthusiasm, prognostication became a flourishing business.

In this regard, the full title of Leonard Digges's 1555 treatise is worth quoting here: *A prognostication of right good effect fructfully augmented, contayninge playne, briefe, pleasant, chosen rules, to iudge the wether for euer, by the sunne, moone, sterres, cometes, raynbowe, thunder, cloudes, with other extraordinarie tokens, not omitting the aspectes of planets*. A good example of successful scientific vulgarisation, the book incorporated Aristotelian views about meteorology, common-sense remarks on the weather, as well as specific indications on how to decipher and interpret signs from the 'fated sky'. Thomas Hill (c. 1528–c. 1574), then mostly 'known as a translator of popular books on science and the supernatural',[53] followed suit and, in the early 1570s, his compilation entitled *A contemplation of mysteries* gathered meteorological observations as well as various superstitions rephrased in a series of simple and striking formulae.[54]

The many early modern weather plays, including John Rastell's proto-scientific *Interlude of the Four Elements* (c. 1511), John Heywood's *Play of the Weather* (1533) and the above-mentioned *Summer's Last Will and Testament* by Thomas Nashe, testify to the attention which many sixteenth-century writers paid to England's atmospheric conditions and to the popular treatises on the subject.

Elizabethan playwrights were directly affected by meteorological phenomena, as poor weather could be responsible for low attendance in the public playhouses. In his 1612 preface to *The White Devil*, John Webster famously discarded the possibility of audience distaste and instead attributed the failure of his play, acted at the Red Bull, to the miserable winter weather ('it was acted in so dull a time of winter, presented in so open and black a theatre, that it wanted [. . .] a full and understanding auditory').[55] Even though Webster may have wanted to cheer himself up, it is nonetheless true that not everyone would attend a play in such conditions and, as a consequence, one should face the idea that the London playhouses were far from being systematically packed then.[56]

Now, contrary to farmers or playgoers, who were more anxious about the immediate consequences of the weather than concerned by the long-lasting impact of climate, those whom Leonard Digges called 'meteoroscoper[s]'[57] tried hard to make sense of the erratic weather patterns in order to place them into broader climatic cycles. But being still deprived of the measuring instruments which could have enabled them to collect reliable data and adopt a truly scientific method,[58] they focused less on the sky proper than on its presumed effects on the human body and mind. They ascribed differences in race, ethnicity and national character to climate influence so that, as a consequence, the idea that the weather interfered with men and women's habits, moods and humours and that it affected their behaviour – including their sexuality – was then quite familiar. In sum, bodily humours were thought to depend on climate and environment, whose external influences were somehow transmitted through the blood as well as through the skin. In Spenser's *Shepheardes Calender* (1579) for example, '[e]very Aeglogue refers to the monthly seasonal changes and the effects they have on the shepherds, with the use of pathetic fallacy ensuring that meteorological and emotional concerns become entwined'.[59] The idea that climate had serious effects on one's temperament and personality was then current among the poets and playwrights who borrowed from classical writers as well as from contemporary foreign texts translated into English. If, according to Mary Floyd-Wilson, 'climatological theory' was regaining ground in early modern England,[60] it must be recalled that not everyone agreed on how climate really worked on human temper as several systems coexisted at the time. Two acquired special prominence.[61] In *Airs, Waters, Places*,[62] Hippocrates stipulated a form of continuity between human physiology and its natural environment, so that the phlegmatic humours of the English were supposedly due to the rainy climate of northern Europe. But, according to Aristotle's

Problems (*Problemata*)[63] – possibly a pseudo-Aristotelian treatise – and to his *Politics* (*Politika*),[64] the body's complexion was determined in reaction to the surrounding climate, internal heat then resulting from external cold, a theory of contraries often used to explain the English settlers' adaptation to the Virginia environment. In 'Problem XIV', one reads for instance that '[t]hose who live in a hot climate are chilled, but those who live in a cold climate have a hot nature'. Yet Aristotle also remarks that opposite situations sometimes lead to similar results: 'Both classes are big,' he writes, 'those in cold climates because of the natural heat existing in them, those in hot climates owing to the heat of the place; for growth is due to hot climates and heat.'[65] This classical climatology was then adopted and adapted by early modern writers so as to fit their readers' tastes and expectations.

Giovanni Botero's *The worlde, or an historicall description of the most famous kingdomes and common-weales*, which first appeared in four volumes published between 1591 and 1596, was apparently a best-seller in England as it went through six editions between 1601 and 1630. Botero essentially follows Hippocrates' views, and in his compendium, he describes how climate affects the living conditions of human beings and determines their behaviour. By contrast, Jean Bodin's erudite *Method for the Easy Comprehension of History* (*Methodus at facilem historiarum cognitionem*, 1566) is more openly Aristotelian and goes much further than Botero's observations. Where the Italian writer simply reflects on the influence of climatic conditions on history, Bodin foregrounds the role played by climate in politics and aims 'at establishing a correlation of culture with climate and topography',[66] a correlation further developed in his *Six livres de la République* (1576), translated into English by Richard Knolles in 1606. The programmatic title of the chapter opening the fifth book of the treatise is particularly representative of the role played by climate in the field of politics: 'What order and course is to be taken, to apply the forme of a Commonweale to the diuersities of mens humors, and the meanes how to discover the nature and disposition of the people.'[67] Bodin insists on the necessity for cunning politicians to adapt their strategies to the dispositions of the peoples placed under their rule. According to him, for example, 'eloquent discourses agree not with the grosse wits of the Northerne people', whose intellectual ineptitude is offset by their sheer physical strength, while the same discourses are often 'too base for them of the South', much more clever than their northern counterparts, but physically weaker.[68] Such observations on climate were popular in England especially because of their practical implications that opened up new perspectives in the working out of political strategy.[69] The Catholic essayist Pierre Charron, another of the French disciples of Montaigne, who clearly drew upon Bodin's theory of climate and classifications, was to become famous in England. In a treatise published in 1601 and translated into English seven years later under the title *Of wisdome*, he explains that

> [t]he first most notable and vniuersall distinction of men, which concerneth the soule and body, and whole essence of man, is taken and drawne from the diuers site [*sic*] of the world, according to which the aspect and influence of heauen, and the sunne, the aire, the climate, the countrie, are diuers. So likewise not only the colour, the feature, the complexion, the countenance, the manners, are diuers, but also the faculties of the soule.[70]

With their vapours, humours and fluids, men and women's bodies were therefore comparable to small independent weather systems.[71] Human passions were liquids saturating the body and in need of control, a little like torrential rains threatening to flood the land.

All in all, by the turn of the seventeenth century, the observation of the sky had established 'a point of contact' between 'experts' (natural philosophers, astrologers) and 'laypersons' (seamen, shepherds, artisans).[72] As to theologians, whose views circulated in both the popular and the learned spheres of English society, they helped bridge the gap between these two spheres. Playwrights who, like Shakespeare, used weather and climate imagery in their works in order to represent and question men's behaviour in connection not just with the natural world but also with religion, power and politics efficiently amalgamated ideas spread by popular and learned *milieux* alike. As a result, early modern plays somehow imbibed the ambient climatic concerns and conveyed them through tropes that, more often than not, suggested how tenuous were in those days the boundaries between religious belief and science, on the one hand, as well as between theology and literature, on the other.

Greek influences

Open to a variety of influences, the stage first and foremost absorbed classical precepts which, once digested and rehashed, often became Renaissance commonplaces. If much has been written about the influence of Latin writers such as Virgil, Ovid, Pliny and the like,[73] that of Greek writers on early modern climatic views has been rather neglected so far, not least because of T. W. Baldwin's long-lasting influence on Shakespeare studies.[74] We have just seen that, in sixteenth-century England, the attention paid to geo-humoralism testified to a certain familiarity with the Hippocratic and Aristotelian traditions. If they followed different paths of interpretation, both stressed the importance of humours and of the atmosphere with its four 'elements' (*Twelfth Night*, 2.3.9), a word which, according to Feste, had by then become 'over-worn' (ibid. 3.1.58).[75] On the one hand, the Greek physician Galen developed Hippocrates' findings while remaining loyal to his theory of humoralism,[76] and his work on human passions and bodily processes not only shaped English medicine at large but also influenced the whole rhetoric of healing and disease in Shakespeare's time. On the other hand, Aristotle's teachings also durably posited the influence of weather and habitat on human beings. Written in 350 BC, his *Meteorology* (*Meteorologica*) enhanced the irregularity of meteorological phenomena as well as their 'inaccessibility to the senses',[77] promoted a non-teleological view of the science of meteorology, and defined the sun as the moving cause acting upon things and beings in the sublunary world. Such ideas were developed in parallel to the Hippocratic tradition and were spread by a number of followers. For example, a treatise traditionally attributed to Aristotle's student Theophrastus of Eresus, composed around 320 BC and entitled *Concerning Weather Signs* (*De signis tempestates*),[78] contains a catalogue of signs said to help people forecast the weather. 'Some of the rules' developed in this book 'were based on the position of celestial bodies, but most were based on observations of the skies and of natural beings'.[79] In 1592, the antiquarian and Protestant thinker Isaac Casaubon 'brought out the first full-dress early-modern edition of Theophrastus',[80] and he issued a second part seven years later, both being published in Lyons. One should not either underestimate the serious impact of the Ptolemaic tradition as, in the second century AD, Claudius Ptolemy also established methods for meteorological forecasts in his *Four Books* (*Tetrabiblos*), a treatise in which he explains that the heavenly bodies affect events on earth. In the sixteenth century, the work still had its adepts, as Shakespeare ironically shows in *King Lear*, and it circulated in various manuscripts before a Latin translation was issued in Nuremberg in 1535.

But this far from exhaustive list of Greek thinkers preoccupied with weather issues and still fairly influential in early modern Europe would be quite incomplete without the mentioning of Epicure, one of the first to reject the traditional explanations about the weather and the promoter of ataraxy, that is, a state of mind characterised by the absence of troubles. His views were transmitted through the Latin poet and philosopher Lucretius who, after being rediscovered in January 1417 in a German monastery by the Florentine humanist Poggio Bracciolini, was widely discussed along the whole Renaissance period.[81] Shakespeare probably came across his views on nature, climate and the cosmos in Montaigne's *Essays* but also, in all likelihood, in a number of earlier texts that consistently engaged with materialism. The pervasive influence of the Latin poet is more than palpable in *As You Like It* and *King Lear*.[82] Yet, for a number of reasons, many of them religious, there was absolutely no consensus on Epicureanism in Shakespeare's time, and not a few among his literary rivals raised their eyebrows over a philosophy that was wrongly imagined to promote the pursuit of unbridled sensuality. It was also dismissed for its refusal to take seriously the messages conveyed by the stars. In his *Planetomachia* ('War of the Planets'), Robert Greene summarises the general suspicion concerning Epicure:

> such is the perverse nature of man, that hee wholy forgetteth the cause of his Creation [. . .] And with the Epicures placeth his content in his Vineyards, and granaries, gaping after pelfe, and thinking his contemplation high enough in gazing after the painted shadowes of fading riches[83]

Based on a gross misunderstanding or, quite simply, on ignorance, since Greene may after all merely indulge here in name-dropping, such a denunciation suggests that Epicureanism actually began to find its way among the learned circles of society, especially among those interested in scientific and environmental matters. A few years later, Helkiah Crooke wrote an immensely popular anatomy manual in which he remarked:

> Let that beastly Epicure now lay his hands upon his mouth, & keepe silence, who was not ashamed to affirme, that the bodies of men were made by chance and fortune, out of a turbulent concourse (forsooth) of a number of Atomies or Motes, such as we see in the Sunne.[84]

To writers like him whose views were dictated by Christian assumptions, Epicure seemed dangerous as he was becoming a reference in the world of early modern science. It would be fair to say, therefore, that Greek meteorological thinking, transmitted through a number of indirect channels, informed popular opinions about climate and weather issues in the early modern era, and that its impact was far from being restricted to the literati. So, the much-ingrained idea that, according to Ben Jonson's famous phrase, Shakespeare's 'small Latin and less Greek' made these important Greek sources inaccessible to him should now be seriously reconsidered as, through a number of second-hand Latin or English sources, these texts have in fact deeply influenced his representation of weather and climate as well as the very writing of his plays by way of his imagery and dramatic language.

Shakespeare's skies

The recent advent of ecocriticism in Shakespeare studies has made us increasingly aware that the representation of climate in early modern drama intersected with discourses on misrule, politics, science, animals, food and crops, as well as on race, emotions and humours.[85] Much has already been written about the last three topics in particular as climatic conditions have often been used of late as a key factor helping to explain racial considerations and conduct assessment in Elizabethan and Jacobean England. However, in the approach worked out in this book, climate, or rather the idea of climate, is to be considered less as a means to understand or enhance other types of phenomena than as an end in itself. Indeed, as climate resists being represented, few critics have been interested in considering its role per se,[86] a gap which the present study aims to bridge. Furthermore, if, as Jennifer Munroe posits, ecocriticism proceeds from the need to focus on the 'nonhuman world' and to 'concentrate on its material qualities and relations',[87] this book does not exclusively focus on actual climate to the detriment of its metaphorical representations. On the contrary, it seeks to reconcile the material and the metaphoric, the actual and the imagined. The seven Shakespeare plays addressed in the successive chapters of this book are all, in various ways, affected by weather and climate issues and by what Gwilym Jones has called 'environmental irony', that is, by the unnerving intrusion of real weather into the playwright's fictional world.[88] *A Midsummer Night's Dream*, *Romeo and Juliet*, *As You Like It*, *Othello*, *King Lear*, *Anthony and Cleopatra* and *The Tempest* indeed illustrate various types of interaction between men and the weather. They present in turn the consequences of an excess of rain on cattle and crops, of freezing temperatures on men's morale, or of tempests on their shipwrecked victims. They probe the way prayers may affect the state of the 'airy region' (*Romeo and Juliet*, 2.1.63), the way magic may calm down or raise the winds, and they examine how men's misbehaviour produces disquieting seasonal variations. Each play has its own specificity and resorts to particular climatic myths (e.g. Phaeton, Noah's ark, Danaë or Iris), but in those seven works, the dramatic plot is grounded in, framed by and focused on natural phenomena. In fact, the plot incidents take place in the sky as much as on the earth thanks to a series of images and analogies that 'hold, as it were, the mirror up to nature' (*Hamlet*, 3.2.22–3). Human disasters develop in the here and now, but the zodiacal and climatic imagery used by the playwright gives them an extraordinary extension. In particular, they allow the plays to reverberate with heavenly signs, cosmological allusions and seasonal references at large, thus enriching the text by multiplying its layers of meaning.

The two so-called 'festive comedies' studied in this book, namely *A Midsummer Night's Dream* and *As You Like It*, appear far less festive if we approach them from a climatic angle, which inevitably reveals some of their darker facets. Shakespeare's three love tragedies, *Romeo and Juliet*, *Othello* and *Anthony and Cleopatra*, offer interesting insights into the way the playwright associates heavens and humours, climate and the planets, and how this particular combination in turn colours or affects the dramatic events and prepares the audience for an inevitable outcome. *King Lear* certainly prolongs this vision by making the storm a mirror of the old king's disturbed psyche, but it adds to it a complexity of its own as well as a radical pessimism inspired by Lucretius' materialistic views. If, up to a point, the tragedy takes up and considerably worsens the darker sides of *As You Like It* (the comedy is often referred to as a 'dark pastoral'), the carnivalesque traits of its arsy-versy world bring it closer to a sombre and frightening feast of fools than to the temporary misrule and madness of the romantic comedies. As to *The Tempest*, its very title suggests Shakespeare's renewed interest in climatic issues at the end of his career. His last romance summarises the whole canon and makes it clear that, at this particular stage of his dramatic production, environmental and weather issues now mostly matter as theatrical illusions and climatic constructions.

In order to explore Shakespeare's apprehension of the early modern 'fated sky', these plays provide a number of interesting case studies and form a coherent ensemble. In *A Midsummer Night's Dream* (c. 1595–6), wetness informs the play as a whole. The moon spreads humidity in Athens while the weather turns rainy and cataclysmic, due to the unruly behaviour of Oberon and Titania who are the source of the general confusion turning the world upside down. Their quarrel over the little Indian boy alters the cycle of the seasons and, as a result, the would-be paradise of the forest is 'filled up with mud' (2.1.91). If, on the one hand, Titania's lines on climatic 'distemperature' (2.1.109) certainly have some sort of topical relevance, on the other, reducing them to a mere commentary on the vagaries of the English weather in the 1590s would hardly do justice to the richness and complexity of this comedy. For the young playwright, the *Dream* and its ever-shifting environment serve in fact as an experimental ground to challenge medieval beliefs and to test fresh hypotheses, such as the idea that people's attitudes may in fact be responsible for climatic imbalance.

In *Romeo and Juliet* (c. 1595–6), weather and humoral determinism play a central role with the background references to the dog days that may, up to a certain extent, be held responsible for both plague and misrule so that, beyond bad luck or misfortune, the influence of the stars turns out to be preponderant in the lovers' fate. In such a context, the play's heavenly signs take on an importance almost equal to that of the earthly events, to the point that heat may be considered as a major actor in the tragedy. As an anagram of 'hate', 'heat' indeed overdetermines the climate of the play. Both words foreshadow the flare-up of violence in Verona, leading Romeo and Juliet to be trapped in an overall astronomical, humoral and climatic pattern giving them virtually no chance to escape the stifling air of the city. Omnipresent in the tragedy, light and lightning emphasise the violence of passions and reinforce the inevitability of the lovers' final death march inscribed in the sonnet prologue.

The third chapter turns to *As You Like It* (1599), a comedy of the cold with its exiled Duke and his court having to make the best of a bad bargain in a freezing forest. Marked by a saturnalian atmosphere, which favours melancholy and bitter-sweet songs to the detriment of the not-so-innocent games of love, the play presents us with a multiple temporality in which Shakespeare alludes to several ritual times, themselves associated with various types of weather. But jolly and festive as the comedy may sometimes be, Arden's air remains desperately frosty – a frostiness synonymous with exile and tyranny. If springtime, 'the only pretty ring-time' (5.3.16), is duly announced, it never fully materialises at the end. As a result, even though the multiple marriages about to be celebrated apparently point to a satisfying resolution of the plot, the characters' tirades, laden with clichés, still suggest frozen and sterile thoughts.

Othello (1604) is another play obsessed with breath and wind, a cosmological piece in which climate and air – aptly imaged as 'the heavens' breath' in *Macbeth* (1.6.5) – coalesce to make the Moor the victim of his own humours as much as of the satanic Iago. The importance given to cosmic elements as well as to the planets and their influence on men and women's behaviour serves to elevate and magnify a play sometimes wrongly reduced to the genre of the domestic tragedy.[89] Besides, the recurring imagery related to *pneuma* turns the scene into a dark carnival with its frightening disaster at the end epitomised by the image of the 'tragic loading of [the] bed' (5.2.374). If a providential tempest preserves Cyprus from the assaults of the Turkish fleet, Othello and Desdemona's love then becomes a highly tempestuous affair that ends in tragic suffocation.

King Lear (1605–6), where the vehemence of the old king's defiant speeches is matched by the raging storm striking the heath, goes several steps further as far as violence and darkness are concerned. If, in *Of the lavves of ecclesiasticall politie* (1593), Richard Hooker assumed that natural phenomena coincide with the voice of God, the playwright here directly questions the alleged divine origin of climatic manifestations in a dark and nihilistic vision of life. As he fights against the storm, superbly staging his own distress, Lear proceeds to an inverted exorcism justifying his mad dream to have all forms of human life destroyed on earth rather than having his reason and mental sanity restored. Influenced by Lucretius' atomism, the play makes an extensive dramatic use of the traditional humoral and cosmological interplay of the four elements. Eventually, as gall invades Lear's heart and eradicates both hope and tenderness, a disquietingly grotesque tonality pervades the tragedy which forces us to look at the title character's internal turmoil and to try and an*atom*ise his folly.

Anthony and Cleopatra (1606–7) follows up on *Lear*'s Lucretian philosophy as the play proposes yet another reflection on climatic issues which bridges the gap between humoral theories and atomist patterns. Shakespeare's 'Epicurean' (2.1.24) vision of the weather, which operates in the play's subtext, suggests that the atomism already displayed in previous works underpins the playwright's obvious concerns about climate and the environment. Questions such as the infinite, the void and the flux in particular are major concerns in this Roman play obsessed with shifting shapes and the dissolution of living beings. The destructive yet paradoxically creative power of the Nile provides a rich background for the tragedy, and it is in this hot and moist context that the playwright calls attention to Cleopatra's intimate meteorology and bodily humours. While climate is mainly present in a number of geo-humoral features characterising the lovers as well as in a dense metaphoric network (tears, jealousy, cruelty and burning pain are conveyed through imagery of gales, hail and tempests in a deliberately hyperbolic, mock-Petrarchan stance), a rich system of cosmological references runs through the text, thus suggesting that the title characters are not just lovers but also explorers of sky and heavens as it were. It is no mere coincidence that Anthony's famous tirade on clouds, representative as it is of the centrality of celestial issues in this tragedy, should coincide with the poetic and dramatic climax of the play.

In the last chapter, similar concerns with the shiftiness of clouds resurface in *The Tempest* (1611), a play often quoted for its ecological significance in which Shakespeare once again addresses the question of climate and the four elements in his revisiting of the travel narratives of his time with their recurring fear of sea-storms. In this rewriting of Virgil's *Aeneid*, not entirely devoid of Homeric reminiscences, the playwright returns to the initial questions of the *Dream*: can men and women rule the elements? If we trigger off a climatic disorder, can it be mended? And if we lose control, what may then ensue? The playwright thus reassesses the role of man's 'potent art' (5.1.50) in the ordering of nature. A clear precedent for the tempest raised by Ariel, Prospero's airy spirit, the storm caused by Aeolus on Juno's order in her attempt to revenge herself against Venus by eliminating her son, Aeneas, is at the origin of the foundation of Rome, the city that will later destroy Dido's Carthage in the wake of the Carthaginians' final defeat at the end of the third Punic war. But Prospero's tempest, which is also meant as a form of revenge against his enemy brother Antonio, is more problematic as it oscillates between the illusory and the real, magic and science, the sublime and the mundane. Providing us with kaleidoscopic views, the play cogently explores the power of the elements and once again reaffirms that, for Shakespeare, what appears in the celestial sphere cannot be dissociated from, and is definitely as crucial as, what happens on earth.

All in all, this study of climate and environment in Shakespeare questions the widely accepted idea of the playwright's opportunistic, one-off approaches to sky-related issues. I argue that, as a man of the theatre acutely aware of his audience's anxieties and eager to address them, the Renaissance dramatist worked out a system of underlying climatic references in order to display the serious incidence of environmental pitfalls on human lives and, conversely, to say something about the disasters resulting from men and women's interference with their natural environment. First seen as a reservoir of metaphors, Shakespeare's skies rapidly become used for their scientific, philosophical or material resonances rather than for their mere hyperbolic significance. In Sonnet 21, for example, the poet ironises against those who write love poems laden with exaggerated conceits involving 'sun and moon' (l. 6) and 'all things rare / That heaven's air in the huge rondure hems' (ll. 7–8). His own airspace evolves from play to play to become increasingly palpable onstage, so that, from the end of the 1590s onwards, the spectator could almost imagine himself or herself walking on 'the floor of heaven / [. . .] thick inlaid with patens of bright gold' (*The Merchant of Venice*, 5.1.58–9).[90] Turning the airy essence of the firmament into the earthy substance of the floor, what we might call an inverted transubstantiation links the heavens to our mundane here and now as, for the playwright, humankind is as firmly rooted in the skies as it is naturally grounded in the earth.

In such a perspective, the heavenly 'vault' so often referred to by Shakespeare (in *Romeo and Juliet*, the word appears no fewer than eight times) is an image that works like Achilles' shield in *The Aeneid*. A beautiful artefact decorated with constellations, the shield is not only a mirror of the cosmological order of the universe but also a form of looking glass, or window disclosing future events to the spectator/reader. At the twilight of his career, the author of *The Tempest* clearly sought to capture something of the abstract, disembodied 'heavens', this inaccessible 'expanse in which the sun, moon, and stars [were] seen',[91] so as to encapsulate them in a theatricalised 'great globe' (*The Tempest*, 4.1.153) encompassing a tangible, circular sky closely connected to the earth, which then seemed to allow people, rich or poor, to peep into the unknown (i.e. nature, God, the cosmos, the afterlife) in order to assuage some of their existential as well as climatic fears. Through this pragmatic rather than programmatic approach, Shakespeare managed to reconcile the ideal and the ethereal with the concrete daily realities as well as with the material resources and properties of the stage, thus giving his audience the blessed though short-lived illusion that the inaccessible was at long last within reach.

Notes

1. William Whittingham, *The Bible and Holy Scriptures conteyned in the Olde and Newe Testament*, 'The Holy Gospel of Jesus Christ, according to Luke', chap. XVII, n.p.
2. In 1783, the British astronomer William Herschel discovered double and multiple nebulae, known today as 'galaxy groups'. See Steven J. Dick, *Discovery and Classification in Astronomy: Controversy and Consensus*, p. 333.

3. In a paper entitled 'On the Modification of Clouds' (presented to the Askesian Society in London in December 1802), the pharmacist Luke Howard introduced three cloud types: the cirrus, the cumulus and the stratus.
4. On the astrological sense of 'influence', see François Laroque, 'The Science of Astrology in Shakespeare's Sonnets, *Romeo and Juliet* and *King Lear*'.
5. This apocalyptic rhetoric in connection with climatic issues has now resurfaced in a number of scientific warnings. Sociologists and climatologists indeed often stress the rise of an apocalyptic global warming discourse which has fuelled strong 'ethical injunctions' since the late 1980s. On this, see for instance Philip Smith, 'Narrating Global Warming', p. 751.
6. This passage has already inspired the title of Benson Bobrick's *The Fated Sky: Astrology in History*.
7. See for instance William Cuningham, *The cosmographical glasse*, fol. 64. On the new attractiveness of the torrid zone for Renaissance travellers and on the enslavement of its inhabitants because of their presumably weak mental capacities, see Nicolás Wey Gómez, *The Tropics of Empire: Why Columbus Sailed South to the Indies*, and Surekha Davies, *Renaissance Ethnography and the Invention of the Human: New Worlds, Maps and Monsters*.
8. See Steven Mullaney, *The Reformation of Emotions in the Age of Shakespeare*, p. 58. As Mullaney makes clear, '[t]he perfect balance of humors [. . .] was found [. . .] in people of a Mediterranean climate – unsurprisingly, since humoral theory and its Greek and Roman physicians were native to such a clime.'
9. Mary Floyd-Wilson, *English Ethnicity and Race in Early Modern Drama*, p. 2. On climate theories in connection with race, see also Jean E. Feerick, *Strangers in Blood: Relocating Race in the Renaissance*.
10. François E. Matthes, 'Report of Committee on Glaciers', p. 520. Matthes originally intended the phrase to describe an 'epoch of renewed but moderate glaciation which followed the warmest part of the Holocene [i.e. the Medieval Climatic Optimum]'. It now applies to a different time period. In his 1967 *Histoire du climat depuis l'an mil*, for instance, the French historian Emmanuel Le Roy Ladurie uses the phrase to describe a cooling period spanning from 1590 to 1850. Le Roy Ladurie's seminal study has been translated into English under the title *Times of Feast, Times of Famine: A History of Climate since the Year 1000*.
11. The quotation refers to the 27 June 1607 entry in the journal of Henry Hudson and the crewman John Playse. See Dagomar Degroot, 'Exploring the North in a Changing Climate: The Little Ice Age and the Journals of Henry Hudson, 1607–1611', p. 78.

12. Brian Fagan, *The Little Ice Age: How Climate Made History, 1300–1850*, p. 90. Recent studies relying on tree-ring dating (i.e. on the study of growth rings to identify changes in the atmospheric conditions over a long period of time) also provide valuable climate information, even though only a few studies actually focus on Great Britain. Two main reasons can be put forward: the exploitation of forest resources in Great Britain has drastically limited the number of ancient trees, while conifers, which provide researchers with the most relevant results, remain relatively rare in this part of the globe. For more general remarks on the cooling trend characterising Shakespeare's period, see Rob Wilson et al., 'Last Millennium Northern Hemisphere Summer Temperatures from Tree Rings, Part I: The Long Term Context'. Thanks to the in-depth study of tree rings, the authors observe a 'coherent picture of warm conditions around the end of the 10th century, a cooling trend for about 500 years, a prolonged cool period from ca. 1450–1850 and steep warming until present' (p. 10).
13. Mike Hulme, 'Climate', p. 33.
14. On this, see H. H. Lamb, *Climate, History and the Modern World*, p. 211.
15. Philip Armstrong, 'Preposterous Nature in Shakespeare's Tragedies', p. 107.
16. Thomas Nashe, *Summer's Last Will and Testament*, p. 134.
17. Helkiah Crooke, *Mikrokosmographia a description of the body of man*, Book 1, chap. 4, p. 10. The verb 'infuse' is used by Graziano in a similar sense in *The Merchant of Venice* when he tells Shylock that '[the] souls of animals infuse themselves / Into the trunks of men' (4.1.131–2).
18. Marsilio Ficino, *Platonic Theology*, p. 235.
19. Carla Mazzio, 'Coda: Scepticism and the Spectacular: On Shakespeare in an Age of Science', p. 242. Mazzio relies on Crane to develop her argument.
20. See Mary Thomas Crane's *Losing Touch with Nature: Literature and the New Science in Sixteenth-Century England*, p. 17. Crane notably quotes Lewis Wolpert who argues in *The Unnatural Nature of Science* that science is deeply 'unnatural', i.e. 'counter-intuitive'.
21. For a dramatic description of the 'heavens', see *Hamlet*, 2.2.301–5, when the Prince gestures at the constellations represented on the roof jutting out over the stage: 'This most excellent canopy the air, look you, this brave o'erhanging, this majestical roof fretted with golden fire – why, it appears no other thing to me than a foul and pestilent congregation of vapours.'
22. Samuel Schoenbaum, *William Shakespeare: A Documentary Life*, p. 7.
23. Still according to the *OED*, the first use of the word dates back to 1603 and is found in Philemon Holland's translation of Plutarch's *Moralia*.

24. Originally, the Anglo-Norman *avirounement* served to denote proximity, a sense which has been kept in the current usage of 'environment'.
25. On the early modern enthusiasm for the exploration of nature, see Alix Cooper, *Inventing the Indigenous: Local Knowledge and Natural History in Early Modern Europe*. For environmental issues in Renaissance literature and their connection with an overall sense of loss (caused by deforestation, receding woods, withdrawing fields, pollution problems, etc.), see Ken Hiltner, *What Else Is Pastoral? Renaissance Literature and the Environment*.
26. On these perennial conceptions still used, today, to respond to ecological problems, see David Macauley, *Elemental Philosophy: Earth, Air, Fire, and Water as Environmental Ideas*.
27. England's fairly harsh climate did not necessarily entail poor harvests since, according to some historians, the economic consequences of the 'Little Ice Age' may have been exaggerated. I am just arguing that the weather had an undeniable influence on the population's daily morale. On the reassessed impact of the 'Little Ice Age', see for example Morgan Kelly and Cormac Ó Gráda, 'The Waning of the Little Ice Age: Climate Change in Early Modern Europe', p. 319.
28. Mike Hulme, *Weathered Cultures of Climate*, p. 3.
29. John Florio, 'Tempo', in *A Worlde of Wordes*, p. 415.
30. Mike Hulme, *Weathered Cultures of Climate*, p. 5.
31. Mike Hulme, 'Climate', p. 30.
32. This is also the first definition supplied by the *OED*, in which the word 'climate' primarily designates 'each of the bands or belts of the earth's surface stretching from west to east and associated with specific parallels of latitude' (1.a).
33. John Florio, 'Clima', in *A Worlde of Wordes*, p. 75.
34. In 1613, Samuel Purchas still understood 'The Climes or Climates' as 'the spaces betwixt two parallels'. See Samuel Purchas, *Purchas his pilgrimage*, The First Booke, chap. IX, p. 43.
35. Thomas Blundeville, *M. Blundevile his exercises containing sixe treatises*, The Second Booke, chap. 18, fol. 193v.
36. On this, see Toni Francis, 'New Directions: Imperialism as Devilry: A Postcolonial Reading of *Doctor Faustus*', p. 120.
37. Christopher Marlowe, *Doctor Faustus. A- and B-Texts* (1604, 1616), p. 183.
38. Mary C. Fuller, *Remembering the Early Modern Voyage: English Narratives in the Age of European Expansion*, p. 124.
39. Edmund Spenser, *The Faerie Queene*, VII.vii.23.
40. Ibid.: 'Rayne, hayle, and snowe do pay them [human creatures] sad *penance*, / And dreadfull thunder-claps (that make them quake) / With flames and flashing lughts that thousand changes make' (my emphasis).
41. Keith Thomas, *Religion and the Decline of Magic: Studies in Popular Beliefs in Sixteenth and Seventeenth Century England*, p. 41.

42. Ibid. p. 31.
43. Wolfgang Behringer, *A Cultural History of Climate*, p. 89.
44. On Luther's stance against weather superstitions and his condemnation of idolatry, see Euan Cameron, *Enchanted Europe: Superstition, Reason, and Religion 1250–1750*, p. 160.
45. On literacy in early modern England, see for instance Tessa Watt, *Cheap Print and Popular Piety, 1550–1640*, p. 7. Watt shows that rates have been probably underestimated, and that '[a]mongst husbandmen and labourers especially, there may be a large number who attended school up to the age of six, and learnt the primary skill of reading'.
46. In connection with weather predictions based on birds, see Banquo's speech in *Macbeth*, 1.6.3–6: 'This guest of summer, / The temple-haunting martlet, does approve / By his loved mansionry that the heavens' breath / Smells wooingly here.'
47. On weather prognostications, see Raymond Gillespie, *Devoted People: Belief and Religion in Early Modern England*, p. 128.
48. See Louis B. Wright, *Middle-Class Culture in Elizabethan England*, p. 84.
49. Keith Thomas, *Religion and the Decline of Magic*, p. 284. On the popularity of astronomy, see Francis R. Johnson, *Astronomical Thought in Renaissance England: A Study of English Scientific Writings from 1500 to 1645*. See also William Aspenhall (ed.), *The Correspondence of Philip Sidney and Hubert Languet*, p. 23. In a letter to Sidney dated 1 January 1574, Languet writes: 'You are right to pay attention to astronomy; without some knowledge of it, it is impossible to understand cosmography; and he who reads history without a knowledge of this, is very like a man who makes a journey in the dark.'
50. Keith Thomas, *Religion and the Decline of Magic*, p. 284.
51. Ibid. p. 285.
52. Ibid. p. 286.
53. John Considine, 'Hill, Thomas (*c.*1528–*c.*1574)'.
54. Throughout this book, I will rely on the 1574 edition printed by Henry Denham. Presumably, this copy is not the first edition of Hill's treatise, which was probably issued around 1571 if we are to believe the Stationers' Register, where it is entered as '*Rayre wonders and feyrefull syghtes* 1570–71'. I thank the Rare Books and Music Reference Team of the British Library for their help.
55. 'To the Reader', in *The Works of John Webster*, vol. 1, pp. 140–1.
56. On attendance figures, see Ann Jennalie Cook, *The Privileged Playgoers of Shakespeare's London, 1575–1642*, p. 190.
57. Leonard Digges, *A prognostication of right good effect fructfully augmented*, sig. B.iij.
58. The invention of the barometer is ascribed to Evangelista Torricelli and dates back to 1644. The first thermoscope (the ancestor of the thermometer) was probably perfected by Galileo himself at the end of the

sixteenth century. In 1662–3, Christopher Wren invented the tipping-bucket mechanism which served to measure the amount of rainfall, and in the 1670s, Robert Hooke devised the mechanical weather clock. On this, see Stephen Burt, *The Weather Observer's Handbook*.

59. Abigail Shinn, '"Extraordinary discourse of vnnecessarie matter": Spenser's *Shepheardes Calender* and the Almanac Tradition', p. 144.
60. Mary Floyd-Wilson, *English Ethnicity and Race in Early Modern Drama*, p. 35.
61. For an overview of these theories, see Mary Floyd-Wilson and Garrett A. Sullivan Jr (eds), *Environment and Embodiment in Early Modern England*, p. 5. For them, the Hippocratic tradition promotes a form of continuity between the subject and his environment while the Aristotelian one favours a 'counteractive' model.
62. Hippocrates' views began to circulate among English literati after his treatise of *Airs, Waters, Places* had been translated into Latin in 1570 with a commentary by Girolamo Cardano.
63. The treatise was published in the original Greek tongue in the sixteenth century, but the Latin translation of the *Problemata* completed by Theodora Gaza and published in 1474 had made the text more easily accessible. It was eventually translated into English in 1595 as *The problemes of Aristotle*.
64. Leonardo Bruni's Latin version, written around 1436, of Aristotle's *Politics* updated the thirteenth-century translation of William of Moerbeck. An anonymous English translation, *Politiques, or Discourses of gouernment*, was published in London in 1598.
65. Aristotle, *Problems*, vol. 1, 'Problem XIV', pp. 319–21. Also quoted in Jean E. Feerick, 'A "Nation . . . Now Degenerate": Shakespeare's *Cymbeline*, Nova Britannia, and the Role of Diet and Climate in Reproducing Races', p. 38.
66. R. W. Desai, '"What means Sicilia? He something seems unsettled": Sicily, Russia, and Bohemia in *The Winter's Tale*', p. 320.
67. Jean Bodin, *The six bookes of a common-weale*, The Fifth Booke, chap. 1, p. 545.
68. Ibid. p. 559.
69. Robert Burton was to testify to this long-standing influence in *The Anatomy of Melancholy*, p. 423.
70. Pierre Charron, *Of wisdome*, The First Booke, chap. XLII, p. 163.
71. On the humoral system in early modern England, see Gail Kern Paster's pioneering study, *The Body Embarrassed: Drama and the Disciplines of Shame in Early Modern England*.
72. Katharine Anderson, 'Looking at the Sky: The Visual Context of Victorian Meteorology', p. 303.
73. On this, see Louis B. Wright, *Middle-Class Culture in Elizabethan England*.
74. T. W. Baldwin, *William Shakspere's Small Latine and Lesse Greeke*. Baldwin, however, acknowledges (albeit reluctantly) the place of Greek

grammar in Shakespeare's education (vol. 2, p. 619). He concludes that '[i]f Shakespere reached any Greek author at all, it is a safe assumption that he began and continued on the Greek testament, with English and Latin translations available' (p. 627).

75. On the atmosphere, I refer in particular to Aristotle's *Meteorology* (I.2, pp. 4–5). In *Twelfth Night*, through Feste, Shakespeare scoffs at those systematically using an affected term like 'element' instead of the more basic 'sky', as Keir Elam notices in his edition of the play (n. to 3.1.57, p. 254).
76. According to the Greek physician, and as Mark Grant explains in *Galen on Food and Diet* (p. 14), the humours and the seasons may 'agree' or 'disagree with each other'. Basically, for Galen, 'blood, air and spring are moist and hot [. . .]; yellow bile, summer and fire are hot and dry, whilst black bile and earth and autumn are dry and cold; phlegm, water and winter are cold and moist.'
77. Craig Martin, 'Conjecture, Probabilism, and Provisional Knowledge in Renaissance Meteorology', p. 271.
78. *Concerning Weather Signs* was probably consulted by Aratus, the author of *Phaenomena*, a didactic poem where Greek astral myths are presented and in which the weather is dependent on the positions of the stars. The latter text was one of the chief sources of Virgil's *Georgics* (37 BC) – a work recommended for its weather lore in Digges's *Prognostication of right good effect fructfully augmented* (sig. Biii.v).
79. Craig Martin, *Renaissance Meteorology: Pomponazzi to Descartes*, p. 12.
80. James Dougal Fleming, *The Mirror of Information in Early Modern England: John Wilkins and the Universal Character*, p. 102.
81. On this, see Stephen Greenblatt, *The Swerve: How the Renaissance Began*, and Gerard Passannante, *The Lucretian Renaissance: Philology and the Afterlife of Tradition*. For a reassessment of Lucretius' influence throughout the ages, see Pierre Vesperini, *Lucrèce. Archéologie d'un classique européen*. Vesperini notably posits that Epicurean philosophy, far from foreshadowing modernity, 'was the most archaic of Athenian schools' (p. 15, my translation).
82. On Lucretius as a major influence on Shakespeare's weather imagery, see Jonathan Pollock, 'Of Mites and Motes: Shakespearean Readings of Epicurean Science'.
83. Nandini Das (ed.), *Robert Greene's Planetomachia (1585)*, p. 9. For Greene, the 'celestiall influences' of the planets were 'predominant in terrestrial creatures' (p. 21).
84. Helkiah Crooke, *Mikrokosmographia*, Book 1, chap. 3, p. 8.
85. See for example Gabriel Egan, *Green Shakespeare: From Ecopolitics to Ecocriticism* and, by the same author, *Shakespeare and Ecocritical Theory*; Simon C. Estok, *Ecocriticism and Shakespeare: Reading Ecophobia*; Dan Brayton, *Shakespeare's Ocean: An Ecocritical Exploration*; Bruce Boehrer, *Environmental Degradation in Jacobean Drama*; Greta

Gaard, Simon C. Estok and Serpil Oppermann, *International Perspectives in Feminist Ecocriticism*; Lynne Bruckner and Dan Brayton (eds), *Ecocritical Shakespeare*; and Jennifer Munroe, Edward J. Geisweidt and Lynne Bruckner, *Ecological Approaches to Early Modern English Texts: A Field Guide to Reading and Teaching*. For a critique of ecocritical theories, see Timothy Morton, *Ecology Without Nature: Rethinking Environmental Aesthetics*. For a survey of the field, see Jennifer Munroe, 'Shakespeare and Ecocriticism Reconsidered'.

86. See, however, Gwilym Jones, *Shakespeare's Storms*. As its title indicates, Jones's book focuses on one particular kind of natural phenomenon to the exclusion of all others.
87. Jennifer Munroe, 'Shakespeare and Ecocriticism Reconsidered', p. 462.
88. Gwilym Jones, *Shakespeare's Storms*, p. 9.
89. On this, see Sean Benson, *Shakespeare, Othello and Domestic Tragedy*.
90. Cf. *The Reign of King Edward the Third*, a play only partly written by Shakespeare, 13.32–4: 'With their approach there came this sudden fog / Which now hath hid the airy floor of heaven, / And made at noon a night unnatural.'
91. I here quote from the first definition of 'heaven' supplied by the *OED*. See also the second meaning of the word: 'The part of the atmosphere above and closest to the earth's surface, within which humans observe terrestrial weather systems, flight, and other activity in the sky local to earth.'

Chapter 1

'We see / The seasons alter': Climate Change in *A Midsummer Night's Dream*

> [. . .] the winds, piping to us in vain,
> As in revenge, have sucked up from the sea
> Contagious fogs which, falling in the land
> Have every pelting river made so proud
> That they have overborne their continents. (2.1.81–5)

A Midsummer Night's Dream takes its title from a famous seasonal celebration of Celtic origin linked to the advent of summer in the British Isles. At the time when the play was presented at the Theatre and, presumably, also at court,[1] the Elizabethan year was divided into two halves, namely winter, from Christmas to 24 June, and summer, which went from 25 June to Christmas. In this early modern calendar, 24 June corresponds to Midsummer, a day likely to prove 'very hott',[2] as well as to the latest possible date for Corpus Christi. Poised as it is between the sacred (corresponding to the winter half of the calendar) and the profane (the summer half marked by rural celebrations), Shakespeare's comedy appears as a threshold play, whose delicate balance is exemplified by its double timescape and double setting.

The play's double timescape entails the alternation of day and night while its intriguing double setting takes us back and forth between the comedy's main locale, Athens, under the rule of Duke Theseus, and the forest's green and damp atmosphere under the rule of Oberon and Titania. The wood near Athens actually recalls a folk England whose chilliness is also repeatedly evoked in *As You Like It* ('Blow, blow, thou winter wind', 2.7.174). While southernmost Greece was hot and dry, the English climate was fairly cold and rainy, so that it seems that Shakespeare fashioned his personal geography and exotic settings with his immediate surroundings in mind. In his plays, 'the far becomes the near' and it sometimes 'becomes so dangerously near that it invariably proves necessary to recreate familiar distances to separate the far-off world from the world of Christian civilisation'.[3]

The role of the moon

As noted by Philippa Berry, a tragedy such as *Romeo and Juliet* is obsessed with 'the sun's passage in late July',[4] while *A Midsummer Night's Dream*, by contrast, is mainly concerned with the shining of the moon. The comedy starts with a reference to the moon ('how slow / This old moon wanes', 1.1.3–4), a crucial weather correlate described as 'cold' and 'fruitless' (1.1.73) by Duke Theseus, who is about to marry the Amazon Hippolyta, the doppelgänger of Titania, the queen of fairies in the play. 'In the air, the moon causes sometimes lightning, thunder, cloud mists, winds, rain, and storms [. . .] nor has it lesser powers over the earth, over fields, crops, and trees [. . .] And, what's more, with its great humidity, the moon is extremely harmful to our physical health.'[5] Moderata Fonte's observations in *The Worth of Women* (1600) correspond to the contemporary commonplaces in Shakespeare's comedy and it is probably no coincidence that, in Ovid's *Metamorphoses* (3.173), the moon goddess is referred to as Titania and, in Arthur Golding's translation, as one of Titan's daughters. In calling his queen Titania, Shakespeare not only provides his preternatural heroine with the ability to influence the weather but he also endows her with the fertility of Lucina, the goddess of childbirth in Roman mythology, while Phoebe's chastity seems reserved for the Amazon Hippolyta. As a result, one of the main tasks of the moon goddess in the *Dream* consists in '[d]ecking with liquid pearls the bladed grass' (1.1.211), that is, in supplying Athens and its green surroundings with showers of fertilising rain.[6]

Titania's fecundity is only nominal since the little changeling boy she is so happy to raise is actually the child of her late Indian votaress. In act 2, she nostalgically reminisces how she and her female friend 'laughed to see the sails conceive / And grow big-bellied with the wanton wind' (2.1.128–9).[7] Curiously, through the implicit analogy between the ships' sails and the female uterus, the 'wanton wind' seems to have impregnated the womb of Titania's mortal friend with the child now jealously kept by the fairy queen.[8] Strange as this idea may seem, it was then widespread among classical writers such as Aristotle, who mentions it in Book VI of his *History of Animals*,[9] while in Pliny's *Natural History*, part of a chapter is devoted to wind-begotten horses:

> In Portugall, along the river Tagus, and about Lisbon, certaine it is, that when the West wind bloweth, the mares set up their tailes, and turne them full against it, and so conceive that genitall aire insteed of the naturall seed: in such sort, as they become great withall, and quicken in their time, and bring foorth foles as swift as the wind, but they live not above three yeares.[10]

It is doubtful whether Shakespeare's contemporaries still believed in the seminal powers of the wind,[11] but the 'wanton wind' mentioned by Titania does seem to trace us back to this age-old tradition. The playtext more specifically evokes a pregnant woman whose swinging gait imitates the movements of trading ships riding the windy sea with ballooning sails and, as the sea was then thought to be chiefly governed by the moon,[12] Titania, the doting moon goddess, suggests that she has been an active agent and a conniving witness of her votaress's impregnation. Having 'erase[d] the reproductive agency of a biological father', the fairy queen no longer appears as a mother, but as a father figure having appropriated the codes of 'patriarchal fatherhood'.[13] Therefore, as Valerie Traub observes, Titania 'usurps patriarchal power' since '[t]he child is the manifest link of a prior affection between women that is associated with their shared fecundity and maternal largess'.[14]

In the Ovidian story of Pyramus and Thisbe performed by Bottom and his fellow amateur players, the moon also plays a crucial role. Disappointingly, in Arthur Golding's translation, the emphasis is first and foremost put on coolness ('by a cool and trickling spring', The Fourth Book, l. 111) and darkness ('As soon as darkness once was come, straight Thisbe did devise / A shift to wind her out of doors', The Fourth Book, ll. 114–15) rather than on moonlight, an element only mentioned in passing when Thisbe makes out a lioness in the dark ('Whom Thisbe spying first / Afar by moonlight, thereupon with fearful steps gan fly', The Fourth Book, ll. 122–3).[15] Yet this unobtrusive detail was to be emphasised by Ovid's followers. In *De mulieribus claris* (Cap. XIII), Boccacio makes it clear, for example, that Thisbe found her way to the woods thanks to the natural light that enabled her to see where she was going ('Ardentior forte Tisbes prima suo fefellit et amicta pallion intempesta nocte, sola patriam domum exivit et, luna monstrante viam, in nemus intrepida abiit').[16] In Shakespeare's comedy, the celestial orb is an inescapable force (de) regulating the sublunary world. It even influences the low-born artisans who, having perfectly understood its pre-eminence in the Pyramus and Thisbe story, ask for an almanac (3.1.50) showing the lunar phases of the year so as to know whether or not the moon will shine on the night of their performance at Duke Theseus's court. Indulging in a fit of naive realism, they decide to add the role of Moonshine in order to emphasise the veracity of their representation ('one must come in with a bush of thorns and a lantern and say he comes to disfigure, or to present, the person of Moonshine', 3.1.55–6).

The almanac consulted by the mechanicals in the play may not have been very different from *The Kalender of sheephards*, a translation of *Le compost et kalendrier des bergiers* reprinted umpteen times since its first publication in 1502 and in which Luna is depicted as an oversexed woman whose genitals are concealed behind a human-faced moon.

Fig. 3 Luna in *The Kalender of sheephards*, 1585, sig. LIIIIv. By courtesy of the Bodleian Library.

Since Aristotle's time, the phases of the moon had been thought to coincide with women's periods,[17] a traditional correlation between the moon's phases and the female cycle also present in *The Kalender*. Unsurprisingly, in *A Midsummer Night's Dream*, the moon is once again conspicuously related to female coldness and (in)fertility and to women's 'monthlie tearmes and flowres',[18] explicitly symbolised in the comedy by the white pansy turned 'purple with love's wound' (2.1.166–7) after being hit by Cupid's arrow.

Besides, the anonymous author of *The Kalender* portrays the celestial orb as 'the Lady of the night' as he acknowledges that 'the chiefe light and clearinesse that is by night' is due to the 'presence of the Moon, for the moone is much more nearer approached unto us then any starre is'. Readers eventually learn that the 'moone is colde and moyst of nature and her colour is much fayrer then silver, and her chiefe house is Cancer'.[19] Such popular beliefs appear in Shakespeare's comedy where the moon is described as 'cold' (1.1.73; 2.1.156) and 'wat'ry' (2.1.162; 3.1.188). For Theophrastus, this sort of 'dusky' moon is a sign of wet weather[20] while, in the play, the feminised landscape of the Athenian forest becomes fuzzy and moist like the moon. Far from being an ideal natural space favouring freedom and all sorts of erotic possibilities, the main setting appears as a place where the harsh weather conditions dwarf rather than elevate men and women.

Such a deflation of the forest habitat is reinforced by an intriguing discrepancy in the lunar phases that punctuate the *Dream*. Though '[f]our happy days' are supposed to 'bring in / Another moon' (1.1.2–3), Theseus and Hippolyta's wedding finally takes place the day after. Richard Wilson notes that the 'old moon' (1.1.4) – which may also be read as a derogatory allusion to Queen Elizabeth – 'is still up in the sky on Midsummer night, and is again seen, by Puck, after midnight at the end of the second day [5.2.2]'.[21] All this means that the new moon, traditionally believed to be 'beneficial for sowing crops' and for 'human conception' as well,[22] never actually shows up. Therefore, beyond its political implications, the truncated lunar cycle of the *Dream* entails an absence of menstruation and an impossible regeneration, two major deficiencies that lay bare the imperfection of the Athenian green world. These circumstances may explain why the milk-white flower swollen red when hit by Cupid's arrow, minutely described by Oberon (2.1.166–8) and mischievously used by Puck, is both central to the plot and never visible onstage: subservient to the whims of the moon, female sexuality, though violently longed for, keeps being deferred.

Admittedly, this fairly gloomy vision of the forest in Shakespeare's play anticipates our contemporary vision of a world marked by eco-pessimism and disillusion about climate and partly belies Northrop Frye's optimistic analysis which, in 1957, presented the forest comedies as places of harmony and restoration. In *Shakespeare Our Contemporary* (1961), the Polish actor and critic Jan Kott had already opted in favour of a somewhat darker perspective in his insistence on the image of a 'bitter Arcadia' poles apart from Frye's festive world. Yet, both Frye and Kott delineate Shakespeare's green world 'in female terms': Frye's description evokes 'specific aspects of female anatomy' and is 'resonant of the nurturing presence of the "good mother"', while Kott's more provocative vision gives pride of place to 'the seducer and the "wicked stepmother"'.[23] The fact is that this gendered topography is present in the *Dream* and it is after all for the spectator to decide whether, with her Indian boy, Titania's role is closer to that of a wicked stepmother or to that of a comforting maternal presence.

India, the native country of the 'little changeling boy' (2.1.120), is an important exotic locale and Titania's long-lasting friendship with an Indian 'vot'ress of [her] order' (2.1.123) suggests that she spends part of her life in India. In Chaucer's *The Merchant's Tale*, one of Shakespeare's main sources for the *Dream*,[24] the Pluto-Proserpina background is part and parcel of the tale's structure, which incidentally happens to be related to the myth of the seasons. In Chaucer's text, Proserpina is portrayed as 'queene of fayërie' (l. 2316) and she takes sides with the young wife May, while Pluto, the 'king of fayërie' (l. 2227), is associated with the old husband January.[25] Shakespeare's set of fairies reformulates the symbolism of Chaucer's tale in a rather obvious way since Oberon, the fairy king, corresponds to Pluto's alter ego, while Titania matches Proserpina's role and function. According to David Wiles, '[j]ust as Proserpina resides in the underworld during the winter, so apparently Titania resides in India, making an unscheduled return to honour Theseus' wedding.'[26] Wiles's hypothesis implicitly equates the events in the woods near Athens with the advent of springtime. However, as a consequence of Hermia's poetic trope expressing her amorous despair ('[Lysander] hath turned a heaven unto a hell', 1.1.207) and under the effects of global cooling, these barren woods have become something of a hell on earth. So, if Wiles's association of India with hell conveniently allows him to fuse 'the Fairy Queen of English folklore' with the 'classical goddess of Spring' in the figure of Titania,[27] Shakespeare's play lends itself to a markedly different reading, one which associates the Athenian woods with a syncretic, postlapsarian universe afflicted with spells of bad weather. I would rather suggest that India, in the play, represents spring and fertility, a moist, maternal place of female intimacy where children are born, while the forest would correspond to a bleak wintry season[28] when Titania-Proserpina must live and fight with Oberon-Pluto and illustrate the traditional war of the sexes. Moreover, since the Middle Ages, paradise had traditionally been located in the east,[29] so that Titania's nostalgic evocation of India certainly makes sense when replaced in such a world picture. As a result, the Midsummer festival present in the play's title at best signals a time 'of mistakes and wandering wits, popularly attributed more to the influence of the moon than to that of the summer solstice'.[30]

Material conditions

The unseasonable Athenian place in which '[t]he spring, the summer, / The childing autumn' and the 'angry winter change / Their wonted liveries' (2.1.111–13) is what the play actually proposes to explore. Tellingly enough, it offers a number of al fresco scenes rather appropriate for open-air performances of the play and it emphasises the influence of natural, and more specifically climatic, phenomena on the human mind.

We know that the *Dream* was first published in 1600 (Q1) and was reprinted years later in a slightly different version in the 1623 Folio. Unfortunately, these two authoritative versions do not tell us whether Shakespeare first wrote his comedy for a public or a court performance when he started composing it or whether he had both in mind.[31] However, if we accept to be guided by the text, either Q1 or F1, things become clearer. As Richard Dutton puts it, *A Midsummer Night's Dream* 'is a play which is quite substantially *about* the way in which public drama might become courtly drama and [. . .] about the nature of the "allowance" that might make it possible'.[32] So, if we are to believe Shakespeare himself, the *Dream* was probably first thought of as a play written for an open stage[33] and eminently adaptable to a courtly venue. In other words, if the comedy may have been presented for the occasion of an aristocratic marriage (and therefore for a private family ceremony),[34] its printed text reveals that it was first and foremost written for 'an outdoor physical structure rather than a hall in a private house'.[35]

Among the clues scattered in the text and suggesting that the *Dream* was initially planned for a public playhouse, the play's numerous exits and entrances are certainly worth noticing. Martin White observes for instance that, at the start of act 2, scene 1, a fairy enters 'at one door and Robin Goodfellow at another' (2.1.0 s.d.).[36] More problematically, he also points out that 'the central recess in the rear wall (if there was one [. . .]), covered with a hanging, might have been employed for Titania's bower [. . .] while the frequent references to the moon might have involved acknowledgement of the decorated roof above the players' heads'.[37] True, most public stages of the 1599 period onwards – including the Globe and the Fortune – were the product of a potent 'cosmographic imagination'[38] and the playhouse where Shakespeare's *Dream* was performed may have had a roof protecting the actors from bad weather with images of the zodiac and of planets painted on its underside. Yet, if this place was the Theatre in Shoreditch, it probably had no real cover above its apron stage (the Rose theatre was presumably one of the first playhouses to be provided with a protective cover)[39] and in this it was similar to the Curtain. As a result, if the *Dream* was first played at the Theatre, the mechanicals' recurring allusions to the moon did not necessarily allude to the materiality of the playhouse but were more surely meant to recreate particular atmospheric conditions that tried hard to suggest darkness in the traditional daylight performances.

The emphasis put on night and darkness is probably one of the main characteristics of the play and it is all the more significant as darkness was symbolically related to outbursts of temper as well as to love and its mysteries. Under the influence of their 'spleen' (1.1.146), Hermia and Lysander both rebel against the injustice of the old father Egeus and of the law of Athens. The young man compares their frail love to 'the lightning in the collied night' likely to 'unfol[d] both heaven and earth' (1.1.145–6), thus imaging their emotions in terms of the most intense manifestation of stormy weather. Shakespeare then inverts the usual dark/bright paradigms[40] insofar as the lovers' subsequent nocturnal stay in the forest becomes an epiphany that corresponds to the brief and intense flash of lightning[41] mentioned by Lysander. This eerie moment suspended in time is framed by what must be understood as a form of diurnal blackness: indeed, before their elopement, the Athenians describe themselves as engulfed in a dreary daylight assimilated to 'jaws of darkness' that 'devour' (1.1.148) their dream and, at the end of the play, when the couples finally wake up at dawn, what they make out around them is all fuzzy and blurred:

Demetrius: These things seem small and undistinguishable
Like far-off mountains turned into cloud.
Hermia: Methinks I see these things with parted eye
When everything seems double. (4.1.186–9)

Ironically, the morning daylight prevents the lovers from seeing things as clearly as in the dark forest, so that they once again associate their new state of mind – uncertainty and doubt – with the weather, this time misty or cloudy.[42] Contrary to the brief and intense vision provided by 'the lightning in the collied night', their current confused apprehension of 'mountains turned into cloud' reinforces the scepticism[43] of the young couples apparently unable to suspend their disbelief at this stage: was the Athenian forest a locus for radical love experiments? Was it only a dream? Did it all really happen?

If we accept that the wood near Athens is a fantasised image of Shakespeare's England, whose skies were rarely clear in those days,[44] the often harsh weather conditions that prevailed during many a public performance must have had an enduring influence on the spectators. Indeed, average temperatures in Europe had begun to drop around the beginning of the fourteenth century (after a 400-year warm medieval period)[45] and the 1590s were a notoriously cold period[46] with average temperatures about two degrees Fahrenheit lower than in the previous centuries.[47] These difficult conditions must then have found a striking echo in Titania's complaints about the rainy weather of fairyland. As the *Dream* repeatedly depicts poor meteorological conditions, the persisting rain or cool temperatures that were likely to attend its performances certainly contributed to arouse, among the audience, real feelings of empathy with the plight of the young lovers plodding along in the damp Athenian night.

A pessimistic cosmology

Theologians were then keen to supply religious explanations for natural weather disturbances. Luther, for example, argued that the Fall of Adam affected not only man but also nature in general. The whole of God's creation was said to have degenerated into a corrupt nature (*natura corrupta*) and, in his reading of Genesis (3:1), he explains that '[a]ll these things are deformed after the (original) sin, so that all creatures, even the sun and the moon looked as if they were put in a sack'.[48] With the subsequent destruction of the world by the Flood, the situation became even worse. Seeing the deluge as a catalyst for speeding up the world's decay was a fairly common view in Renaissance thinking, especially in the tradition of the pessimistic cosmologies that presented the earth as an ageing world.[49]

In *A Midsummer Night's Dream*, this belief is hinted at in Titania's complaint about the poor climate and the natural catastrophes that plague the play's environment. In this monologue – often seen, today, as the foreshadowing of a vibrant ecological creed[50] – she itemises a catalogue of disasters, including fog, flooding, failed crops, miserable flocks, polluted air, bad temperatures and rheumatic diseases, all due to the wet climate:

> Therefore the winds, piping to us in vain,
> As in revenge, have sucked up from the sea
> Contagious fogs which, falling in the land
> Have every pelting river made so proud
> That they have overborne their continents.
> The ox hath therefore stretched his yoke in vain,
> The ploughman lost his sweat, and the green corn
> Hath rotted ere his youth attained a beard.
> The fold stands empty in the drownèd field,
> And crows are fatted with the murrain flock.
> The nine-men's-morris is filled up with mud.
> [. . .]
> Therefore the moon, the governess of floods,
> Pale in her anger washes all the air,
> That rheumatic diseases do abound;
> And thorough this distemperature we see
> The seasons alter: hoary-headed frosts
> Fall in the fresh lap of the crimson rose,
> And on old Hiems' thin and icy crown
> An odorous chaplet of sweet summer buds
> Is, as in mockery, set. The spring, the summer,
> The childing autumn, angry winter change
> Their wonted liveries, and the mazèd world,
> By their increase, now knows not which is which; (2.1.88–114)

An early annotator – perhaps William Johnstoune, but the evidence is scant – of the 1623 First Folio noted in the margins: 'Change of seasons of the yeere to the preiudice of mens health and food.'[51] The passage must have caught the attention of many a reader. Indeed, in this resonant foul-weather speech shot through with allusions to

Edmund Spenser's *Shepheardes Calender*,[52] one finds a number of ideas congruent with the learned views of the time as well as topical lines whose immediacy must have struck the groundlings who were then going through a period of particularly bad weather. The 'pale' celestial body described here and which is correlated to torrential rains ties in with Leonard Digges's 1555 observations on the colours of the moon. 'The *Moone* pale, or somewhat inclined to blacke, obscure or thick, threatnith rayne,' he explains.[53] To the first audiences of the play who were used to the cold and wet summers of the mid-1590s and to the bad harvests that plagued the country from 1591 to 1597, this cataclysmic description was probably no surprise.[54] If we look at Simon Forman's account in his diary, the wet weather of the year 1594 seems to have proved particularly depressing:

> This moneths of June and July were very wet and wonderfull cold like winter, that the 10 dae of Julii many did syt by the fyer, yt was so cold; and soe was yt in Maye and June; and scarce too fair dais together all thay tyme, but y rayened every day more or lesse. Yf yt did not raine, then yt was cold and cloudye.[55]

In his *Annals*, John Stow also reports that, in 1594, it rained almost continually throughout the realm from May until St James's Day (25 July):

> This yeere [1594] in the month of May, fell many great showres of raine, but in the moneths of June and July, much more: for it commonly rained everie day, or night, till S. James Day, and two daies after together most extreamly, notwithstanding in the moneth of August there followed a faire harvest, but in the moneth of September fell great raines, which raised high waters, such as staied the carriages, and bare downe bridges[56]

Things worsened the following year, at the time when the *Dream* was presumably composed, as Stow notes that, in 1595, 'bitterly cold temperatures prevailed from April 20 until the end of May',[57] ruining the harvests and causing dearth and public unrest:

> The 20. of Aprill being Easter day in the afternoone about five of the clock, it thundered and lightened sore, with some raine, notwithstanding the weather very colde, and so continued the rest of that moneth, and also the moneth of May.

> This yéere by meanes of the late transporting of graine into forraine countries, the same was here growen to an excessive price, as in some places from fouretéene shillings to foure markes the quarter, and more, as the poore did féele, for all thinges else, whatsoever was sustenance for man, was likewise raised without all conscience and reason.[58]

A similar dearth is suggested in the *Dream*, when Titania bemoans the desolation of the 'drownèd field' (2.1.96). It is, of course, up to the spectator to decide whether her complaint is sincere and thus exonerates her from any sort of interference with the weather, or if it is nothing but the hypocritical posture of 'a spirit of no common rate' (3.1.145), as she portrays herself. Her lines actually capitalise on the 'trope of the evil north', a theme 'ubiquitous in the literature of medieval England'[59] and still found in a number of early modern texts.

In connection with this trope, the *Dream*'s ambiguous queen must have called to mind some evil fairy not unrelated to the witches often suspected of practising weather magic[60] – a belief still firmly held by many of Shakespeare's contemporaries, including educated ones such as the Swiss traveller Thomas Platter who, in 1599, remained convinced that witches were 'still the cause of much hail, thunderstorm &etc'.[61] Throughout the comedy, Titania seems to play with the meteorological conditions near Athens by altering them at the drop of a hat: if she first triumphantly emphasises the bitterness of the climate in 2.1, she proudly asserts a few moments later that '[t]he summer still doth tend upon [her] state' (3.1.146). This shows that, if she personally delights in perfect summer weather, she can also, like the Weird Sisters in *Macbeth*, conjure up fogs and floods in order to blight the earth. In this perspective, Shakespeare's queen of fairies could be regarded as a female Faustus figure, commanding at her will '[a]ll things that move between the quiet poles' (*Doctor Faustus*, A-Text, 1.1.58), thereby bringing about 'whirlwinds, tempests, thunder, and lightning' (ibid. 2.2.165). In line with this, her soliloquy looks like a blasphemous anti-prayer, for if she perfectly depicts the Athenians' pathetic situation, she neither begs for clemency nor hopes to have her sins absolved. On the contrary, a few lines further down, she insists on her right to keep the little changeling boy for herself, thus cynically accepting the possibility of another spell of foul weather in retaliation against her refusal to obey Oberon.

Her attitude must of course be seen in the specific context of the period. In Shakespeare's England, church prayers were often meant to assuage the effects of the harsh weather. One typical instance is found in *Certayne Godly exarcises meditacions and prayers*, an anonymous pamphlet dating from 1565 and which contains 'A prayer for fayre weather':

> In thy handes (O lord) are both heaven and earth: So gouerne them therfore, we besech thee for thy names sake, that we maye haue seasonable weather to receaue the fruites of the earth. Stay the immoderate plentie of rayne and vnseasonable weather, and powre rather vpon vs the plentyfull dewe of thy spyrite, that we may daily grow to the perfection of christen menne. Take awaye from vs our stony hartes, & geue vnto vs in the stede therof, hartes of fleshe, that we may cease at the length thus to prouoke thy wrathe to brynge vpon vs eyther this so great a plague as the losse of the fruites of the earthe, eyther elles other plagues of thy iuste iudgement.[62]

The unreformed English Church already had masses said for fair weather and, in the *Book of Common Prayer* (1549), this tradition was maintained, since one prayer for rain – rather short – and one for fair weather – significantly longer – are included in it. According to Brian Cummings, the two compositions turn out to be 'quintessential' ones in the *Book of Common Prayer*.[63] The second one reads as follows:

> O Lorde God, whiche for the sinne of manne, didst once drowne all the worlde, except eight persons, and afterward of thy great mercye, didste promise never to destroy it so agayn: We humbly beseeche thee, that although we for oure iniquities have woorthelye deserved this plague of rayne and waters, yet upon our true repentaunce, thou wilt sende us such wether wherby we may receive the fruites of the earth in due season, and learne both by thy punishment to amende our lives[64]

At the turn of the seventeenth century, country parishioners probably heard similar prayers several times a month, which prompted John Earle to satirise those who could recite no more than two prayers, namely those for 'Rayne or faire weather'.[65] In them, as might be expected, harsh weather was regarded as a just punishment for sinners, and the improvement of meteorological conditions was presented as depending on their willingness to repent. However, in the pagan world of the *Dream*, the characters never express the faintest wish to amend their lives, least of all the strong-willed fairy queen, Titania.

Encompassing religious, magical, archetypal and topical concerns, Titania's lines in act 2, scene 1 superimpose several layers of meaning which not only provide the spectators with realistic details depicting a threatening and threatened microcosm but also invite them to adopt a broader perspective on the macrocosm at large. As a result, the fairy queen offers a syncretic description reminiscent of the worldwide inundation described in Genesis (6–9) and paving the way for Francis Bacon's depiction of Tartarus in the *Wisdom of the Ancients* as 'a place signifying perturbation [. . .] between the lower superficies of heaven and the centre of the earth, in which all perturbation and fragility and mortality or corruption are frequent'.[66] Away from the Garden of Eden, Shakespeare's Athenian green world works as an inverted pastoral marred by chaos and bad weather.

The upside-down world depicted by Titania resurfaces later on in the play when Hermia, trying to make sense of Lysander's sudden desertion, alludes to the antipodes:[67]

> The sun was not so true unto the day
> As he to me. Would he have stolen away
> From sleeping Hermia? I'll believe as soon
> This whole earth may be bored, and that the moon
> May through the centre creep, and so displease
> Her brother's noontide with th'Antipodes. (3.2.50–5)

Depicting an anthropomorphic – and thereby chaotic – universe which the pathetic fallacy helps to endow with humours and human feelings, Hermia's lines are not simply amusing here: they also have disquieting undertones as they imply the possibility of night at midday when they conjure up the idea of a terrestrial globe through whose centre the moon could creep. This image, absurd as it may seem, is a graphic way of alluding to cosmic (and comic) disorder and to a form of apocalypse. The young woman, however, hardly believes that such a topsy-turvy situation will ever occur. By contrast, Titania is fully alert to the power of natural forces. Her speech shows that, contrary to the incredulous Hermia, she has already taken stock of the situation, aware as she is of the role she has played so far, and she therefore lucidly blames herself and Oberon for being the causes of all these climatic evils: '[w]e are their parents and original' (2.1.117).

In other plays of the period, by contrast, characters enduring particularly unseasonable weather often present themselves as passive, powerless subjects. In Samuel Daniel's pastoral drama *The Queenes*

Arcadia (c. 1606), for instance, two shepherds voice their concerns about climate change and they clearly see themselves as the main victims of the harsh weather affecting England:

> And me thinkes too our very aire is changed
> Our wholesome climate grown more maladive
> The fogges and syrene offend us more
> (Or we made think so) than they did before.[68]

Daniel's infectious 'fogges' remind us here of the 'contagious fogs' (2.1.83) that are being sucked from the sea in the *Dream*.[69] Shakespeare's comedy also makes it clear that Robin himself is able to produce fog to lead the lovers astray ('The starry welkin cover thou anon / With drooping fog as black as Acheron', Oberon orders him in 3.2.356–7). These repeated allusions to fog in the drama of the period probably owe something to the Londoners' obsession with the thick fog and smoke which was then partly due to the Tudor shipping of coal down the east coast and up the Thames which, at the turn of the seventeenth century, increasingly hindered the vision of the city dwellers and constricted their daily movements.[70] In the case of *A Midsummer Night's Dream*, it is likely that, during some outdoor performances of the play, the fog scene, which did not necessarily require the use of special effects but must have involved some particular stage business with actors pretending not to see one another, may have coincided with a foggy afternoon.

In the passage from *The Queenes Arcadia* quoted above, the word 'fog' is followed by 'syrene', a term now obsolete but which then referred to a 'light fall of moisture or fine rain after sunset in hot countries [. . .] formerly regarded as a noxious dew or mist' (*OED*, 'serene', n.)[71] and which is similarly evocative of a potentially harmful climate likely to weaken even the strongest of men. Underlining the fallen state of mankind, this lament about the poor English weather distantly echoes Titania's complaint in the *Dream*, except that the queen of fairies, far from incriminating the bad weather conditions for the topsy-turvy state of her green realm, claims joint responsibility with her husband for the disruption of natural cycles. As in John Heywood's 1533 *Play of the Weather*, where the rough weather results from a general disagreement between Saturn (the god of frost and snow), Phebus (the god of the sun), Aeolus (the god of winds) and Phebe (the goddess of rain), Shakespeare's two main fairies/deities blame themselves for bringing about environmental destruction.[72] They are, as Titania acknowledges, the sole and direct cause of the disturbance of the seasons and of the floods and storms that have ruined the agrarian economy.

The bad weather is thus not the source but the aftermath of marital discord. Up to a point, it is Titania's lack of discipline and self-restraint[73] that has generated the climatic chaos that she describes. In the *Symposium*, Plato emphasises such causal links as he associates ideal love with a temperate climate:

> When the elements to which I have already referred – hot and cold, wet and dry – are animated by the proper species of Love, they are in harmony with one another: their mixture is temperate, and so is the climate. Harvests are plentiful; men and all other living things are in good health.[74]

From a Platonic perspective, the absence of harmonious love in *A Midsummer Night's Dream* may well be the direct cause of the disruption of climate in the Athenian woods. The mad pranks of Puck further illustrate the impossibility of concord in such an unstable environment. Instead of restoring harmony, he wreaks havoc among the couples and makes them err in the labyrinthine forest near Athens. In an intriguing way, Puck prefigures the Cupid described by Francis Bacon in the *Wisdom of the Ancients*, a text that presents the ever-shifting god of love as an atom.[75] Cupid – and Puck, for that matter – is always 'elegantly described with perpetual infancy or childhood' and, similarly, 'the first seeds of things or atoms, they are little and diminute, and always in their infancy'.[76] In other words, like the atoms in Lucretius' first-century poem *De rerum natura*,[77] Puck seems to move around in a vacuum where he freely exerts his power on the outside world. Ironically, when she finds herself under the influence of the love juice, Titania wishes that her ass-headed lover were purged of his 'mortal grossness' and that he could 'like an airy spirit go' (3.1.151–2). She imagines the impossible and comic metamorphosis of the bulky, bumbling, earth-bound Bottom into a tiny atom. So, anticipating *King Lear*, another anti-pastoral play, albeit much darker than the *Dream*, Shakespeare's early forest comedy may tentatively be read as something of an atomist text *avant la lettre*.[78]

Melting, dissolving, creating

Humidity and coldness are prevalent not only in the not-so-idyllic Athenian woods, but in the world at large. In the west, a flood has occurred, 'fill[ing] up with mud' the 'nine men's morris' (2.1.99), while in the east, where Titania's Indian votaress used to live, the 'wanton wind' (2.2.129) keeps blowing. Moisture seeps into every little corner of the comedy's world and the metaphoric language of the play is replete with allusions to humidity and to bodily fluids. Amusingly, when he hears that Pyramus is '[a] lover, that kills himself, most gallant, for love' (1.2.20), Bottom concludes that the role 'will ask some tears in the true performing of it' (1.2.21–2). Saddened by her father's hostility towards Lysander, Hermia mentions 'the tempest of [her] eyes' (1.1.129), and Helena laments the behaviour of her beloved Demetrius, who 'hailed down oaths that he was hers' (1.1.243)[79] before suddenly changing his mind:

> And when this hail some heat from Hermia felt,
> So he dissolved, and showers of oaths did melt. (1.1.244–5)

This general wetness turns the forest into a female locus (dis)ordered by Titania since, as 'leaky vessels',[80] women's moist, lunar physiology accounted for their alleged lack of reason. In *The Secret Miracles of Nature*, originally published in 1561, the Dutch physician Levinus Lemnius writes that

> women are subject to all passions and perturbations [. . .] a woman enraged, is besides her selfe, and hath not power over her self, so that she cannot rule her passions, or bridle her disturbed affections, or stand against them with force of reason and judgement[81]

Even though it was only translated into English during the second half of the seventeenth century, Lemnius's Latin work began to circulate in the 1560s and its views became highly influential among scholars and men of science. We do not know whether Shakespeare was aware of this particular treatise but, in the *Dream*, nature clearly represents female agency. It is in essence a feminine and a destructive force. Even Hiems, the male Winter King embodying old age, seems mocked and unmanned by Nature:

> And on old Hiems' thin and icy crown
> An odorous chaplet of sweet summer buds
> Is, as in mockery, set. (2.1.109–11)

So these disorders, being linked to women's misrule,[82] are also part and parcel of the upside-down world of comedy and, as such, they trigger the carnivalesque pranks of the lovers and the mechanicals' unwitting buffoonery later on in the play.

The characters all tend to describe their feelings in geographic and meteorological terms. While Titania depicts earthly disorders before alluding to an eastern seascape, Puck says that he can girdle the earth '[i]n forty minutes' (2.1.175–6), thus telescoping time and space. The goblin's reference to forty minutes remains unexplained but one may notice that the figure forty coincides with the number of days of abstinence during Lent. We learn further down that Oberon and Titania are able to 'compass the globe' (4.1.96–7) and that 'the moon / May through the centre [of the earth] creep' (3.2.53–4). Such references to a purely poetic geography are intertwined with multiple allusions to weather elements. When Demetrius is under the spell of the flower's juice, he mentions '[t]hat pure congealed white – high Taurus's snow / Fanned with the eastern wind' (3.2.142–3) and, once he has awaked to his former self, he can only say that his 'love to Hermia, / Melted as the snow' (4.1.164–5), subsequently comparing the events of the night to 'far-off mountains turnèd into clouds' (4.1.187). The image of the melting snow used to render the loss of one's name and identity is also found in *Richard II* when the king declares:

> [. . .] Alack the heavy day,
> That I have worn so many winters out
> And know not now what name to call myself!
> O, that I were a mockery king of snow,
> Standing before the sun of Bolingbroke
> To melt myself away in water-drops! (4.1.247–52)

Such allusions to the evaporating memories of the night show that 'we are such stuff / As dreams are made on' and that 'our little life / Is rounded with a sleep', as Prospero puts it in *The Tempest* when he explains that the actors of the masque are now 'melted into air, into thin air' (4.1.148–58). Here Demetrius suggests that maturity and self-knowledge may be reached after fantasy and passion have faded away and, like Shakespeare's magus, an expert in weather magic in front of Ferdinand and Miranda, he assimilates the play to a cloudy, dream-like vision.[83]

Together with *The Tempest*, *A Midsummer Night's Dream* is certainly Shakespeare's most ostentatiously climatic play. With its focus on the power of the moon, on female agency, on seasonal disturbance as well as on rain and wetness, it presents us with a dense network of meteorological themes that reveal how much, in the period, weather issues could be put to use in order to reflect a gendered vision of life and of the world. In the *Dream*, it is Titania's surprisingly human behaviour and, more generally, the violence of female impulses that account for the harshness of the climate.[84] Yet if Shakespeare does generally stick to gender stereotypes in the comedy, he nonetheless reverses the traditional modes of thinking about climate. By presenting the weather as a *consequence* rather than as a *cause* of chaos, he deconstructs the traditional determinist perspective, so as to make the sky the principal source of human troubles. His rearticulated framework is by no means definitive though, as the playwright here tests a simple hypothesis only to present opposite arguments in his other plays. The *Dream* thus works as a laboratory of ideas and as a catalyst for theatrical freedom. Theseus's soliloquy, at the beginning of act 5, makes the audience aware that this is all a comedy and certainly not a meteorological treatise. Resorting to cosmological imagery enhanced by a chiasmic structure, the Duke of Athens turns the poet into a demiurge of sorts:

> The poet's eye, in a fine frenzy rolling,
> Doth glance from heaven to earth, from earth to heaven,
> And as imagination bodies forth
> The forms of things unknown, the poet's pen
> Turns them to shapes and gives to airy nothing
> A local habitation and a name (5.1.12–17)

Here, it is no longer the sky regulating men's behaviour or men's behaviour disturbing the natural order that is given pride of place, but the poet's ability to fathom the sky and fashion the elements according to what his imagination suggests. The vertical up and down movement from earth to heaven and from heaven to earth allows him to create things *ex nihilo* as it were, so that 'airy nothing' presents itself as a blank page on which he can pen whatever fantasy pleases him. Far from the current contemporary meteorological determinism and free from any teleological perspective, Shakespeare vicariously reclaims a form of creative liberty armed with the Adamic power to give 'things unknown' a 'local habitation and a name'. Once this has been accepted by the audience, he can explore a variety of contexts, situations and atmospheres. As a result, if in *A Midsummer Night's Dream*, a comedy under the aegis of the moon, the playwright calls attention to a particularly moist climate in connection with female power, he provides us with another vision of summer in *Romeo and Juliet*. This time, he foregrounds Verona's blazing heat in difficult climatic conditions that result in a crescendo of male aggressiveness as well as in the general flare-up of violence.

Notes

1. See David Wiles, *Shakespeare's Almanac: A Midsummer Night's Dream, Marriage and the Elizabethan Calendar.*
2. John Rainoldes, *Th'Overthrow of Stage-Playes*, p. 118: 'Where, when at midsummer, in very hott weather, Andromeda (a Tragedie of *Euripides*) being played, manie brought home a burning ague from the theater.'
3. Richard Marienstras, *New Perspectives on the Shakespearean World*, p. 163.
4. Philippa Berry, 'Between Idolatry and Astrology: Modes of Temporal Repetition in *Romeo and Juliet*', p. 380. For more on the role of the sun in *Romeo and Juliet*, see below, Chapter 2, pp. 57–79.
5. Quoted in Chris Laoutaris, *Shakespearean Maternities: Crises of Conception in Early Modern England*, p. 136.
6. Cf. 2.1.14–15, when the fairy declares: 'I must go seek some dewdrops here, / And hang a pearl in every cowslip's ear.'
7. On the presumed bawdiness of the early modern wind, see below, Chapter 4, n. 129, p. 148.
8. Up to a certain point, then, the '*vot'ress* of [Titania's] order' (2.1.123, my emphasis) is described as a 'mother' (2.1.123) but one who died in childbirth (2.1.135). She is provided with a paradoxical form of fertility vaguely reminiscent of that of the Virgin Mary in her 'immaculate conception' through the breath of the Holy Ghost – which is also the case of Titania, the moon goddess. Besides, the word which Shakespeare uses to refer to Elizabeth I in Oberon's speech also serves to designate Titania's friend. The Indian votaress and the 'imperial vot'ress' (2.1.163) thus become almost interchangeable figures endowed with some sort of lofty virginal power. I would suggest that Queen Elizabeth I, the Indian votaress and Titania actually form a lunar trinity reminiscent of 'the triple Hecate's team' mentioned by Robin at the end (5.1.375). In this perspective, the late Indian votaress would be the equivalent of Proserpina in Hades, the imperial votaress the counterpart of Diana on earth, while Titania could be identified with Cynthia in heaven.
9. Aristotle, *History of Animals* (*Historia animalium*), vol. 9, p. 98: 'The mare is said also about this time to get wind-impregnated if not impregnated by the stallion, and for this reason in Crete they never remove the stallion from the mares.' For further details on the subject, see Conway Zirkle, 'Animal Impregnated by the Wind'.
10. Philemon Holland (trans.), *The historie of the vvorld*, 'The eighth Booke', chap. LXII, p. 222. In the third century AD, similar arguments were used by Christian writers such as Lactantius to explain that the womb of the Holy Virgin 'was made fruitful' thanks to 'the wind and the breeze'. See Lactantius, *The Divine Institutes*, Book IV, chap. XII, in *The Ante-Nicene Fathers: The Writings of the Fathers Down to A.D. 325*, vol. 7, p. 110.

11. In John Webster's *The Duchess of Malfi*, Ferdinand, like Pliny, imagines that Castruccio's 'Spanish jennet' (1.1.112) 'was begot by the wind' (1.2.114–15). See René Weis (ed.), *The Duchess of Malfi and Other Plays*, p. 112.
12. On this, see for instance George Chapman's 1607 *Bussy d'Ambois*: 'His great heart will not down, 'tis like the sea / That partly by his own internal heat, / Partly the stars' daily and nightly motion, / Ardour and light, and partly of the place / The divers frames, and chiefly by the Moon, / Bristl'd with surges, never will be won' (1.2.138–43).
13. Mario DiGangi, *Sexual Types: Embodiment, Agency, and Dramatic Character from Shakespeare to Shirley*, p. 81. This analysis also applies to Cleopatra whose ambivalence is suggested by her being compared, in Shakespeare's *Anthony and Cleopatra*, to Zeus impregnating Danaë with a golden shower (2.5.46–7). See below, Chapter 6, p. 182.
14. Valerie Traub, *The Renaissance of Lesbianism in Early Modern England*, p. 68.
15. Ovid, *Metamorphoses*, trans. Arthur Golding, p. 124.
16. 'Thisbe, being perhaps the more ardent, was the first to slip away from her parents. Wrapped in a cloak, she left her father's home alone in the dead of the night and, with the moon lighting her way, went fearlessly into the grove.' See Giovanni Boccacio, *Famous Women* (*De mulieribus claris*, c. 1375), chap. XIII, pp. 56–7.
17. See *The problemes of Aristotle*, 'The first Probleme', chap. 'Of womens monthly termes', n.p.: 'Question. *Why haue not women their flowers all at one time of the moneth* [. . .]*? / Answer.* By reason of their diuers complexions [. . .] A woman which is of a sanguine complexion, hath her flowers in the first quarter, a cholerike in the second, a melancholy in the third, and so in the rest.'
18. Ibid. 'The first Probleme', [no chap. indicated], Answer to the question '*Why haue women longer hayre then men?*', n.p.
19. *Heere beginneth the Kalender of sheephards*, 'Of the Moone', [sig. LV].
20. Theophrastus, *Concerning Weather Signs* (*De signis tempestates*), in *Enquiry into Plants*, vol. 2, p. 397.
21. Richard Wilson, *Shakespeare in French Theory: King of Shadows*, p. 153. See also Jay L. Halio, *Shakespeare in Performance: A Midsummer Night's Dream*, p. 6.
22. Helen Hackett, '*A Midsummer Night's Dream*', p. 348.
23. Jeanne Addison Roberts, *The Shakespearean Wild: Geography, Genus and Gender*, p. 24.
24. David Wiles, *Shakespeare's Almanac*, p. 73.
25. Ibid. p. 75.
26. Ibid.
27. Ibid.
28. In *Measure for Measure*, Claudio imagines hell as a 'thrilling region of thick-ribbèd ice' (3.1.123). On the origins of hell as a frozen place, see Jacques Le Goff, *Birth of Purgatory*, p. 50.

29. Isodorus of Seville, in his seventh-century *Etymologiae*, notably placed Eden in the east and, even though no consensus was reached on the exact location of paradise, Asia was still often associated with paradise on early modern maps, as on Abraham Ortelius's 1567 map of India, *Asia orbis partium maximae nova descriptor* (*New description of Asia, the largest part of the earth*).
30. François Laroque, *Shakespeare's Festive World: Elizabethan Seasonal Entertainment and the Professional Stage*, p. 141.
31. On this, see Richard Dutton, *Shakespeare, Court Dramatist*, p. 61. The Folio text contains acts and several stage directions and replaces Philostrate, the Master of the revels, with Egeus. So it is likely that Q1 was written first and revised some time later, between 1608 and 1623.
32. Ibid. p. 63.
33. On this, see also Jay L. Halio, *Shakespeare in Performance: A Midsummer Night's Dream*, p. 10.
34. Peter Holland, 'Introduction', in *A Midsummer Night's Dream*, pp. 111–12.
35. Martin White, *The Shakespeare Handbooks: A Midsummer Night's Dream*, p. 6.
36. Ibid.
37. Ibid.
38. John Gillies, *Shakespeare and the Geography of Difference*, p. 70.
39. It must be acknowledged, however, that 'the only contemporary illustration – an engraved "View of the Cittye of London from the North towards the South", made around 1597–8 – is too unclear to provide any detail of its structure. It has been suggested that – especially if it had no cover – the Theatre's stage might have been portable, to allow the yard to be cleared for "feats of great activity".' See Martin White, 'London Professional Playhouses and Performances', p. 301.
40. A similar inversion of the traditional dark/light opposition occurs in the final couplet of Sonnet 65 ('O none, unless this miracle have might: / That in black ink my love may shine bright') and in *Anthony and Cleopatra*, when Lepidus says of Anthony that '[h]is faults in him seem as the spots of heaven, / More fiery by night's blackness' (1.4.12–13). Here the character's vices are associated with light and his better qualities and normal self with night.
41. Interestingly, the word 'lightning' is first used in the canon in a purely metaphorical sense. In early plays like *Titus Andronicus* (2.1.3) and *Love's Labour's Lost* (4.2.116) it is associated with Jupiter, while in *Venus and Adonis* (l. 348), *A Midsummer Night's Dream* and *Romeo and Juliet* (where it crops up no fewer than four times) it suggests the idea of quickness. But as the material conditions of early modern performance evolved, the word was increasingly resorted to for its connections with the storm. In later plays such as *King Lear* and *The Tempest*, 'lightning' must then be taken literally as referring to the sudden frightening flashes of light across an angry sky.

42. On clouds, see below, Chapter 6, pp. 192–7, and Chapter 7, pp. 242–4.
43. On the association of clouds with scepticism, see Donne's 1622 Easter sermon preached 'from his newly appointed post at St Paul's'. For his use of 'clouds' to 'cast doubt on human knowledge', see Melissa M. Caldwell's *Skepticism and Belief in Early Modern England: The Reformation of Moral Value*, p. 170.
44. As H. H. Lamb points out in *Climate, History and the Modern World*, p. 249, '[s]urveys of the cloud cover in European representational style paintings of various periods have shown averages of nearly 80 per cent cover in pictures from the period 1550–1700, 50–75 per cent at various times in the eighteenth century, 70–75 per cent in Constable's and Turner's time (1790–1840) and 55–70 per cent in the twentieth century.'
45. As already noted earlier, the world would only start warming again in the early 1800s.
46. Brian Fagan, *The Little Ice Age: How Climate Made History, 1300–1850*, p. 90.
47. This drop in temperature was significant enough to leave Iceland completely surrounded by ice and to freeze the Thames in England, while the canals in Holland were routinely frozen. On this, see Emily Oster, 'Witchcraft, Weather and Economic Growth in Renaissance Europe', p. 218. See also the numerous Dutch paintings dating back to the late sixteenth and early seventeenth centuries that represent skating figures on frozen canals (one of the most famous being the *Winter Landscape with Iceskaters* painted around 1608 by Hendrick Avercamp).
48. 'Haec omnia post peccatum deformata sunt, ita ut creaturae omnes, etiam Sol et Luna quasi saccum induisse videantur.' Martin Luther, 'Vorlesungen über 1. Mose', p. 68. Quoted in Michael Kempe, 'Noah's Flood: The Genesis Story and Natural Disasters in Early Modern Times', p. 152.
49. Michael Kempe, 'Noah's Flood', p. 152.
50. Henry S. Turner, *Shakespeare's Double Helix*, p. 34.
51. Akihiro Yamada (ed.), *The First Folio of Shakespeare, A Transcript of Contemporary Marginalia in a Copy of the Kodama Memorial Library of Meisei University*, p. 45, N3, a (through-line-numbers: 473–538). Online edition at <http://shakes.meisei-u.ac.jp/ALL.html#45> (last accessed 18 April 2018).
52. On Spenser's influence on the *Dream*, see Harold Brooks (ed.), *A Midsummer Night's Dream*, pp. xxxiv–xxxix and pp. lviii–lxii. The 'hoary-headed frosts' lamented by Titania seem to echo Spenser's 'My head besprent with hoary frost' (December, l. 135), as James P. Bednarz remarks in 'Imitations of Spenser in *A Midsummer Night's Dream*', p. 90.
53. Leonard Digges, *A prognostication of right good effect fructfully augmented*, sig. B.
54. Brian Fagan, *The Little Ice Age*, p. 94.
55. Quoted by Margrethe Jolly in *The First Two Quartos of Hamlet: A New View of the Origins and Relationship of the Texts*, p. 162.

56. John Stow, *The annales of England* [. . .] *continued*, p. 1278. This passage is summarised in Todd A. Borlik, *Ecocriticism and Early Modern English Literature: Green Pastures*, p. 120.
57. Todd A. Borlik, *Ecocriticism and Early Modern English Literature*, p. 121.
58. John Stow, *The annales of England*, p. 1279.
59. Colleen Franklin, '"An Habitation of Devils, a Domicill for Unclean Spirits, and a Den of Goblings": The Marvellous North in Early Modern English Literature', p. 28.
60. In this regard, one should notice how influential Reginald Scot's 1584 *Discouerie of witchcraft* must have been on Shakespeare. In this treatise, Scot emphasises (and dismisses) the connection between harsh weather and witchcraft practices (The Second Booke, chap. 9, p. 33: '[witches] ride and flie in the aire, bring stormes, make tempests, etc.'). On weather-working witchcraft, see below, Chapter 7, pp. XX.
61. In his diary (*Thomas Platter's Travels in England*, p. 189), Thomas Platter observes that there are numerous witches in England, before adding (somewhat erroneously, since a second witchcraft act increasing penalties was passed in 1563) that 'report goes that [the English] do not punish them with death there, because the queen was once on the water, and a number of witches had planned her destruction in a storm, but another witch prayed for her and held off the tempest, as she herself confessed' (25 September 1599, London). Platter alludes here to the North Berwick witches whose trials took place in 1590 and 1591.
62. *Certayne Godly exarcises meditacions and prayers*, sig. Dd. r–v.
63. Brian Cummings (ed.), *The Book of Common Prayer: The Texts of 1549, 1559, and 1662*, 'Notes to Communion', p. 704.
64. Ibid. 'For fayre whether', p. 38.
65. Quoted in John Craig, 'Bodies at Prayer in Early Modern England', p. 179.
66. Francis Bacon, *Wisdom of the Ancients*, chap. XII, p. 232. The English version of Bacon's treatise dates from 1619, but the book was initially published in Latin in 1609 as *De sapientia veterum*.
67. The 'antipodes' were discovered at the beginning of the sixteenth century (Amerigo Vespucci's 1503 *Mundus novus* made it clear that the southern antipodes did exist) and the word referred to the regions situated directly opposite England, on the other side of the globe.
68. Quoted by Todd A. Borlik, *Ecocriticism and Early Modern English Literature*, p. 128. Borlik compares this excerpt from Daniel's play to 'Lear's harangue imploring [. . .] the rains to deluge the earth' which, according to him, may have sounded like 'an inversion of the Rogation liturgy'.
69. Cf. Thomas Middleton's toxic 'fog or mist raised from Error' in *The Triumphs of Truth* (1613), in *The Collected Works*, p. 973, ll. 495–6. In Middleton's pageant, however, fog is a distinctively urban phenomenon.

70. On the filthy air and fogs already blighting the fossil-fuelled city of London in the early modern era, see William M. Cavert, *The Smoke of London: Energy and Environment in the Early Modern City*. Further information on fog (then frequently associated with deception) in connection with the early modern stage may be found in Alan Dessen's 'Mist and Fog on the Elizabethan and Jacobean Stage'.
71. The first use of the word quoted by the *OED* dates back to 1591 as John Florio mentioned the 'Seren' in his *Second Frutes*: 'The Seren neuer hurts a man in these colde countries.'
72. For a study of Heywood's interlude from an ecocritical perspective, see Jennifer L. Ailles, 'Ecocritical Heywood and *The Play of the Weather*'.
73. On women's supposed lack of self-control, see Michael C. Schoenfeldt, *Bodies and Selves in Early Modern England: Physiology and Inwardness in Spenser, Shakespeare, Herbert, and Milton*, p. 36.
74. Plato, *Symposium*, p. 471.
75. Francis Bacon was familiar with the atomistic philosophy of Democritus, as is confirmed by his quotations from Lucretius in several of his works. In chap. XVII of his *Wisdom of the Ancients*, this is indeed how he tries to make sense of the mythical allegory of Cupid: '[Love's] virtue is such that it [i.e. archery] works upon a distant object, because that whatsoever operates afar off seems to shoot, as it were, an arrow. Wherefore whosoever holds the being both of atoms and vacuity must needs infer that the virtue of the atom reacheth to a distant object; for if it were not so, there could be no motion at all, by reason of the interposition of vacuity, but all things would stand stone still, and remain immovable' (p. 241).
76. Ibid.
77. Shakespeare probably became acquainted with this philosophical poem thanks to a number of early modern texts, among which John Florio's translation of Montaigne's *Essays* was the most important one. For his quotations from the *De rerum natura*, Montaigne himself relied on the 1563 edition of Dionysius Lambinus, *Titi Lucretii Cari De rerum natura libri sex*.
78. *As You Like It* is yet another Shakespearean comedy promoting Lucretian philosophy and paving the way for *King Lear*. See below, Chapter 3, pp. 87–92.
79. Here, as elsewhere in Shakespeare, 'hail' connotes exacerbated violence. In the Pseudo-Aristotelian treatise *On the Cosmos*, hail is said to occur 'when a snow-storm condenses and gains weight from the compression so that it comes down faster. According to the size of the pieces broken off [from the clouds] the masses become larger and their motions more violent.' See *Cosmic Order and Divine Power: Pseudo-Aristotle, On the Cosmos*, p. 31. On hail, see also below, Chapter 6, pp. 182–3.
80. For an influential discussion of this notion, see Gail Kern Paster's article 'Leaky Vessels: The Incontinent Women of City Comedy', as well as

her book *The Body Embarrassed: Drama and the Disciplines of Shame in Early Modern England.*

81. Levinus Lemnius, *The Secret Miracles of Nature*, pp. 273–4.
82. Admittedly, misrule also characterises the male figures in the play, since Puck, in particular, causes the general confusion by misapplying the love-in-idleness potion. Yet had Titania behaved differently, he would never have needed to resort to this love charm. By contrast, in the daylight world of the Athenian court, this logic is reversed since misrule then coincides with Theseus, who ironically becomes the 'frantic' lover he himself depicts in 5.1.10. As to the submissive Hippolyta, she is the only one able to detect reason and 'constancy' (5.1.26) in the strange events of the previous night.
83. See below, Chapter 7, pp. 242–4.
84. There are other examples in the Shakespearean canon of the causal links between human misrule and weather disturbances. A case in point is *Julius Caesar*, where the conspirators seemingly trigger a series of strange climatic events ('Nor heaven nor earth have been at peace tonight', Caesar worryingly notices in 2.2.1). One may also think of *Macbeth*, in which a number of disquieting meteorological phenomena are rendered in terms of metatheatrical images (see in particular Ross's remark to the Old Man, 'Thou seest the heavens, as troubled with man's act, / Threatens his bloody stage', 2.4.5–6).

Chapter 2

'[T]he fire is grown too hot!': *Romeo and Juliet* and the Dog Days

These violent delights have violent ends
And in their triumph die like fire and powder,
Which as they kiss consume. (2.5.9–11)

In *A Midsummer Night's Dream*, the playwright blatantly questions the traditional ideas about climate by showing how free will and ill omens may sometimes combine and how far humankind may be blamed for the various climatic disasters plaguing the lives of peasants as well as of the population as a whole. *Romeo and Juliet*, often considered as the *Dream*'s companion piece, develops its own approach to climate, one that incorporates traditional calendar references as well as astrological beliefs. As Carla Mazzio puts it, '[i]t is fascinating to reflect upon *Romeo and Juliet* [. . .] as a crushing tragedy of astrological determinism at a time in England when predictive astrology was being seriously challenged if not denounced outright.'[1] That the play lends itself to such astrological readings seems confirmed by Paris's remark that 'Venus smiles not in a house of tears' (4.1.8), a phrase which alludes both to his ill-timed marriage with Juliet[2] and to the particular influence of the planet Venus in the zodiac.[3] This does say something about the pre-eminence of the 'fated sky' in Shakespeare's first love tragedy, all the more so as the line comes after Romeo's passionate speech about Juliet's eyes, which he compares to '[t]wo of the fairest stars in all the heaven' (2.1.57).

The astrological and climatic determinism of the play works in such a way that its various incidents first seem to take place on a vast celestial stage before they even unfold on the earth. Consequently, Juliet's life is as dependent on Verona's 'fated sky' as it is on the calendar of its earthly events. If we accept the idea that the seasons are the 'climatic presentation of the passage of time',[4] the month of July here takes on a particular significance through images that associate July with Juliet and underline the stifling summer heat which paralyses the city of Verona. Shakespeare's plot emphasises *courte durée* in opposition to the *longue durée* prevailing in late romances like *The Winter's Tale*, so that, in this early tragedy, summer appears as a climactic time of the year. According to the mythographer Stephen Batman, this season 'is hot and drye, and beginneth when the sunne is in the first part of the signe that is called Cancer: and then the sonne is most hight above us, and beginneth to passe downe little and little, even to the ende of the signe, that is called Virgo'.[5] These characteristics make up the play's climatic background, which probably upset Elizabethan playgoers as '[e]xtremely dry seasons were generally believed to be more sickly than wet [ones]', as Mary J. Dobson reminds us in her study of death, disease and environmental influences in early modern England.[6] Yet the scorching heat and the drought that earmark Romeo and Juliet's birthplace are more than simple weather-related features; they create a breeding ground and a poetic reservoir for the play's flamboyant rhetoric, hysterical sentiments and ardent passions.

Summer and winter seasons

Being reduced to just a few hot days, the tragedy dismisses the cyclical time of nature and fertility and replaces it with standstill action and sterility. In this connection, the ghostly presence of the full moon, 'pale with envy' because Juliet is 'far more fair than she' (2.1.48), is certainly not a good omen, while Juliet herself acts rather cautiously when she calls for the advent of 'cloudy night' (3.2.4), more propitious for 'amorous rites' (3.2.8). Complete darkness allowed lovers to hide and an apparently moonless night was then commonly regarded as synonymous with the defeat of Cynthia, the goddess of chastity.

The first attempt at giving the play a broader scope is found in act 1, scene 2, when Capulet must moderate Paris's enthusiasm after the latter has expressed his desire to marry Juliet immediately. Juliet's father had first asked the impatient suitor to '[l]et two more summers wither in their pride / Ere we may think [Juliet] ripe to be a bride' (1.2.10–11), before proceeding with other seasonal references:

Earth has swallowed all my hopes but she;
She's the hopeful lady of my earth.
But woo her, gentle Paris, get her heart,
My will to her consent is but a part;
[. . .]
At my poor house look to behold this night
Earth-trading stars that make dark heaven light.
Such comfort as do lusty young men feel
When well-apparelled April on the heel
Of limping winter treads, even such delight
Among fresh fennel buds shall you this night
Inherit at my house. (1.2.14–30)

Framed by an epanalepsis repeating the word 'earth', the first two lines imply that Juliet had siblings but that she alone has survived. Significantly, she is depicted as the 'hopeful lady' of her father's 'earth', an anagram of 'heart', a word which could just as well make sense in Capulet's sentence (l. 15). Right from the start, the play deals with a situation where the 'heart' ominously gives way to the 'earth', that is, to the tomb. In Capulet's lines, young Juliet is said to be in the spring, or April, of her life, while her ripening years will correspond to the summer[7] of her hypothetical wedding. Capulet's discourse thus points to a discrepancy between his personal timeline and the 'real' timescape of the play, that is, the high summer season. He believes his daughter to be still immature, unaware that she is already in the summer of her life, being ready to fall in love. Curiously, if Capulet evokes '[l]imping winter' the better to dismiss its influence, its mention is enough to cast a shadow on his entire speech. However, winter here belongs to the past and Capulet is determined to put all his hopes in the more clement months of the year.

But as far as they are concerned, the sources of *Romeo and Juliet* all rely on snow and ice imagery. In Luigi Da Porto's 1530 *Istoria novellamente ritrovata di due nobili amanti*, it is snowing outside[8] and the freezing climate seems to contaminate the characters' physiology, since, in this text, Marcuccio/Mercutio's hands are cold both in winter and in summer[9] – a detail taken up by Arthur Brooke but dropped by Shakespeare. In *Le Novelle del Bandello* (1554) and Brooke's *The Tragicall Historye of Romeus and Juliet* (1562), the lovesick Romeo, still unaware of Juliet's beauty and enamoured of the disdainful Rosaline, is so depressed that he is wasting away like snow melting under the heat of the sun:[10]

He can not staye his steppe, but forth still must he ronne,
He languisheth and melts awaye, as snow against the sonne,
His kindred and al[l]yes do wonder what he ayles,
And eche of them in friendly wise, his heavy hap bewayles. (97–100)[11]

Brooke makes it clear that his imagery is grounded in 'real' cold when he specifies that the ball organised by old Capulet takes place in winter:

The wery winter nightes restore the Christmas games,
And now the season doth invite to banquet townish dames. (155–6)[12]

The cold atmosphere prevails all along, so much so that the lovers' sorry plight is compared to a perpetual winter where night predominates: 'The dawning they shall see, ne sommer any more, / But black-faced night with winter rough, (ah) beaten over sore' (1727–8).[13] Contrary to this, the summer imagery predominates in Shakespeare's play. Yet Brooke does not entirely neglect summer symbolism. In the lines just quoted, summer is mentioned as a conspicuously absent season ('ne sommer any more') and, from time to time, Shakespeare's predecessor provides an indirect insight into Juliet's tortured mind by comparing it to a stormy summer sky:

As oft in summer tide, blacke cloudes do dimme the sonne,
And straight againe in clearest skye his restless steeded do ronne,
So Julet wandring mynd yclowded is with woe,
And by and by her hasty thought the woes doth overgoe. (1253–6)[14]

The playwright may have taken his cue from this particular passage when he decided to situate his love tragedy in a hot season which literally stifles Verona's inhabitants. Although his reasons for doing so remain unknown, one can certainly figure out some of them. First, by switching the cold background of his source for a particularly hot environment, Shakespeare dramatised the consequences of a scorching heat unfamiliar to his English audiences[15] to associate it with the violence of the male world. Secondly, in his retelling of a well-known story, he added a touch of novelty in his much deeper exploration of the passions of choler and hatred traditionally associated with hot blood and hot weather. Thirdly, he thus availed himself of strong effects of contrast thanks to the juxtaposition of heat and fire with other elements linked to winter and whiteness. In the wake of Petrarch's description of the poet-lover who feels 'icy cold in warm weather and burning hot in cool', a means to 'encompass the extremes that the lover must embrace in order to comprehend the infinite variety of love', the playwright resorts to 'freezing and frying' imagery enabling him to both depict and question the aftermath of passion.[16] As a result, the references to the snow become particularly striking in the play. In the naive discourse of the two young lovers ('So shows a snowy dove trooping with crows', Romeo tells Juliet (1.5.47); 'Whiter than new snow upon a raven's black', Juliet later echoes (3.2.19)), they forcefully convey their common ideal of purity. What would have looked like simple clichés in a different context is now endowed with a new significance against the background of Verona's hot summer months. Because the spectators are constantly reminded of this stifling summer, they become more alert to the winter images foreshadowing paralysis and death. After Juliet has drunk the Friar's potion and faked death in act 4, scene 4, Capulet once again alludes to the untimely presence of winter in the middle of summer: 'Death lies on her like an untimely frost / Upon the sweetest flower of all the field' (4.4.54–5). This paves the way for images of whiteness and stillness that are all the more striking as the sun usually parches the inhabitants of Verona – a fact made obvious early on in the play, when Benvolio declares to Romeo that 'one fire burns out another's burning, / One pain is lessened by another's anguish' (1.2.45–6). This adumbrates a rather gloomy vision of Romeo and Juliet's milieu: it is both a dry and hot hell on earth, where love becomes a burning pain with no way out but death that freezes them stiff.

'Shake-scene'/Shake-sky: The Nurse's earthquake

The ancient Greeks thought that excessively hot and dry weather could induce violent earthquakes, a belief apparently not unfamiliar to the inhabitants of Shakespeare's Verona where a heat spell seems to stifle all and sundry. The Nurse, for instance, refers to an earthquake that took place eleven years before. In her helter-skelter narrative, its violence was such that she was forced to run away from a dovehouse as the earth suddenly started to shake:[17]

> 'Tis since the earthquake now eleven years,
> And she was weaned – I shall never forget it –
> Of all the days of the year upon that day,
> For I had then laid wormwood on my dug,
> Sitting in the sun under the dovehouse wall. (1.3.25–9)

Earthquakes were then reputedly caused by subterranean winds trapped in the earth which suddenly struggled to get out (hence considered as meteorological phenomena having nothing to do with plate tectonics),[18] when they were not viewed as 'occasioned by the power and influence of the Planets'[19] or simply as wholly unnatural events. As such, they were associated with the frightening images of John's Book of Revelation (6:12–17; 8:5; 11:13, 19; 16:16–21).

The earthquake mentioned by the Nurse and narrowly correlated to climatic disorders is of particular significance here as it points to both disruption and unruliness. Described in a purely comic way, it may appear, as Hotspur mockingly puts it in *1 Henry IV*, as 'a kind of colic pinched and vexed / By the imprisoning of unruly wind' (3.1.27–8); but in *Romeo and Juliet*, by hinting at Verona's organic disorders, it also creates an ominous context that prepares us for the tragic events to follow. These spasms being already part of the Nurse's distant past, the spectators are led to see the tragic events of the play as the more or less direct aftermath of the shaking of the earth in Juliet's early childhood. An event that first and foremost serves to revive past events in the Nurse's memory, the earthquake allows the garrulous old woman to exaggerate the importance of a trivial incident in Juliet's early life and to insist on the infant's strong personality since earthquakes were traditionally associated with passion.[20] The possibility that baby Juliet replaced the Nurse's dead daughter Susan (1.3.20–3) lies at the root of her stream of consciousness and it probably accounts for her fond, irrational 'remembrance of things past' (Sonnet 30, l. 2).

Linked as it is to the earthquake in quasi-Rabelaisian fashion, the small, intimate moment of Juliet's weaning literally takes on cataclysmic connotations as the shaking of the earth is oddly put in connection with the peevish bodily convulsions of the child suddenly having to give up her nurse's breast-feeding.[21] The trauma of Mother Earth being ripped open seems just as important as the Nurse's round mound of flesh suddenly becoming inaccessible to the baby girl because of its bitter taste. So, for all the amusing elements in the situation depicted here, it would seem that, right from her infancy, the child's growth has been familiarly associated with both grotesque and traumatic events.[22] In other words, the Nurse's recollection is marked as much by a terrifying natural event as by an ostentatious physicality conveyed through multiple references to an open, grotesque female body.[23] Providing a moment of comic relief, these physical details could be disregarded insofar as they might downplay the importance of the passage. Still, because it so powerfully foreshadows the final tragedy and the seismic essence of Romeo and Juliet's passion, this particular episode is often highlighted and has led to a myriad of interpretations.

Oddly enough, the impact of classical beliefs is rarely considered as such. Yet, ever since antiquity, seisms had terrified people, making them feel vulnerable and insecure.[24] '[T]he disaster of an earthquake', Seneca writes, 'extends far and wide, is inevitable, insatiable, deadly for the entire state. It gulps down not only homes or families or individual cities; it inters entire nations and regions.'[25] Most of Shakespeare's contemporaries seem to have been horror-struck by its destructive power as much as by its foreshadowing of frightening upheavals and disasters.[26] Early modern spectators must therefore have immediately grasped the sinister implications of the seismic waves evoked by the Nurse, whose unconscious prophetic and proleptic role verges upon divination or witchcraft.

While the influence of classical writings on Shakespeare seems undeniable here, that of contemporary events is also possible, albeit more questionable. Indeed, in order to date the play as exactly as possible, critics have traced the earthly convulsions mentioned by the Nurse back to the real earth tremors which took place in England on 6 April 1580, for instance, or to the 'terrible Earth-quake' mentioned in William Covell's 1595 *Polimenteia* which occurred on 1 March 1584.[27] There were also landslips at Blackmore, in Dorset, on 13 January 1583 and at Mottingham, in Kent, on 4 August 1585 – the latter causing 'intense local excitement'.[28]

However, beyond any sort of topical allusions, what matters most is the symbolism of this saynette which reminds us of an archetypal earthquake, namely the one reported in the New Testament, when Christ, whose thirst on the cross has become unbearable, asks the Roman soldiers for some water to allay his agony[29] – instead of which he is sadistically offered a sponge full of gall (Matthew 27:34). In little Juliet's case, gall has been applied on the Nurse's nipple in order to wean her. This comic detail is indeed quite remote from the divine tragedy of the crucifixion but, taken together with the Nurse's mention of doves and a dovecote, 'which also belonged to meditations on the Passion and the mystery plays',[30] it may then have rung a bell in the spectators' minds. In their delineation of a comic, cameo version of Christ's agony, such allusions work like a filigree in the text or a hidden miniature inside the greater picture that does say something about the child's destiny: passion, in its Christian as well as its etymological sense (i.e. 'suffering, enduring'), is certainly a brand marking Juliet's flesh from her very early childhood.

To passion, ill-luck (or star-crossing) should certainly be added, since the weaning episode is framed between the two sad events of the passing away of the Nurse's husband and the death of her baby daughter Susan. So 'Shake-scene', according to the nickname allegedly given him by the dying Robert Greene,[31] resorts here to the intermittent natural disorders that happen to shake the earth in order to shape and shake the face of the sky and turn it into a shaky 'fated sky'. However, in his first love tragedy, the playwright is not yet the defiant 'Shake-sky' he will later become. He carefully relies on reported speech ('"Shake", quoth the dovehouse', 1.3.35), on repetition and on evocative suggestions, using watered-down colours to paint this small climatic tableau. Only in *King Lear* will he depict a broader picture of a major cosmic upheaval to represent the terrifying wrath of the elements in the midst of the deafening silence of the gods.

'What's in a name?'

A little further down in the Nurse's speech, we learn that Juliet, Capulet's only surviving child, owes her name to her being born on 31 July, on the eve of Lammastide (i.e. the harvest festival which took place on 1 August).[32] When asked to give Juliet's age, the Nurse repeatedly quibbles on the specific acoustics of Juliet's name which, in her own rambling style, she parses as 'Jule'/'Ay':

> 'Thou wilt fall backward when thou hast more wit,
> Wilt thou not, *Jule*?' And, by my halidom,
> The pretty wretch left crying and said '*Ay*'.
> [. . .]
> 'Wilt thou not, *Jule*?' quoth he,
> And pretty fool, it stinted and said '*Ay*'.
> [. . .]
> 'Will thou not, *Jule*?' It stinted and said '*Ay*'. (1.3.44–59, my emphasis)

The name 'Juliet' is clearly derived from 'July', the month of her birth, even though she is already called this in Brooke's *Tragicall Historye of Romeus and Juliet* and 'Julietta' in Bandello's earlier story. In order to probe the limits of nominalism, Shakespeare keeps the original name and then creates a rich background in which its meanings and connotations may be multiplied. He deliberately changed the season in his sources to make his heroine the victim of climatic excess, while in paying renewed attention to the medieval doctrine of 'ephemeral elections',[33] the playwright efficiently called attention to Juliet's fate.[34]

By the way he insists on the signifier 'Juliet', the playwright shows that he understands the importance of the name (*nomen omen*, according to the Latin motto) in a deeper and more acute way than his predecessors. In the ball scene, for instance, Juliet is said to 'han[g] upon the cheek of night / As a rich jewel in an Ethiop's ear' (1.4.158–9). The word 'jewel' here functions as the acoustic double of 'Jule' and it makes Juliet's name reverberate throughout the play.[35] These lines form an erotic blazon, an astrological emblem and a case of *discordia concors*. Such complex imagery is not gratuitous; it ties in with the references to summer and heat that pervade the play. Furthermore, the allusion to the 'Ethiop's ear' does not simply connote sensuality but is also a subliminal reference to the myth of Phaeton in Ovid's *Metamorphoses*.[36] This myth indeed expressed early anxieties about growing heat and climate change and it allowed Shakespeare to convey something of the popular fears often associated with the apocalyptic visions of a world destroyed by fire.[37] In Arthur Golding's translation, one of the marginal notes of the Phaeton episode refers to the following quotation as '[t]he burning of the world':[38]

> The restless horses of the sun, began to neigh so high
> With flaming breath that all the heavens might hear them perfectly.
> [. . .]
> [. . .] they left the beaten way
> And, taking bridle in their teeth, began to run astray.
> [. . .]
> Then wheresoever Phaëton did chance to cast his view,
> The world was all on flaring fire. The breath the which he drew
> Came smoking from his scalding mouth as from a seething pot;
> His chariot also under him began to wax red-hot.
> He would no lenger dure the sparks and cinders flying out.
> Again, the culm and smouldering smoke did wrap him round about,
> The pitchy darkness of the which so wholly had him hent
> As that he wist not where he was nor yet which way he went.
> The wingèd horses forcibly did draw him where they would.
> The *Ethiopians* at that time (as men for truth uphold),
> The blood by force of that same heat drawn to the outer part
> And there adust from that time forth, became so black and swart.
> (The Second Book, ll. 203–301, my emphasis)[39]

As it provides an explanation for the apparition of the black race,[40] this climatic myth surfaces in a number of Elizabethan and Jacobean plays. In Ben Jonson's *Masque of Blackness* (1605), the burning sun is also seen as the origin of Niger's daughters' black skin:

> As of one Phaëton, that fired the world,
> And that before his heedless flames were hurled
> About the globe, the Ethiops were as fair
> As other dames, now black with black despair, (ll. 137–40)[41]

But while, in Jonson, the myth mainly serves as a political ornament,[42] it provides a necessary complement and context to the summer imagery in *Romeo and Juliet*. Set during the dog days in Verona, the tragedy completely reverses the pattern established in *A Midsummer Night's Dream*, unless it is the *Dream* that turns upside down the weather pattern informing *Romeo and Juliet*. If, in the comedy, misrule affects the weather, misrule in the tragedy is caused by the weather and climate acts as an agent of fate.

Canicular intervals and carnal intercourse

High temperatures are repeatedly mentioned in the play and they tend to contaminate its dramatic language. Probably due to the combined effects of anger and heat, Capulet becomes 'too hot' (3.5.174), while Mercutio criticises Romeo and Benvolio for 'burn[ing] daylight' (1.4.41), that is, for wasting time. As to Friar Laurence, who is seen collecting herbs in the small hours, he anxiously anticipates the sun's 'burning eye' (2.2.4), while Benvolio warns Mercutio that, in 'these hot days, the mad blood is stirring' (3.1.4). More significantly, perhaps, the Nurse, rather than associate the letter 'R' with Romeo, links it with 'the dog's name' (2.3.198), thereby indirectly alluding to the dog days plaguing the city of Verona at that time of year.[43] The sound /r/ is further associated with 'rage', a recurring word used in the play to designate anger or choler and which also refers to rabies, the dog's disease.[44]

On top of being considered the hottest ones in the year, the canicular days were also reputed as rather risky when it came to sexual intercourse. Old Capulet explicitly associates the words 'summers' and 'pride' (1.2.10), the last one probably alluding to Juliet's sexual desire.[45] Similarly, according to Mercutio's obscene jest 'the bawdy hand of the dial is now on the prick of noon' (2.3.106–7), midday heat supposedly led to sexual arousal. The word 'prick', which originally designated the small puncture indicating noontime on the dial, here also refers to the male sexual organ. Although she immediately grasps Mercutio's double entendre, the no-nonsense Nurse feigns to be offended by such an off-colour joke ('Out upon you, what a man are you', 2.3.106).[46]

Yet, beyond its comic and erotic implications, the heatwave that parches the fair city of Verona has tragic consequences. In Friar Laurence's words, such heat, which is synonymous with uncontrolled passion, is bound to flare up and provoke a final explosion:

> These violent delights have violent ends,
> And in their triumph die like fire and powder,
> Which as they kiss consume. (2.5.9–11)

For Romeo's confessor, the lovers' kiss is no better than the fuse or spark that will provoke a most dangerous explosion, so that the kiss of life at the ball reads as an anticipation of the final kiss of death in the Capulets' monument. Romeo and Juliet therefore unwittingly put themselves at risk, all the more so as the dry summer season turns heat into a hate that proves especially noxious for the young couple. Bernard Capp notes that, in early modern England,

> [a]bstinence during August may help to explain the dramatic fall in baptisms in May and June. There is clearly a summer trough in the baptismal figures, usually ascribed (on no clear evidence) to abstinence during harvest. A seasonal taboo during the heat of summer, linked through the dog days with some kind of astrological rationale, may help explain the low figures for the early summer.[47]

According to Tudor astrologers, the 'position of the stars' in the months of July and August 'made sex dangerous',[48] and a number of almanacs and popular treatises warned people against the dire consequences of sexual intercourse at this time of year.[49] In his health manual published in 1612, William Vaughan strongly recommends avoiding 'carnal copulation' in summer:

> It is best to vse carnal copulation in Winter, and in Spring time, when nature is desirous without the help of Arts dregs, and at night, when the stomacke is full, and the body somewhat warme, that sleepe immediately after it, may lenifie the lassitude caused through the action therof. In Sommer in Iune and Iuly when the spettle thickens on the ground, it cannot be good.[50]

This, however, does not deter Juliet from evoking in explicit, if not blasphemous, terms the sexual orgasm that Romeo, her beloved pilgrim turned secret husband, will hopefully give her in the course of their wedding night:

> Come, gentle night; come, loving, black-browed night,
> Give me my Romeo, and when I shall die
> Take him and cut him out in little stars (3.2.20–2)[51]

Even though Shakespeare's heroine seems aware of the role played by stars and planets in her own destiny, the sexual ecstasy to which she refers in these cosmic images actually delineates the dis-*astrous* course of their love story. In Juliet's poignant words, the dismembered Romeo, metamorphosed into an Ovidian constellation, undergoes a metaphoric *sparagmos*. In the sonnet prologue, physical desire was already presented as doomed since the 'parents' loins' are described as 'fatal' (l. 4), a phrase ironically reformulated at the end when Romeo makes a significant analogy between the female womb and the cannon's mouth (5.1.65). If, at the beginning, Romeo somewhat naively hopes that his and Juliet's 'bud of love by summer's ripening breath / May prove a beauteous flower when next [they] meet' (2.1.163–4), he is gradually led to discover that, far from favouring the flowers' growth, the summer's breath/breeze will instead cause them to fade away. Romeo and Juliet's love story sadly buds in the wrong season, when, because of the traditional abstinence required during the dog days, Eros is to be replaced by his darker brother Thanatos.

To cap it all, July's hot spell was also thought of as the unhealthiest period in the year.[52] Stephen Batman explains that these 'Canicular daies' are particularly evil ones. During this month, he affirms, 'all hot passions [and] evils increate, [and] that time is most disconvenient [and] unaccording to medicine'.[53] Unsurprisingly, then, the play contains several references to plague, conspicuously absent in the sources.[54] In his Queen Mab speech, Mercutio briefly alludes to the miasma theory which used to account for the spreading of the disease when he explains that the fairy 'gallops night by night' (1.4.68) '[o]'er ladies' lips, who straight on kisses dream, / Which oft the angry Mab with blisters plagues, / Because their breath with sweetmeats tainted are' (1.4.72–4). Having been mortally wounded by Tybalt, Romeo's friend curses both Montagues and Capulets and exclaims 'A plague a both houses' (3.1.91) before giving up the ghost. In the last act, the reference to pestilence is, by contrast, unmetaphored.[55] The audience realises that, just like London, Verona is a plague-ridden city, something which has a crucial impact on the play's tragic outcome. Most importantly, Friar John informs Friar Laurence that he failed to deliver his letter to Romeo because of the 'pestilence' that had broken out there and the resulting quarantine aimed at 'staying' the infection (5.2.5–16). By preventing the delivering of Laurence's letter to Romeo, this bout of plague speeds up the catastrophe and, as a result, the two title characters become the indirect victims of the epidemic.

The links between climate and health were well established in early modern England and epidemic illnesses such as plague were thought to be partly due to environmental factors. Plague treatises of the period commonly attributed contagion to punishment by God while they simultaneously blamed astrological factors and the foul air breathed by the city inhabitants. Kristy Wilson Bowers summarises these beliefs by stating that, '[a]ccording to miasmatic theory, air became polluted or poisoned by any number of factors, including a dangerous alignment of planets, unusual natural phenomena such as earthquakes, and local factors, such as accumulated garbage that created noxious (and therefore unhealthy) smells.'[56] Early modern treatises unquestionably embrace such explanations, and among many others, Thomas Brasbridge's book, *The poore mans ieuuel, that is to say, A treatise of the pestilence* (1578), supplies a useful illustration of what they looked like. A reliable author by contemporary standards, Brasbridge was a Church of England clergyman who had probably gone through a period of plague during his Banbury childhood in the mid-1540s.[57] To him, the disease is first of all a consequence of man's sins. 'The second cause', he writes, 'is giuen by Astronomers, and is called an euill constellation.'[58] The third one is purely climatic:

> The thirde cause [. . .] is the corruption of the ayre: whiche beyng corrupted, is apte to infect mans body. For euery man that liueth, draweth the breath which we haue, of the aire round about vs. If it be stinking, venomous, and corrupt, the body of man [which] is neare to it, is in danger of corruption, wherof oft times is ingendered the Pestilence. The aire is corrupt, either generally in diuers Cities, or Countries: or particularly, in some one place.[59]

In Verona, self-destructive hate is rampant and corruption commonplace. In effect, the 'star-crossed lovers' suffer from an unfavourable astronomical situation, and to cap it all, the Nurse reminds us that a major earthquake had shaken up the earth a few years before. For an English playgoer not necessarily aware of Brasbridge's treatise but fully conscious of the threat represented by plague, such signs were correlated to the presence of miasma, that is, of the foul air that infected the population and put their lives in danger.

Summer and hot weather were then generally associated with the plague bacillus.[60] So, when Old Capulet asks his servants to 'quench the fire' and complains that 'the room is grown too hot' (1.4.141) before the ball, he may express a form of anxiety regarding the disease, as heat and crowded places then appeared as two threatening catalysts of plague. Juliet's father may also deliver a proleptic observation on the erotically loaded atmosphere of the ball, an observation paving the way for the play's tragic outcome. In fact, the 'fire' mentioned by Capulet, a rather unexpected element in a feast given during a very hot summer night, works simultaneously as an intertextual clue pointing to the Christmas feast in Shakespeare's main sources and as a symbolic undertone foreshadowing the consuming passion of the 'star-crossed lovers'.

In such circumstances, Romeo and Juliet indulge in their ardent feelings of love while Verona consumes itself with hate/heat and with the long-lasting rivalries between the two great families of the city. In this context, death is seen as a hoped-for relief from the pangs of passion. The lovers welcome the early signs of an impending storm as the precursors of heavenly light and intense happiness, and after killing Paris before the Capulets' monument, Romeo cannot help exclaiming:

> How oft, when men are at the point of death,
> Have they been merry, which their keepers call
> A lightning before death! O, how may I
> Call this a lightning? (5.3.88–91)

If, in the first scenes of the tragedy, an anxious Juliet realises the danger of hasty contracts and expresses her doubts about 'the lightning which doth cease to be / Ere one can say it lightens' (2.1.161–2), the lightning in Romeo's speech takes on a contrary connotation which proves more comforting than really unnerving. Joy being now firmly paired with death, the 'lightning' mentioned by the young man cannot be reduced to the proverbial expression of 'the lightning before death' or of the one 'referring to the merry-making of the condemned criminal'.[61] It also refers to dawn which, in the play, appears as a fragile, transitory moment between light and darkness and as a time of lightening/lightning. The threshold of afterlife is yet another form of dawn, when the lovers, be they Paris, Romeo or Juliet, are allowed to experience a brief but intense happiness. Romeo's reference to the 'lightning' of death also ties in with his own demise as his last words associate death with the advent of rough weather:

> Come, bitter conduct, come, unsavoury guide,
> Thou desperate pilot, now at once run on
> The dashing rocks they sea-sick weary bark. (5.3.116–18)

These lines echo Capulet's comparing his daughter to a 'bark', a 'tempest-tossed body' (3.5.130, 136). Romeo is ready to imitate Juliet for better and for worse. Here, one detects the hellish vision of the 'pilot' Charon who ferried souls across the Styx. This dreamy vision of the ocean at such a crucial moment in the young man's life may also be connected to Clarence's dream in *Richard III*. But while Clarence concentrates on a static macabre vision, Romeo focuses on the intense energy required to cross the frontier between life and death; what emerges from his dying speech is not an image of burial, but a climatic reference to stormy, tempestuous weather: Romeo's last moments of life are those of a mariner caught in a tempest.[62]

Transported across the river/ocean of the underworld, Juliet's husband fulfils the prophetic statement of the Prologue, so that no real suspense is left as to the hero's fated end. The lovers are indeed 'star-crossed', an idea which Shakespeare reiterates in introducing a revealing confusion about the state of Verona's sky. He uses the word 'vault' and its cognate 'vaulty' nine times in his tragedy, all of them in reference to the Capulet tomb or monument.[63] Yet, in the morning of the wedding night, at dawn, when Romeo refers to 'the *vaulty* heaven so high above our heads' (3.5.22, my emphasis), the playwright subtly allows high and low, heaven and Hades to communicate and coincide. Like the 'Ethiop' image at the ball (1.4.159), later implicitly taken up in Juliet's invocation of Phaeton in her prothalamion ('Gallop apace, you fiery-footed steeds', 3.2.1), the word 'vault' and its adjectival compound are marked by a dire ambivalence foreshadowing disaster and death.

In connection with the sense of doom pervading the play, the dog days should be replaced within a broader context. The phrase 'dog day' itself stems from astrology and climatology as it designates Sirius, the brightest star in the Canis Major (or Great Dog) constellation which rises about the same time as the sun in the month of August.[64] Stephen Batman, among other mythographers, explains the origin of the *dies caniculares* in his 1582 encyclopaedia.[65] In

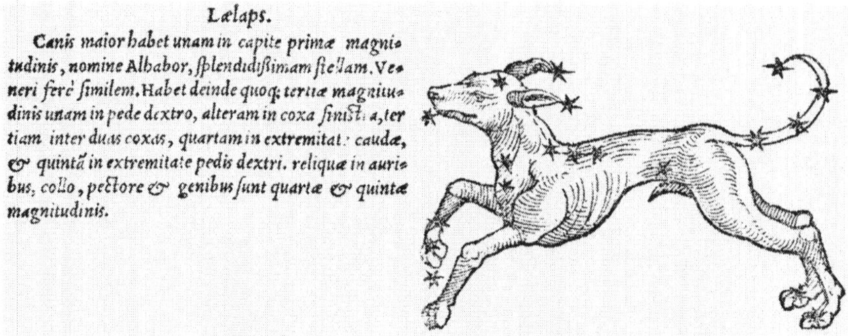

Lælaps.

Canis maior habet unam in capite primæ magnitudinis, nomine Albabor, ſplendidiſſimam ſtellam. Veneri ferè ſimilem. Habet deinde quoq; tertiæ magnitudinis unam in pede dextro, alteram in coxa ſiniſtra, tertiam inter duas coxas, quartam in extremitate caudæ, & quintã in extremitate pedis dextri. reliquæ in auribus, collo, pectore & genibus ſunt quartæ & quintæ magnitudinis.

Fig. 4 Caius Julius Hyginus, 'Canis Major', in *Fabularum liber*, 1535, p. 100. By courtesy of the Bibliothèque Nationale de France.

popular stellar calendars such as *Hyginus fabularum liber* (Basel, 1535), the end of July and the early days of August were depicted as 'canicular' because these months concluded with the rising of Sirius.[66] Sirius was placed at the tip of the constellation of Canis Major and was thus visually associated 'with a descent into the jaws, or mouth, of death'.[67]

The lives of the 'star-crossed lovers' and their passionate revolt against their parents and the adult world in general may be partly explained by the fact that Sirius was traditionally associated with extremely hot temperatures. For the physicians of the time, there was no doubt that atmospheric heat was the main cause of an increase of choleric humour in the body. In 1590, William Clever explained that if winter usually stifled angry feelings, the hot months of the year triggered man's choler:

> Cornelius Celsus saith, that like as the sunne is lodged vnder the darke vaines of the clouds, in the night season, so heat is hidde and shrowded in the secrete vaines of all thinges in winter, and waxeth more outward and stronger with the increase of the yeare, vntill by the temperature of the sunne it bee powerfull and mightie.
>
> Arnoldus de noua villa affyrmeth, that the complexion of man ariseth and falleth, with the temperature or distemperature of the yeare: so may we gather hereby that choller rageth and super aboundeth more in sommer then winter, only by an intentiue qualitie.[68]

In Shakespeare's tragedy, yellow bile[69] and ardour prompt the actions of the main characters who are presented as the victims of an aforeknown destiny. Romeo is depicted by Friar Laurence as a 'distempered head' (2.2.33), a phrase in which the past participle points to 'an imbalance in the bodily "humours" of blood, phlegm, choler and melancholy'.[70] In the same vein, Samson and Gregory exchange a series of quibbles on the word 'colliers' that designated 'those who carry coals' as well as 'dog collars'. 'Colliers' are thus associated with 'choler' and 'dogs' (1.1.2–7), all three being keywords that encapsulate the metaphoric, symbolic and climatic resonances of the play's imagery and mythical patterns.[71]

So, if the *Dream* examined innovative modes of thought, *Romeo and Juliet* returns to a Greek and medieval world picture, according to which the four elements of *ignis*, *aer*, *aqua* and *terra* organise, operate within and influence vegetal, animal and human bodies alike. More importantly, perhaps, *Romeo and Juliet* enhances the hate/heat metathesis illustrated in the family rivalry, which is both revived and aggravated by Verona's parching sun. It reveals how interrelated climate and sexuality were in Shakespeare's time. If heat was believed to be responsible for passionate feelings, we have seen that it was also perceived as a dangerous moment for sexual intercourse. Yet, Shakespeare insists, if religious devotion could be regarded as compatible with chaste love, adolescent passion, on the contrary, seldom served Petrarchan ideals and the abstinence traditionally required during the hot season of the year. As a result, the two title characters are made to experience moments of fierce tension and of terrible frustration. No wonder then that, in *Romeo and Juliet*, the violence of despair and the energy of youth turn out to be the two sides of the same coin. The numerous allusions to light and lightning in the play make it clear that love at first sight, or *coup de foudre*, does illuminate the night of ingrained tradition and blind hatred, while it also prefigures the inevitable dis-aster (etymologically, an adverse event caused by an unpropitious star) and retribution that follow.

As Maurice Blanchot put it, '[t]he disaster ruins everything, all the while leaving everything intact.'[72] Though it is uncertain whether the French essayist had Shakespeare's tragedies in mind when he started to write his fragmentary book, his observation is not irrelevant in the case of a play such as *Romeo and Juliet*, where local mores and tradition are such that they plague and paralyse a 'fair Verona' (Prologue, l. 2) which is subsequently identified as a site of inertia, suspicion and superstition. It is precisely because nothing ever changes under the suffocating sky of the Italian city-state (one may incidentally notice that, in the Shakespearean canon, the rule of metamorphosis seems much more coterminous with sylvan landscapes – think of *A Midsummer Night's Dream* and *As You*

Like It – than with a stunted urban environment) that the disaster brutally strikes its very centre and, out of the blue, stifles some of its best young lives. Because it is strongly denied from the start, violence becomes worse and worse as the plot unfolds and leads to the moment when, under the influence of the sultry weather, the exacerbated hatred of the two warring clans can only add fuel to the fire and bring about the explosive passions that make the play a truly clima(c)tic and seismic love tragedy.

Notes

1. Carla Mazzio, 'Coda: Scepticism and the Spectacular: On Shakespeare in an Age of Science', p. 240.
2. On Venus's connection with 'urgent amorous affections' and with marriage, see Alison Findlay, *Women in Shakespeare: A Dictionary*, p. 412.
3. In *The Fated Sky: Astrology in History*, Benson Bobrick explains that 'the twelve houses of the horoscope were divisions of the heavens relative to birthplace, as connected to the rotation of the Earth on its axis as it revolved. In chart interpretation, they encompassed every human aspect of life' (p. 25).
4. Tripthi Pillai, 'Constructing Experiences and Charting Narratives: The Future in/of *A Midsummer Night's Dream*', p. 153.
5. Stephen Batman, *Batman vppon Bartholome his booke De proprietatibus rerum, newly corrected, enlarged and amended*, 'De temporum divisione', Book 9, chap. 6, p. 144. See also William Vaughan's 1600 *Naturall and artificial directions for health*, p. 62: 'Summer begins, when the sun entreth the signe of *Cancer*, which is the twelfth daie of *Iune*. In this time choler is predommant, heat increaseth, the winds are silent, the sea calme, fruites doe ripen, and Bees doe make honey [. . .] It is daungerous taking of Physick, and speciallie in the dog dayes.'
6. Mary J. Dobson, *Contours of Death and Disease in Early Modern England*, p. 20.
7. On early modern marriage and seasonal patterns, see Keith Wrightson, *Earthly Necessities: Economic Lives in Early Modern Britain*, p. 232. Wrightson explains that '[i]n pastoral regions spring or early summer marriages were more common. In predominantly industrial parishes, or in towns, however, there was no pronounced seasonal pattern.'
8. Luigi Da Porto, *Istoria novellamente ritrovata di due nobili amanti* (1530), in *Romeo and Juliet: Original Text of Masuccio Salernitano, Luigi Da Porto, Matteo Bandello, William Shakespeare*, p. 25.
9. '[H]is hands were always freezing', ibid. p. 27.
10. Quoted in Nichole DeWall, '"A Plague 'o Both Your Houses": Shakespeare and Early Modern Plague', p. 69.

11. Arthur Brooke, *The Tragicall Historye of Romeus and Juliet*, in *Narrative and Dramatic Sources of Shakespeare*, vol. 1, p. 288.
12. Ibid. p. 290. See, too, ll. 949–50, p. 310: 'The summer of their blisse, doth last a month or twayne / But winters blast with spedy foote doth bring the fall agayne.'
13. Ibid. p. 330.
14. Ibid. p. 318.
15. Heat, however, was not completely foreign to England. In his *Six bookes of a common-weale*, Jean Bodin reports, for instance, that the year 1556 was an especially hot one: 'And the like chaunced in England, in the yeare 1556, as I haue seene by letters from M. de Nouailles ambassador in England for the French king: in the which he doth assure, That the heat had bene so vehement, as the flame kindled by the sunne, burnt the fruits and villages throughout a whole countrey' (The Fifth Booke, chap. 1, p. 548).
16. S. K. Heninger Jr, 'Sequences, Systems, Models: Sidney and the Secularization of Sonnets', p. 73.
17. On this episode, see Barbara Everett, '*Romeo and Juliet*: The Nurse's Story'.
18. Largely ingrained in the Elizabethan era, this explanation was first given by Aristotle in the fourth century BC. See his *Meteorology*, II.7–8. See also Thomas Hill, *A contemplation of mysteries*, fol. 65v: 'An Earthquake, is [. . .] a breaking forth or trembling, a smyting or agitation, when an exhalation is included in the bowels of the same, séeking a violent issuing forth: in the which, this so smyteth & shaketh the earth. And of the same smiting, is the Earthquake named. Or it is a certaine working, by which the earth is moued, caused especially of winde or drie vapours, inclosed within the hollownesse of the earth, and through the heate of the sunne, and other starres heated and enforced, by which motion and enforcement, this so shaketh, that oftentymes houses, yea whole Cities, and townes, are throwne downe, and suncke with the same, that the sinnes of men may be punished, and others called vnto repentaunce.'
19. Philemon Holland (trans.), *The historie of the vvorld*, The Second Booke, chap. LXXIX, p. 37.
20. On this, see Thomas Hill, *A contemplation of mysteries*, fol. 65v: 'An Earthquake, is a passion or affection of the earth.'
21. François Rabelais similarly focuses on a nurse's nipples in his 1532 *Gargantua* (in *Les Cinq Livres de faits et dits de Gargantua et Pantagruel*, chap. XXXX, p. 336).
22. On the grotesque body, see Mikhail Bakhtin, *Rabelais and his World*, p. 26. Interestingly, in his 1601 translation of Pliny's *Natural History*, Philemon Holland describes the consequences of earthquakes in a very Rabelaisian way which assimilates the earth to the human body: 'As for the gaping chinke, sometimes it remained wide open, and sheweth what

it hath swallowed: otherwhiles it closeth up the mouth and hideth all' (*The historie of the vvorld*, The Second Booke, chap. LXXX, p. 38).

23. On the specific connections between the female grotesque and the Nurse, see Lina Perkins Wilder, *Shakespeare's Memory Theatre: Recollection, Properties, and Character*, p. 67.
24. On the terror caused by earthquakes, see Virgil, *Georgics*, in *Eclogues, Georgics, Aeneid 1–6*, Book 1, l. 461: 'Germany heard the noise of battle sweep across the sky and, even without precedent, the Alps rocked with earthquakes [. . .] Horror beyond words, beasts uttered human speech; rivers stood still, the earth gaped upon.'
25. Seneca, *Natural Questions*, VI.1, p. 131.
26. On earthquakes as portents, see Cicero, *On Divination, Book 1*, p. 57 (35): 'often have earthquakes given true predictions of many serious events to our commonwealth and many to every other state'. See also p. 71 (78).
27. William Covell, *Polimenteia*, sig. Y2r–v. For further details, see Sidney Thomas, 'The Earthquake in *Romeo and Juliet*'.
28. Sarah Dodson, 'Notes on the Earthquake in *Romeo and Juliet*'.
29. Beatrice Groves, *Texts and Traditions: Religion in Shakespeare 1592–1604*, pp. 63–4.
30. Ibid. p. 64: 'An early seventeenth-century sermon connects the spirit which Christ gave up when he died with the dove released from the ark [. . .] In mystery plays too, the moment of Christ's "giving up the ghost" was staged with a dove [. . .] It is perhaps the representation of Christ's spirit as a dove that led Richard Rolle, in his "Meditations on the Passion", in which he imagines Christ's crucified body as infinitely open and permeable, to use the image of a dovecote for the Saviour's body.'
31. Or by the author of *Greene's Groatworth of Wit* (1592), passing himself off as Greene. For Henry Chettle as the author of the pamphlet, see D. Allen Carroll (ed.), *Greene's Groatsworth of Wit Bought with a Million of Repentance*, pp. 1–31.
32. As the Nurse puts it, '[o]n Lammas Eve at night shall she be fourteen' (1.3.23).
33. 'Ephemeral elections' allowed its astrological practitioners to choose appropriate dates for specific events, depending on the aspect of the heavens and the position of the planets.
34. See *King Lear* (and below, Chapter 5, pp. 157–8), where Edmund's view (1.2.109–31) corresponds to the cynical naturalist or materialistic stance, while in *Romeo and Juliet* the influence of the dog days, or Canis Major, and the questions of date, calendar and nativity, as well as conception, are presented as crucial elements that determine the tragic events of the play.
35. On this, see François Laroque, 'Roméo et Juliette, entre violence et jouissance', p. 13.

36. See also Manilius, *Astronomica*, IV, p. 289.
37. Another explanation for the origins of the black race, which became increasingly popular in the Jacobean era, was related to the curse of Ham (Genesis 9:20–7).
38. This phrase is a marginal gloss taken from Ovid, *Metamorphoses*, trans. Arthur Golding, The Second Book, p. 69.
39. Ibid. pp. 66–9.
40. In this connection, see François Rabelais's 1532 *Pantagruel*, in *Les Cinq Livres de faits et dits de Gargantua et Pantagruel*, chap. II, p. 424.
41. Ben Jonson, *The Masque of Blackness*, in *Court Masques: Jacobean and Caroline Entertainments 1605–1640*, p. 4.
42. In Jonson's masque, once cleansed in water and dried by the English sun, i.e. by James I, the black ladies can recover their white skin. Yet, as David Lindley points out, Jonson faced a practical problem: 'at the end of the masque, there could be no immediate transformation of the ladies back into their "real" selves' (Ben Jonson, *The Masque of Blackness*, in *Court Masques*, p. 215).
43. In the play, the dog days, a time of excessive heat from mid-July to mid-August, are also hinted at through Friar Laurence's name. Saint Laurentius, whose feast day fell on 10 August (i.e. during the canicular period), was indeed known as the patron saint of cooks as he was martyred by being roasted alive on a grill. The dog days are also mentioned in *King Henry VIII*: 'There is a fellow somewhat near the door – he should be a brazier by his face, for, o'my conscience, twenty of the dog-days now reign in's nose' (5.3.38–41).
44. See 'Rage', *OED*, 4.†c. 'Rabies in a dog. *Obs.*' It is significant, therefore, that the play often uses the word 'rage' (which crops up four times) rather than that of 'choler' (mentioned twice), for example.
45. See 'Pride', *OED*, II.†11. 'Sexual desire, esp. in a female animal; heat. *Obs.*'
46. See Sophie Chiari, '"Vat is de clock, Jack?": Shakespeare and the Technology of Time' and 'Le temps et ses objets dans l'oeuvre de Shakespeare'. Interestingly, a German satirical print of the 1540s owned by the British Museum (reverse copy after Peter Flötner, Prints & Drawings Department, Registration number: E,8.186) shows a human-shaped sundial with a gnomon running from mouth to anus.
47. Bernard Capp, *Astrology and the Popular Press: English Almanacs 1500–1800*, p. 121.
48. Ibid. p. 120.
49. Ibid. p. 122. Capp reports that Thomas Hill, in a treatise like *The profitable arte of gardening nowe the thirde time set forth*, 'advised readers to take rue and herb-grace regularly as an anti-aphrodisiac during July, a period for which he urged sexual abstinence' (see pp. 42, 45, 78 in the 1574 edition of Hill's book).

50. William Vaughan's 1612 *Approved directions for health*, chap. 4, p. 69. An abridged version of this passage is already present in the 1600 edition of the same treatise, chap. 7, p. 46.
51. The stellar metaphor triggered by the fantasised presence of the beloved gives way to earthly driven impulses when the two lovers realise that they are separated: Romeo compares himself to 'dull earth' (2.1.2), while Juliet sees herself as 'vile earth' (3.2.59).
52. As part of climatic disturbance, hot temperatures could be an indirect cause of death insofar as it 'delay[ed] planting or harvesting', which had 'dire consequences for much of a population in a given region'. See Robert Markley, 'Summer's Lease: Shakespeare in the Little Ice Age', p. 136.
53. Stephen Batman, *Batman vppon Bartholome his booke De proprietatibus rerum*, 'De temporum divisione', Book 9, chap. 15, p. 146.
54. If plague is not mentioned in the sources, the disease imagery is nevertheless present in them. For instance, in Brooke's *Tragicall Historye of Romeus and Juliet* (p. 356), as Juliet awakes in the tomb, she discovers Romeo's corpse, much to her dismay, and she describes her immediate surroundings as 'this fowle *infected* place' (my emphasis).
55. On the 'unmetaphoring' process, see Rosalie L. Colie, *Shakespeare's Living Art*, p. 11.
56. Kristy Wilson Bowers, *Plague and Public Health in Early Modern Seville*, p. 33.
57. Julian Lock, 'Brasbridge, Thomas (1536/7–1593)'.
58. Thomas Brasbridge, *The poore mans ieuuel, that is to say, A treatise of the pestilence*, 'The second Chapter Sheweth the second cause of the Pestilence', n.p.
59. Ibid. 'The third Chapter Sheweth the third cause of the Pestilence', n.p.
60. On the contrary, cold climates were considered as 'responsible for scrofula, which partly explained why it was so prevalent in Britain'. See Stephen Brogan, *The Royal Touch in Early Modern England: Politics, Medicine and Sin*, p. 20.
61. Neil Rhodes, 'Time', p. 292.
62. See Ruth Nero, 'Tragic Form in *Romeo and Juliet*', p. 248.
63. See 4.1.11, 4.3.32, 4.3.38, 5.1.20, 5.3.86, 5.3.131, 5.3.254, 5.3.276 and 5.3.290.
64. On this constellation, see Manilius, *Astronomica*, I, pp. 35–7. On 'The power of the Dog Starre', see Philemon Holland (trans.), *The historie of the vvorld*, The Second Booke, chap. XL, pp. 19–20.
65. Stephen Batman, *Batman vppon Bartholome his booke De proprietatibus rerum*, Book 8, chap. 36, p. 137, and more specifically chap. 39, p. 138.
66. Caius Julius Hyginus, *Fabularum liber*, p. 100. Sirius rose at a time which was approximately situated on 11 August. Hippocrates explains that '[o]ne must [. . .] guard against the rising of the stars, especially of

the Dog Star [. . .] For it is especially at these times that diseases come to a crisis. Some prove fatal, some come to an end, all others change to another form and another constitution.' See Hippocrates, *Airs, Waters, Places*, in *Hippocrates*, vol. 1, p. 105.

67. Philippa Berry, *Shakespeare's Feminine Endings: Disfiguring Death in the Tragedies*, p. 37.
68. William Clever, *The flower of phisicke*, p. 96.
69. By the early sixteenth century, yellow bile was the most common term for the choleric humour.
70. René Weis (ed.), *Romeo and Juliet*, p. 203, n. to l. 29.
71. Later on in the play, in 1.4, Tybalt mentions once againt choler in connection with 'patience perforce' (1.4.202).
72. Maurice Blanchot, *The Writing of the Disaster*, p. 1.

Chapter 3

'Winter and rough weather': Arden's Sterile Climate

When the ennui of winter's sterility gleamed.[1]

Where *Romeo and Juliet* showed an urban milieu peopled by heat-stricken characters, *As You Like It* turns to a more rural and windy environment. As in Stéphane Mallarmé's 'The Swan', *As You Like It* is a comedy where Shakespeare somewhat perversely transforms the *locus amoenus* of Arden's green locale into a wasteland of ice and idleness, thus making 'the boredom of sterile winter' sound really 'resplendent'. All the same, the forest paradoxically provides its natives and refugees alike with a breathing space that promotes self-discovery and self-knowledge. It is a sacred place, a natural 'temple' ('for here we have no *temple* but the woods,' Touchstone says to Audrey in 3.3.44–5), with its columns of carved trees where lovers and exiles, fools and philosophers come to pause and meditate. Interestingly, the word 'temple' is derived from the Latin *templum*, a term itself probably related to the word *tempus* referring to both time and season. As a matter of fact, the words for weather phenomena were originally the same as those used to designate time and, in a bucolic Arden owing much to the pastoral poetry of Virgil, this triple association between time, weather and the sacred sounds natural enough.[2]

As already stated above, the rather unpredictable nature of the English climate was a real issue for Shakespeare's contemporaries, so much so that the need to allay the fears of meteorological uncertainties encouraged people to rely on the forecasts of popular almanacs. In *As You Like It*, almost everybody seems obsessed with the weather, and if the fool is said to 'bask [. . .] in the sun' (2.7.15),[3] Amiens's two songs (2.5.1–8 and 2.7.174–97) rather point to a bad weather spell afflicting the pastoral world. Both airs are tinged with pessimistic undertones that reflect the difficulties of the Ardenites and their constant efforts to cope with the situation. In his first song, Amiens notices that, in the forest, one 'shall see / No enemy / But winter and rough weather' (2.5.5–7). Depressing enough in itself, this burden actually delights the melancholy Jaques and is repeated later on in the scene (2.5.35–8) by all the characters onstage. Amiens is asked to sing again almost immediately after the arrival of Orlando bearing old Adam on his back, and he is once again keen to depict a postlapsarian world afflicted by constant gusts of wind:

> Blow, blow, thou winter wind,
> Thou art not so unkind
> As man's ingratitude; (2.7.174–6)

Admittedly, the harsh winter repeatedly evoked by Amiens in his songs is somewhat counterbalanced by the pages' song 'It was a lover and his lass' (5.3.13–43), which paves the way for Duke Senior's final 'rustic revelry' (5.4.161) and for the traditional winter vs spring opposition.[4] Yet no spring symbolism radiates through Arden's 'golden world' (1.1.95). It is more of a promise than an actual fact and it will not be fulfilled in the play since the epilogue lays bare the artifice of the marriage ceremony, just in time to remind us that a boy actor simply could not marry another male player.

A Saxon pastoral

In Lodge's *Rosalynde* (1590), the dying John of Bordeaux tells his sons: 'Oh mans life is like lightning that is but a flash.'[5] While the comparison of man's life to a flash of lightning[6] is certainly a vivid one, there is no winter in Lodge's French forest of Ardennes.[7] The few references to rough weather in it are exclusively metaphorical.[8] The birds sing all day long while the sun is seen rising behind the trees: 'The Sunne was no sooner stept from the bed of Aurora, but Aliena was wakened by Ganimede.'[9] As a result, the characters spend the time chatting under sunny skies in weather conditions that are never described as particularly chilly, and as the wedding draws near, no cloud ever obscures the constantly clear sky: 'Thus they past away the day in chat, and when the Sunne began to set, they took their leaves and departed.'[10]

Shakespeare's dramatisation translates the pastoral climate of Lodge's *Rosalynde* into a typical English one with rain, snow and cold temperatures. Arden, where time is at a standstill,[11] overlaps with the geographic reality of his native country and, as in the *Shepheardes Calender*, his shepherds 'feel cold and hunger and articulate the plight of the labouring rural poor'.[12] In Spenser's work, the January eclogue reveals the rather depressing situation of a young man, 'Colin clout a shepheardes boy', in love with a country lass called Rosalinde. Colin notably compares 'his carefull case to the sadde season of the yeare, to the frostie ground, to the frosen trees, and to his owne winter-beaten flocke'.[13] His predicament foretells that of Orlando, Silvius and the like, all caught up in the grips of a frozen forest and in the iron hands of a cold mistress.

The playwright thus elaborates a typically Saxon pastoral that belies the idealised climate of the Latin tradition with its mild, sunny weather. No wonder, then, if coldness also serves to emphasise the sorry plight of Celia and Rosalind as they escape from Duke Frederick's realm. As the two maids are about to leave the court for the forest of Arden, Celia evokes a dimmed weather ('this heaven, now at our sorrows pale', 1.3.94) reflecting the inner torments of their minds. The pathetic fallacy turns the sky depicted by the worried Celia into a mindscape and it functions as a mirror held up to her inner psyche, an image that made sense in the early modern era when the elements responsible for the make-up of climate (i.e. heat, cold, humidity and dryness) were believed to be the same as those present in the make-up of the human body. So while for most Elizabethans the natural habitat obviously influenced human behaviour, many literary works of the time introduced a two-way link between human physiology and climatic conditions, thereby attributing to the sky the inner characteristics of their protagonists.

Such an interaction between humankind and the surrounding atmosphere surfaces when Celia blames Rosalind for not 'loving [her] with the full weight that [she] love[s] [her]' (1.2.6–7): Rosalind's love is not 'righteously tempered' (1.2.10) as she is not capable of forgetting her grief and entirely devoting herself to her cousin. Here, the past participle 'tempered' is the equivalent of 'composed',[14] while 'temper' as a noun also meant '[t]he constitution, character, or quality of a substance or body (originally supposed to depend upon the "temper" or combination of the elements)' (*OED*, II.a†4). In addition, it could refer to '[t]he condition of the atmosphere with regard to heat and cold, dryness and humidity; the prevailing condition of the weather at a place' (*OED*, II.†6).[15] If we take these different meanings into account, what Celia actually implies in her criticism is that Rosalind's love is heavily dependent on the climate and that, like the typical English weather, it changes all the time.

Yet it would seem that, in chilly Arden, the characters remain stuck in a never-ending winter, a season then described as cold and dry, two features aggravated by the northern situation of Shakespeare's forest.[16] Winter therefore often favoured melancholy, a mood correlated to the influence of Saturn, a planet perceived as 'the coldest and driest planetary body'.[17] Jaques, the comedy's proto-environmentalist, is particularly prone to the influence of Arden's climate as he portrays himself as having 'a melancholy of [his] own, compounded of many simples, extracted from many objects' (4.1.13–14). The despondent traveller clearly suffers from the nipping cold which steals in through the trees, an 'impression'[18] of cold shared by most of the characters in the play regardless of their social status, from the high-born Duke exiled in the forest to the simple shepherd Corin, whose philosophy is highly dependent on the bad weather conditions prevalent in Arden: 'the property of rain is to wet and fire to burn' (3.3.12–13).

In describing the forest as a place plagued by tough weather conditions, the playwright reminds us of the meteorological instability characterising England, a country which, on early modern maps, was situated close to 'Hibernia', the name then given to Ireland.[19] Even though Shakespeare's forest encompasses multiple geographical settings, including France and Flanders, it seems reasonable to think that, given its biting air, he probably had in mind his native Warwickshire, since 'Arden' corresponds both to his mother's name and to the vast and familiar woodlands which 'still covered entire areas of the Midlands [. . .] and reached almost up to Wilmcote, a hamlet close to Stratford and the home of Shakespeare's mother'.[20] Like most of his contemporaries, Shakespeare regarded himself as a northerner and he did not hesitate to play with the clichés generally associated with England's northern location on the maps.

This was part of a long-standing tradition as, in the first century BC, Vitruvius had already insisted that 'the northern populations, infused with the thickness of the air and chilled with moisture, have sluggish minds'.[21] This explains why the Ardenites are all so idle and obsessed with songs, sports and celebrations while even the no-nonsense Corin is never seen tending his flock. In his *Natural History*, Pliny declared that 'in the frozen and icie regions, the people have white skins, haire growing long downeward, & yellow; but they be fierce & cruell by reason of the rigorous cold aire [. . .] Here there breed noisome and hurtfull wild beasts of many formes and figures.'[22]

This biased description may explain the presence of a number of savage beasts lurking in Arden, such as the 'green and gilded snake' (4.3.103) and 'the hungry lioness' (4.3.109), both apparently sustained by the moisture of the place. Tacitus, for his part, described Roman Britain in *Agricola* and he observed that there 'the climate [was] miserable, with frequent rains and mists', even though he acknowledged that there was 'no extreme cold' affecting the country.[23] Shakespeare makes Arden's climate decidedly colder than that of Roman Britain, confining rain to plays like *A Midsummer Night's Dream* with its wet Athenian forest, or to *Henry V*, where the English warriors are characterised by their 'rainy marching' (4.3.112). If he was not conversant with Tacitus' writings, the dramatist may have been vaguely familiar with Vegetius, whose works had been translated into English in 1572. In his fifth-century manual devoted to Roman military practices, *On Military Matters* (*De re militari*), the Latin writer confidently states that 'the people of the North, whom the Sunne burneth not so nere, being more rashe and unadvised, yet a great deale better blouded, are most ready of all, and desirous of warre'.[24] In other words, because they are 'remote from the fires of the sun',[25] the northern inhabitants lack moderation and prove irritable, a point ironically corroborated by Shakespeare who revels in the presentation of fairly over-the-top characters such as Orlando, apparently unable to work, but prone to dissention and anger. That the Ardenites should be influenced by the bad climate is therefore not surprising per se. Still, the early modern clichés disseminated by geo-humoralism are here reworked in what quickly appears as a more complex vision of men and women, even though such a view remains connected to the local topography and to its prevailing climate.

Climate theories inherited from classical antiquity through writers and thinkers like Aristotle, Hippocrates and a few others were widespread in England. Shakespeare thus did not have to wait for the translation of Jean Bodin's *Six livres de la République* (1576) or Pierre Charron's *De la sagesse* (1601) in order to read and digest such ideas. Yet a quick glance at Bodin's and Charron's writings enable us to have a better view of the beliefs which circulated at the time and which Shakespeare uses and distorts in *As You Like It*. According to Bodin's *Six bookes of a common-weale*, which appeared in English in 1606, climate determinism affected humankind as a whole.[26] For him, the world was divided into three climatic zones (i.e. the hot, cold and temperate ones), and while the French had like him the undeniable advantage of living in a temperate zone, the less fortunate (and far less clever) English were considered as northerners.[27] A more or less similar classification is presented in Charron's *Of wisdome*, translated in 1608. Like Bodin before him, Charron emphasises the regional and humoral typologies of the time and 'locates the phlegmatic and the sanguine in the north'.[28] He explains that 'Northerne people' turn out to be 'great eaters and drinkers' (contrary to people from the south, who are deemed 'abstinent'), are 'heavy, obtuse, stupid' and 'inconstant' (while the southerners are 'wise and subtile'), are '[l]ittle religious and devout' (contrary to 'Southerne' people, who are '[s]uperstitious, contemplative'), and that their manners are 'cruell and inhumane' (contrary to the southerners, who are '[n]o warriers').[29] While Shakespeare's Ardenites are undeniably 'great eaters and drinkers' and 'inconstant'[30] people, they are not exactly the stereotypes which Charron's classification suggests and they seem to elude such climatic bias, for if they prove rather 'sanguine', they turn out to be more clever and contemplative than one might have expected.

'The winter's wind'

In fact, the play's characters are surprisingly aware of the importance of the weather and of its effects on the way they behave. Having lost everything, they know that the sky is far from being an objective ally unless they simply decide to see it as such. In the Bible, after realising that Paradise has been lost, the first thing Adam can think of is the weather, a stance which Duke Senior seems to repeat in his postlapsarian Arden/Eden when, in his first appearance onstage, he betrays what sounds like an undeniable form of anxiety half-hidden beneath a surface bravado:

> Here feel we not the penalty of Adam,
> The season's difference, as the icy fang
> And churlish chiding of the winter's wind –
> Which when it bites and blows upon my body
> Even till I shrink with cold, I smile and say,
> 'This is no flattery' – these are counsellors
> That feelingly persuade me what I am. (2.1.5–11)[31]

If 'the penalty of Adam' refers to Adam's Fall and to the end of the eternal spring as a punishment for his original transgression of God's commandment, the phrase also ironically serves to anticipate the sudden onstage appearance of old Adam, Oliver and Orlando's aged family servant who has now reached the winter of his life ('Therefore my age is as a lusty winter, / Frosty but kindly', 2.3.52–3). As to Duke Senior's images, they are all about to be unmetaphored in the course of the play as the fangs and bites of the air ('icy fang', 'bites and blows upon my body') to which he alludes are literalised at Orlando's expense, when the young man must fight against a hungry lioness.

In Duke Senior's rambling meditations, one notes the obtrusive presence of the wind, whose importance is underlined by two strongly alliterative and perfectly regular iambic pentameters ('And churlish chiding of the winter's wind – / Which when it bites and blows upon my body', 2.1.7–8): gales are indeed part of Arden's everyday life. According to an early modern belief, Aeolus was 'standing in the Mouth of a Cave', which 'signifieth [that] the Wind [was] engendered' in earthly cavities.[32] In this world of 'caves' (2.7.204; 4.3.140; 5.4.180), the wind often blows hard. The Duke's tirade chimes in with Amiens's 'Blow, blow, thou winter wind', while the bitterness of Arden's frosty air is enhanced by Orlando when he realises that Old Adam loses his strength and blames the 'bleak' (i.e. cold) air for weakening an already debilitated old man: 'Yet thou liest in the bleak air. Come, I will bear thee to some shelter' (2.6.11–12).

Etymologically, Shakespeare's comedy can be defined as a 'tempestuous' (from the Latin *tempestas*, designating either calm or stormy weather) play, focusing on the disturbance or commotion of the wind, which here alternately highlights the exiles' strengths and their weaknesses. Whereas in the stripped-down universe of *King Lear* the wind contributes to the play's tragic outcome (Lear seems to lose his mind in the course of the storm), in *As You Like It* it proves a vital and spiritual force in spite of the dangers it represents for its weaker creatures. Orlando sees the wind as a means of revealing Rosalind's virtue to the world ('Her worth, being mounted on the wind, / Through all the world bears Rosalind', 3.3.67–8) and Duke Senior feels strengthened by winter's cold blasts.

What the Greeks called *pneuma*, or the 'vital spirit', supposedly inhabited the air: this was the assumption found in treatises like those of Hippocrates and Galen.[33] For Duke Senior, the wind is much more than a lyrical, singing presence that transforms Arden's bleak vacuum into a poetic refuge. It is a palpable sign of divinity. God manifests Himself through the breeze, a view later confirmed in the play. Against all odds, it is the sceptical Jaques who associates the wind with a form of religious presence ('I must have liberty, / Withal, as large a charter as the wind, / To blow on whom I please', 2.7.46–8),[34] as he assimilates the all-powerful wind to the Holy Spirit.[35] No wonder then if, in his quest for liberty, he somewhat sacrilegiously aspires to be endowed with the powers of the wind.[36]

One better understands, then, why the pious Duke Senior depicts windy weather whose harshness seems paradoxically 'sweet' to him. Doing so, he echoes a typically English view of the weather, found for instance in Thomas Watson's poetry, a view according to which men used to hard climatic conditions are 'seldom harmed' by the spells of bad weather:

> The man that dwelles farre North, hath seldome harme
> With blast of winters wind or nipping frost:
> The Negro seldome feeles himselfe too warme
> If he abide within his natiue coast;[37]
> So, Loue in me a Second Nature is,
> And custome makes me thinke my Woes are Blisse. (ll. 13–18)[38]

Similarly, Duke Senior's 'woes are blisse'. Such a stance seems closer to the Catholic than to the Protestant faith as, for Luther and his followers, the 'weather was a courier by which God expressed himself. Each type of meteorological event had its own distinctive message that indicated God's providence or anger' and meteorological effects 'could be spurs to conversion or pious living'.[39] So, if fair weather was a sign of God's protection, foul weather could only reflect His wrath and His displeasure at the way men and women behaved. Duke Senior does not follow this Protestant logic. A Catholic exile regretting the 'better days' when the 'holly bell [. . .] knolled [people] to church' (2.7.120–1), he is ready to see 'good'/God in everything, even in the harshest manifestations of the weather as well as in trees, brooks and stones (2.1.16–17).

However, the Duke's stoic vision[40] of his environment not only feeds on his religious faith but also owes much to the approach developed by Lucretius in the first century BC. 'In the beginning', Lucretius explains in his *De rerum natura*, 'Earth brought forth all kinds / Of plants and glowing verdure on hillsides' (Book V, ll. 785–6), and 'The world in its fresh newness would give rise / Neither to rigorous cold nor excessive heats / Nor violent storms of wind, for in a like / Proportion all things grow and gather strength' (Book V, ll. 817–20).[41] However, 'Time transforms / The whole world's nature, and all things must pass / From one condition to another' (Book V, ll. 827–9), 'All is in flux' (Book V, l. 830).[42] The world of *As You Like It* is not the golden world of the 'beginning', it is a world 'almost six thousand years old' (4.1.75–6), where Adamic innocence has been lost. Basically, it is a universe subject to the transformation process depicted by Lucretius, for whom life's ever-changing motion is encapsulated in the quick transformations of the weather:

> [. . .] though the weather of the sky
> Has been but now of limpid purity,
> Quite suddenly it becomes so foul and turbid
> That you might fancy that from every side
> All its darkness had fled from Acheron
> And filled full the great caverns of the sky;
> So rapidly does the hideous night of clouds
> Gather together, and faces of murk horror
> Hang over us on high. (Book IV, ll. 167–75)[43]

In similar fashion, the Ardenites experience the weather changes reflected in the songs at Duke Senior's alternative court, the rough weather of the beginning finally giving way to 'spring-time, / The only pretty ring-time' (5.3.15–16), thus making mutability become one of the chief features of the forest of Arden, a place of role reversals and metamorphoses.

Epicure in Arden

With the benefit of hindsight, Shakespeare's comedy partly situates itself under the influence of Epicure, as various examples and allusions seem to show. For instance, when Celia declares that '[i]t is as easy to count atomies as to resolve the propositions of a lover' (3.3.194), her words hint at Lucretius' atomism,[44] just like Phoebe's mention of eyes that 'shut their coward gates on atomies' (3.6.13). Of course, when the playwright wrote *As You Like It*, 'atom' was a poetic word that frequently referred to 'the particles of dust rendered visible by light' (*OED*, III.7), and the two quotations above certainly rely on this common usage. Yet this word could also be used to designate 'a hypothetical particle, minute and indivisible, held to be one of the ultimate particles of matter' (*OED*, II.3) – a meaning already established, if not widespread, in the Elizabethan era since it dated back to the fifteenth century ('Atom', *OED*, II.3). An open, kaleidoscopic play, *As You Like It* certainly lends itself to a philosophical approach if we accept the fact that the pondered

Celia and the tough Phoebe, undergoing the swirling motion of transformation in the forest, also allude (unwittingly or not) to an evanescent world of atoms, where Christian values are belied by an ironical scepticism. Indeed, if for the Christians God had created the world *ex nihilo*, the atomists staunchly denied any intervention of the gods in the process of creation,[45] since for them everything was the result of a fortuitous gathering of atoms, whose number was infinite and movement perpetual.[46]

It goes without saying that a first-hand reading of Lucretius was not necessary for the expression of such ideas. Shakespeare probably relied on Montaigne's *Essays* in which Lucretius is often quoted.[47] However, when he was writing *As You Like It*, round about 1599, Florio's translation had not yet been published, even though part of the translation may have circulated in manuscript. More convincing to me is the hypothesis that Shakespeare found snatches of Epicurean philosophy elsewhere, for instance in texts such as Marcellus Palingenius's *Zodiacus vitae* (Venice, 1531). In this astronomical poem written in Latin hexameters, the association with Lucretian poetry is made clear in the letter to the reader preceding the poem, which mentions both Lucretius and Epicure. This philosophical work exploring the meanings of the signs of the zodiac and correlating man's behaviour with the course of the planets was used as a schoolbook during Shakespeare's childhood. Interestingly, '[w]hen the Catholic Church placed' Palingenius's poem 'on the Index of proscribed books', it began to attract the notice of Protestant writers.[48] In 1560, the Elizabethan poet and translator Barnabe Googe rendered the first three books into English before extending his work to six books the next year 'and completing the translation as *The Zodiacke of Life* in 1565'.[49] The translation went through another two editions in 1576 and 1588. In these reprints, Googe still combined the description of the ages of man's life with the image of life as a pageant, but he added a marginal gloss of his own, 'The theater or stage of mans life',[50] a phrase that foreshadows Jaques's tirade on the 'Seven ages of man'.

However, the play shows several other interesting parallels with Googe's translation, which suggests that Shakespeare did not simply borrow an isolated passage from *The Zodiacke of Life*, but that he also espoused some of its ideas relating to the mutability of human life and to climate as well. Concerned with the order of the universe, the book 'contains one of the most powerful renaissance reappraisals of Epicurean philosophy, although its Epicureanism, clearly derived from the poem of Lucretius, is curiously blended with a strong neo-Platonic and Hermetic strand of thought' and the poem as a whole 'wishes to establish a principle of pleasure and order within the natural universe'.[51] The most significant extract from 'Capricornus, the tenth Booke', for instance, is the one to which is added the marginal gloss quoted above:

> By this occasion left I al, and from the townes I went,
> Accoumpting it more safe for me to liue in desert place,
> And in this hyll to runne the rest of al my ranging race.
> This hil whereas the altare stands of Syluester the saint,
> Where as the Shrine, and monuments of him his prayses paynt:
> Which place though it seme rough and rude, not meete to dwel vpon
> Yet fitte it is for saincts, and such as liue in peace alone,
> And wholly do apply their mindes the Lord to feare and loue,
> And alwayes seeke to ioyne themselues with saincts in Skies aboue[52]

These lines dealing with exile and solitude remind us of Duke Senior, whose retreat is repeatedly described as a 'desert', a wilderness whose hostility is, however, far from insurmountable since he can see '[s]ermons in stones' and God 'in everything' (2.1.17). In other words, the Duke is poised between the two Latin ideals of *carpe diem* and *contemptus mundi*, an attitude echoed in one of Googe's marginal commentaries: 'Often contemplation of earthly calamitie breedes contempt of the worlde and desire of heauenly joy' ('Capricornus', p. 199). For Duke Senior, though, 'heavenly joy' can wait (the denouement shows him eager to rush back to the court), which makes us realise that a degree of familiarity with Googe's translation certainly contributes to making Shakespeare's caustic irony only more biting and perceptible.

In 'Aquarius, the eleuenth Booke', which presents its readers with the astronomical characteristics of each month, Googe links the month of January to the constellation of 'Ganymedes faire'[53] and he describes winter as a 'frosty' month which sends 'his quaking colde' to the earth.[54] He criticises those who, regardless of God's teachings, 'hunt for praise' – like Duke Senior in *As You Like It* – as well as those who behave as 'Foolosophers',[55] which is certainly the case with Touchstone.[56] And he also insists on the power of the wind, a natural force likely to bring a 'nipping' cold in the blink of an eye, as is the case in Shakespeare's comedy:

> Which sundry sorts of winds do breede for no small powre is dealt
> Unto these windes, and oftentimes their force in ayre is fealt,
> These can in time of Sommer cause the nipping coldes to come

And in the winter season sende abrode the warming Sunne,
These can both giue and take away from vs our corne and grayne,
These can both breede and beare away great sicknesses and payne.[57]

Both Googe and Shakespeare see the wind as a powerful agent which can bring sickness and death as well as relief and joy. It is up to men and women to accept its devastating force and to turn it into an ally. In the same chapter, and in keeping with this, Googe asserts that, even if the climatic conditions are very harsh, men can find a refuge in nature, a place which has the capacity to instil pleasure in spite of its coldness, and which is in fact a mixture of sweetness and sourness:

Nor any clime there is, but that there mortall men may dwell
And finde out places fitte for them and seates that serue them well:
Though colde extreme, or to much heat rayne there continually:
Where nature mischiefes doth permit there plants she pleasure by
And wise she mixeth sowre with sweete, and where diseases raynes
There hath she poynted remedies that can release the paynes.[58]

This blend of 'sowre with sweete' similarly characterises Shakespeare's Arden and, more specifically, the forest's soundscape. We learn, for example, that Phoebe falls in love with the 'sweet' Ganymede not, as in Lodge, because of his/her pretty face, but because of the disagreeable tone of his/her irritated voice ('Sweet youth, I pray you chide a year together, / I had rather hear you chide than this man woo', 3.6.64–5). The verb 'chide', which occurs more frequently in *As You Like It* than in any other Shakespeare play, is part of a motif initiated by the old Duke when he praises the 'churlish *chiding* of the winter's wind' (2.1.7, my emphasis), thus making harsh sounds sweet in the musical forest of Arden.

Ultimately, what Googe's translation posits is that, for all its excesses, nature is a cooperative, even therapeutic force capable of healing men and women and encouraging both charity and humanity. In the comedy, nature's evils and harsh weather similarly generate, if not a form of pleasure, at least a feeling close to the ataraxy or imperturbability so dear to Epicure. In the light of those observations, *As You Like It* and its subliminal Epicureanism pave the way for *King Lear*, another anti-pastoral piece that can be read as a sombre echo to Shakespeare's earlier play.[59] Whereas, in the latter, Duke Senior has to cope with 'the icy fang / And churlish chiding of the winter's wind', Edgar, in *King Lear*, will have to 'outface / The winds and persecutions of the sky' (2.3.11–12), while Lear bemoans that 'the rain came to wet [him] once, and the wind to make [him] chatter' (4.6.101).[60] Taking *As You Like It* as a starting point for his dramatic and artistic investigations, Shakespeare reworked a number of key ideas so as to explore a deeper level of interaction between climate and human beings in an atomist tragedy, one even more strongly influenced by Epicureanism and where the weather does indeed bite and hurt. In *As You Like It*, hope subsists in spite, or because, of the strong wind, whereas in *King Lear*, the metaphoric dangers that informed and underlay the text of the comedy now become real.

Another fundamental difference is that, while the tragedy excludes any possibility of merriment and focuses on the sharp pangs of a stormy winter, *As You Like It* runs into several festive seasons which overlap from one scene to the next. Rewriting Ovidian mythology and Marlowe's poetry, Rosalind-as-Ganymede blames the supposedly hot Mediterranean weather ('a hot midsummer night', 4.1.81) for having caused the death of Leander, who did not drown for love but simply because he was taken with a cramp while swimming in the Hellespont (4.1.81–4). In this vignette, the hardships of climate are one with the sufferings of love and Rosalind's lines blatantly function as an ironic commentary on the play as a whole. *As You Like It* is indeed replete with ill-timed or unseasonable love stories. Touchstone and Audrey run the risk of being ill-married by the incompetent Sir Oliver Martext, who, as Jaques says, 'will but join [them] together as they join wainscoat' and, as a result, 'one of [them] will [. . .] like green timber, warp, warp' (3.4.65–7). Here, the adjective 'green' implies an 'unseasoned'[61] alliance. Another striking instance is provided by Rosalind's tongue, which curvets 'unseasonably' each time she speaks of Orlando.[62]

Frigidity and festivity

A place for all seasons, Arden turns into an unseasonable site and, as Martha Ronk notes, the forest is marred by its anti-pastoral weather. 'Holiday', she writes, 'is the wrong season.'[63] The forest, in sum, is prone to overlong winters that affect the course of human life. Pretending to cure Orlando from his love-sickness, the disguised Rosalind focuses on seasonal change as she glibly tells the young man: '[m]en are April when they woo, December when they wed; maids are May when they are maids, but the sky changes when they are wives' (4.1.117–19). If Rosalind's jibe on 'the squally changeableness of the English spring'[64] amounts to a criticism of the vagaries of love, it first and foremost implies that the return of spring at the end of the play (5.3) is a form of optical, or rather climatic, illusion. No wonder then if Touchstone finds the pages' ditty 'untenable' and meaningless (5.3.45) as he will not be fooled by a pretty song that is mere wishful thinking. So, even if Hymen intends to 'people [. . .] every town' (5.4.126) thanks to the institution of marriage, it is in fact December which permanently settles in the household, instilling coldness and sterility in the couple's relationships. This is corroborated by Hymen's declaration to Touchstone and Audrey, the mirror image of Orlando and Rosalind, at the end of the play: 'You and you are sure together, / As the winter to *foul* weather' (5.4.119–20, my emphasis). The country lass is bound to become the Fool's 'foul', a sluttish woman unable to procreate.[65]

Paired with coldness and sterility, the 'foul' Audrey – who, like Jaques, may well be infected with syphilis[66] – is not the only one associated with barrenness in the play. Phoebe bears a name connecting her with Diana, the moon goddess of chastity, a parallel which is definitely not a good omen for Silvius. As to Orlando, a younger sibling and a brave son, he looks like a possible genitor. Yet when Celia describes him, she focuses on his lips which she compares to Diana's: 'He hath bought a pair of cast lips of Diana. A nun of winter's sisterhood kisses not more religiously: the very ice of chastity is in them' (3.5.13–15). Thus associated with Diana and the winter season, the young man is linked to barrenness, which may foretell a childless and therefore unhappy union with Rosalind.

If it never rains in Arden, the weather there is bitterly cold ('Freeze, freeze, thou bitter sky, / That dost not bite so nigh / As benefits forgot', 2.7.186–8).[67] Pelting rain and flooding are part of Jaques's scriptural joke when he, at the end of the play, ironically compares the lovers paired by Rosalind to the animals rescued in Noah's ark (Genesis 7):[68] 'There is sure another flood toward, and these couples are coming to the Ark' (5.4.35–6).

Of course, the 'Ark' aptly materialises the scenic space of a metadramatic play such as *As You Like It*, particularly in the case of a public performance in a popular playhouse, since the Globe's 'wooden O' could have been perceived as some sort of seabound deck.[69] The 'Ark' carries other resonances though. The excess of water seems to herald a change in the surrounding atmosphere, as the frozen sterility of the Arden forest, so far teeming with bachelors and unmarried pairs, is about to be replaced by the more fertile climate of a new court now peopled by newly wed couples and hopefully expectant wives. On the

Fig. 5 Noah peeping out from his ark. Illustration taken from John Harding's copy of the Bible translated into German by Martin Luther. *Biblia: das ist, die gantze Heilige Schrifft*, Wittemberg, Gedruckt durch Hans Kraffts Erben, 1584. By courtesy of the President and Fellows of Magdalen College, Oxford, Magd.b.7.6.

other hand, the allusion to the Flood is surely a double-edged one here as inundations were then perceived as signs of divine punishment.[70] In the Bible, after mankind as a whole have lost faith in God, they provoke His wrath. In the play, Rosalind's own piety is rather dubious, since she only conjures up Hymen, a mock pagan deity dubiously presiding over the festivities, to have the lovers follow suit, hoping herself to be promptly and happily married to Orlando. The sardonic Jaques is the only one in the forest who sees the need for a proper priest ('Get you to church', 3.4.64). The Ardenites, Jaques implies, have caused the equivalent of the biblical Flood and, like the animals in Noah's ark, they will soon copulate in order to repeople the earth,[71] a cynical vision which can hardly be said to promote love and harmony. Interestingly, in the mystery plays which staged the episode of Noah's ark, Noah's wife, Uxor (who has no name in the Bible), is depicted as a shrew challenging her husband's authority.[72] Chaucer probably took Noah's wife as a model for his Wife of Bath,[73] and early modern audiences may have had this parallel in mind. In *As You Like It*, the cheeky Rosalind shares with Uxor an outspokenness and an authority which suggest that, in the couple she forms with Orlando, the latter, contrary to Petruccio in *The Taming of the Shrew*, may rapidly raise more pity than terror: therefore, the role reversal does not end with the play's multiple weddings but is very much paid attention to at that specific moment.

All in all, this suggests that marriage, in the comedy, is seen more as a threat than as an ideal for the male spectators, a threat not really counterbalanced by any hopes of future progeny since, symbolically at least, the cold weather does not foreshadow a warm-up of the relationships between the lovers. If, in spite of its disquieting effects, rain in *A Midsummer Night's Dream* could still be read as a symbol of fertility, coldness was associated with stillness and death. In 1599, at the time when it was presented at the Globe, the play's obsession with sterility, visible in the figure of Duke Senior (whose difficulty in procreating is enhanced by an absent wife and by his being the father of an only child), also probably served to convey the English people's anxiety about the queen's succession.

Understandably concerned by a similar problem, the Duke, at the end of the play, eagerly celebrates wedding festivities which hopefully look forward to young heirs to come. Arden's mock king seems more clear-sighted and endowed with a better sense of responsibility than Duke Frederick, the real ruler who never reflects on his own power and intentions as head of state. Beyond his apparent cheerfulness, Duke Senior shows a remarkable ability to analyse and master the prerogatives of power from the margins. This is noticeable when he presides over the slaughter of the innocent deer in the middle of nowhere. He is a thinker who seems to share Jaques's taste for meditation.[74] The latter too has acquired form of knowledge in experiencing many cold seasons. In this, he is a good illustration of the association of wisdom and winter, a fairly traditional one in the Anglo-Saxon culture[75] in which coldness and exile are supposed to encourage meditation. Jaques here joins hands with several writers having experienced isolation and freezing temperatures as a source of creativity, Ovid being one of the most eminent, exiled as he was in cold and bleak Tomis.[76] Other classical writers relied on this literary tradition, and if they did not always have to undergo the torments of exile and of an enduring cold, some of their characters are represented as having to go through this kind of ordeal. Virgil, for instance, has in his *Eclogues* the shepherd Meliboeus describe distant Britain as 'penitus toto divisos orbe' ('completely cut off from the world').[77] From the sixth century onwards, it was common to hear of holy men purposely putting themselves into the hands of nature at its coldest, since journeys in biting wintry weather were synonymous with the acceptance of a Christian life extolling the virtues of abnegation and meditation. In the case of *As You Like It*, wintriness is as physical as it is metaphysical but it certainly remains that Duke Senior, 'cast out from the sunlike position' of his dukedom, is one of the many 'frozen exiles who have inhabited English poetry for centuries'.[78]

As a banished king having to endure the hardships of exile, he asks to be constantly entertained by a jolly male court, so that the winter festival he celebrates with those around him is reminiscent of the Saturnalia of ancient Rome. This is corroborated by the fact that lords and shepherds, men and women, do meet and mix together during the wedding celebration at the close, thus harking back to the communal and egalitarian spirit of the Saturnalia which traditionally took place at the time of the winter solstice. Up to a certain point, uncertain as he is whether the return of spring will allow him to recover his former dukedom, Duke Senior plays the part of the mock king of the Saturnalia, whose festive reign was limited to the winter season. In this, he is no Prospero figure since he can avail himself of no magic books and of no power to create storms to trap, test and finally defeat his enemies. He simply tries to make the best of the bad bargain of winter and exile, humbly and bravely claiming that '[s]weet are the uses of adversity' (2.1.12).

If the Arden festivities include role-playing, songs, banquets and dances (during the wedding ceremony), they do not obliterate the savagery of the woods or Arden's harsh climate, and the saturnalian song 'What shall he have that killed the deer?' (4.2.9) more specifically reminds us of the omnipresence of death in the forest. This round proves reminiscent of the 'seasonal masquerades performed by mummers dressed in animal skins and antlers',[79] as in Shakespeare's England Christmas rituals included mummers wearing animal masks. In this connection, the disguised lords singing the deer song recall the end-of-winter maskers of early modern Europe.[80] As a result, the winter festivities of *As You Like It* are also connected to an aggressive form of virility, intrinsically linked to the month of November. In his *Ecclesiastical history*, Bede called November 'blodmonath', that is, the 'blood month',[81] thereby correlating this time of year to the annual slaughter, in early winter, of the livestock that could not be fed and that had to be sacrificed to the gods. This generally took place on the feast of St Martin, also known as Martinmas (or Martlemas), on 11 November, which announced the arrival of winter. In early modern Europe, this ritual was associated with a time of feasting, gluttony and merrymaking.[82] Meats that could be preserved were smoked or salted, and what could not be kept was consumed during the Martinmas feasts. In *The Faerie Queene*, Spenser notably alludes to this tradition as he personifies November as an overstuffed creature, 'full grosse and fat' (VI.vii.40.1). Unsurprisingly, the poet also associates November with Saturn ('For it a dreadfull *Centaure* was in sight, / The seed of *Saturne*, and faire *Nais*, *Chiron* hight', 47–8), turning this month into a saturnalian period of gluttony.[83]

Shakespeare correlates slaughter with laughter as bloody acts lurk beneath the surface of the play. 'This is no place, this house is but a butchery' (2.3.27), Adam sadly remarks as he describes Orlando's devastated house. Allusions to butchery do permeate other Shakespearean plays, and nowhere more blatantly than in *Richard III*, where the past participle 'butchered' is used five times and the plural 'butcheries' twice to refer to the shambles of Shakespeare's zany mass-murderer.[84] Yet in *As You Like It*, the attentive reader/playgoer might well discover a series of anamorphic reflections, as it were, of King Richard III: the playwright stages not one but several winter rulers, including the furious Orlando (the parodic alter ego of Ariosto's *Orlando furioso*), the butcher Oliver, the tyrant Frederick and the exiled Duke Senior. Imbued with a wintry atmosphere, both plays are characterised by a violence that may be more or less repressed.

Performance and rhetoric: Forest frost to frozen words

The cold weather in the play makes the forest a threatening universe, poles apart from the traditional Arcadian world of the Virgilian shepherds. In fact, with its repeated allusions to the harsh weather conditions, the play seems grittily true to its English environment. More generally, partly because of the 'Little Ice Age', plays composed in England during that rather long interval of time are more prone to ice and flood than is the case today.[85] *As You Like It* is no exception, of course, but the specific context of its composition helps us understand more precisely why weather issues are such an obsession in Arden.

During the Christmas holiday of 1598, Shakespeare's troupe dismantled the Theatre and transported its timbers across the nearly frozen Thames in order to settle on the Bankside, where they built the Globe. Incidentally, in late December 1598, 'the weather [. . .] is recorded as having been especially harsh [. . .] and there was heavy snow'.[86] Shakespeare probably remembered this episode when he wrote his comedy. The circular shape of his forest ('the circle of this forest', 5.4.34) tends to suggest that he associated the freezing forest of Arden – incidentally, the words 'frost' and 'forest' are near anagrams – with the world of the stage and, more generally, with the harsh climate that prevailed during the troupe's 'exile' to Southwark on the other bank of the Thames.[87] Of course, in any given play, words have the power to sculpt the air, making it colder or warmer depending on what the actors say: this is what the performative power of language is about.[88] However, when the actor's words do espouse and echo reality, an added touch of magic shows through the performance. This most probably happened with *As You Like It*. Amiens's songs were all the more gripping as the groundlings, standing frozen stiff in the pit, realised that the lyrics did not just describe the fictional setting of the play but also served to conjure up real gusts of icy wind and to recall a recent period of intense coldness. They probably experienced, then, the freedom of the forest thanks to the openness to the sky. Most of the audience had probably come to the Globe by boat, thereby making their theatre-going experience akin to that of leaving the everyday world of the city and escaping into a 'world elsewhere' (*Coriolanus*, 3.3.139), an experience paralleled by Duke Senior, Rosalind and the like in the comedy. As for the actors, they could actually look up at a ceiling painted midnight blue, decorated with stars and signs of the zodiac, and referred to as 'the heavens'.[89] As a man of the theatre, Shakespeare had no particular problem in anticipating this. So, as he drew upon Barnabe Googe's *Zodiacke of Life*, the playwright was necessarily reminded of the zodiacal motifs that traditionally decorated the London playhouses as well as the Globe's heavens, the material counterpart of his source.

All this suggests that the Globe was a particularly fit locale for *As You Like It*, even if Juliet Dusinberre argues that the play may have been performed at court in February 1599, that is, before the opening of the Globe.[90] Hers is a compelling case.[91] This hypothetical performance would have taken place at Shrovetide,[92] when the queen was at the 'rural palace' of Richmond, one of her favourite places – she was to die there on 24 March 1603. Both the Great Hall and the Great Chamber were used for royal entertainments at Richmond, and we have evidence that the Lord Chamberlain's Men were paid for a court performance on 20 February, even though we do not know the title of the play performed there.[93] If this play was *As You Like It*, the invocation of the chilling Arden weather probably held the courtly audience in its grip as they were safely sitting in a fully indoor hall. Among them, especially in wintertime, there must have been an acute sense of the strong outside vs inside opposition or, to put it differently, of the permanence of hospitable surroundings vs the disquieting mutability of the weather.

Dusinberre partly grounds her argument on an epilogue addressed to the queen,[94] which could have been written for a special performance of *As You Like It.*[95] It focuses on time and time-keeping, not only echoing the comedy's contents (Jaques's famous mention of Touchstone's dial in 2.7.20–2, among other things, testifies to his obsession with time in the clockless forest of Arden), but also linking the play to its immediate environment as the gorgeous clock then present in the courtyard of Richmond Palace had just been repaired.

We have seen above that Shakespeare rather loosely inscribes his play within the winter season festivities, which may correspond to the time when he initially wrote the play after the dismantling of the Theatre in December 1598, thus boldly juxtaposing the Christmas revels with the November celebrations. If we accept Dusinberre's hypothesis, a third calendar date coinciding with the festivities of Shrove Tuesday[96] should be added to the two periods already covered by the play. The epilogue ascribed to Shakespeare does invoke the Shrovetide season ('Many a Shrovetide here may bow / To that Empress I do now', ll. 11–12) and we can reasonably surmise that, once the playwright had learnt that his comedy would be performed at court on 20 February 1599, he reworked his text with this specific timing in mind.[97]

Paving the way for another holiday cycle, Shrove Tuesday was the starting point of the Carnival festivities. The omnipresence of the winter gale in Arden can be explained by the gusts of folly blowing through the forest, turning the place into a perpetual Carnival, a feast of fools subverting traditional norms and codes. Unable to control his own bodily functions, Touchstone himself is said to break wind ('Thou loosest thy old smell,' Rosalind tells him in 1.2.85), a remark later corroborated by a tongue-in-cheek Jaques, who professes admiration for the Fool's 'venting' his observations (2.7.41) whenever he chooses to do so and who, as a result, wishes that he too could 'blow on whom [he] please[s]' (2.7.49).[98] Tellingly enough, the word 'folly' was derived from the Latin term *follis*, meaning 'bellows' or 'wind bag'.[99] The circulation of breath, winds and farts was therefore clearly associated with the time of Carnival,[100] and in southern France in particular, Carnival processions included bellows dancers called 'soufflaculs'.

Fig. 6 'Soufflacul', misericord carved by André Sulpice, collegiate church Notre-Dame, Villefranche-de-Rouergue, France, 1473–87. By courtesy of Ciel Bleu.

By and large, Shakespeare's amalgamation of several festivals including Martinmas, Christmas and Shrove Tuesday aligns itself with the general atmosphere of this motley play, created out of a mixture of genres, genders, voices and perspectives. In this context, the chilly coldness of winter actually works as a unifying factor – a civilising factor, so to speak, for better and for worse. It is indeed the harsh weather that prompts the outcasts to gather and look for human warmth. Human assemblies in winter have always symbolised the resistance of civilisation against weather turmoil, and more often than not the roughness of the climate has been regarded as a means of testing one's fortitude.[101] In his *Essays*, Montaigne declared that, harsh as it may have been, the climate never modified his inner self, and that he favoured rainy days over sunny times:

> No weather is to me so contrary as the scorching heat of the parching Sunne [. . .] I love rainy and durty weather as duckes doe. The change either of aire or climate doth nothing distemper mee. All heavens are alike to me.[102]

This, in substance, corresponds to the professed philosophy of Duke Senior, of Jaques and of a few other characters in the play. On the face of it, '[a]ll heavens are alike to [them]', as Montaigne puts it. Shakespeare's travellers are able to cheer up in the face of adversity since the freezing temperatures of the forest do not prevent them from singing and making merry. Yet Shakespeare ironically implies that climate does 'distemper' the Ardenites, who soon realise that the blissful climate of the golden world is desperately absent from the forest. Once in the woods and under the influence of the wind, they are forced to resist, adapt and undergo a deep metamorphosis affecting their inner selves.

Strikingly enough, the playwright deconstructs language as he toys with climatic clichés. Like drama, poetry is repeatedly debunked in the course of multiple scenes, which parody the excesses of Petrarchan images and conventions. Proving unable to proffer much more than commonplaces and stereotypes, the characters suffer from the influence of the bad weather spells that ruin their pastoral ideal. Duke Senior sticks to stoic sayings, Jaques's monologue puts forward hackneyed images on the seven ages of man, Orlando is but a poor sonneteer, Corin's pragmatism verges on simple-mindedness, while the blind Silvius sees in Phoebe another Laura in spite of her brownish figure. By adopting the presumably biased views of the male gender, even the clever, witty Rosalind proves unable to overcome the worst gender prejudices (3.3.290–3). As a matter of fact, both nobles and shepherds utter thawed sentences and 'frozen words' that are strongly reminiscent of Rabelais's *paroles gelées*.[103] Should they begin to melt under the influence of warming weather, they would probably appear as totally meaningless. The title of the play is therefore not the only Rabelaisian echo to be found in Shakespeare's comedy.[104] Yet, if *As You Like It* must surely be reassessed in the wake of its numerous literary antecedents, it should also certainly be perceived as an innovative work whose multiple perspectives and reflection on language anticipate more recent writings, including symbolist ones. As in Mallarmé's swan in 'Le vierge, le vivace et le bel aujourd'hui', the figure of the poet in Shakespeare's play (be it Orlando, Silvius, Phoebe, Jaques or even Touchstone) is trapped in a frozen world and cannot free him/herself from the paralysis of an icy world in order to unbind his or her imagination and become truly creative.

Far from a simple meteorological phenomenon, climate, as it is represented in Shakespeare's second forest comedy, is a spiritual experience that puts to the fore the difficulties linked to fecundity and to the creative act in general. It influences and fashions minds, bodies and thoughts, and while the cold certainly contributes to the anti-pastoralist atmosphere of the forest of Arden, it goes beyond such literary deconstruction by promoting a new paradoxical and disturbing aesthetics, an aesthetics of whiteness, wintriness and windiness, resolutely foreshadowing the chilling atmospheres of *Hamlet* and *King Lear*.[105]

Notes

1. Stéphane Mallarmé, 'The Swan' (1885), trans. Barbara Johnson, in *Lyric Poetry: Beyond New Criticism*, p. 371. The French original reads: 'Quand du stérile hiver a resplendi l'ennui' ('Le vierge, le vivace et le bel aujourd'hui', l. 8).
2. According to Alexandra Harris in *Weatherland: Writers and Artists under English Skies*, '[t]hough some of Shakespeare's plays were commissioned for particular feasts and are associated with certain times of year, all transcend the expectations of the season' (p. 105). The present chapter does not question this but aims at putting to the fore the often-neglected richness of climatic allusions in this particular comedy.
3. Touchstone seems to live in his own temperate climate, apart from the rustic shepherds and the foolish nobles. The word 'touchstone' (*OED*, 1.a) refers to a 'variety of quartz or jasper [. . .] used for testing the quality of gold and silver alloys by the colour of the streak produced by rubbing them upon it'. The forest of Arden, first presented as a 'golden world' (1.1.95), is therefore being tested here by the Fool. This may be why, in the Folio, Touchstone's name only appears in 2.4, i.e. once he has come to live in the forest.
4. In this connection, see the song of the cuckoo and the owl which concludes *Love's Labour's Lost* (5.2.879–912).
5. Thomas Lodge, *Rosalynde. Euphues golden legacies* (1590), in *Narrative and Dramatic Sources of Shakespeare*, vol. 2, p. 163.
6. As the wrestling is about to begin, Celia wishes she had Jupiter's powers to strike Charles the wrestler and make Orlando win the fight: 'If I had a thunderbolt in mine eye, I can tell who should down' (1.2.168).
7. Once in the forest, Rosalynde and Alinda discover the love poems of the unfortunate Montanus: 'First time shall stay his staylesse race, / And winter blesse his browes with corne; / And snow bemoysten Julyes face; / And winter spring, and sommer mourne, / Before my pen by helpe of fame, / Cease to recite thy sacred name.' Ibid. p. 182.
8. Old Adam Spencer, for example, complains that 'the life of man may well be compared to the state of the Ocean seas, that for everie calme hath a thousand stormes'. Ibid. p. 194.
9. Ibid. p. 205.
10. Ibid. p. 246. One week later, Coridon is dressed 'in his holiday sute', as 'Phoebus' Henchman' had just appeared in the sky (p. 247).
11. Winter being the main season in the forest, there is no real place for the development of a full seasonal cycle. This is significant as, for a number of Shakespeare's contemporaries, the seasons were the main marker of the passage of time. See for instance Philippe Duplessis-Mornay, *A woorke concerning the trewnesse of the Christian religion*, trans. Philip Sidney and Arthur Golding, pp. 99–100: 'The Sunne maketh there his naturall course in the Zodiacke betwéene the two

Tropicks or Turnepoynts, so as the Zodiacke is as it were his race, and the Tropicks are his vtmost listes, both the which are so distinguished by degrées and minutes, that hee cannot passe one hearebredth beyond them. The poynts of his two stops are his vtmost bounds, the which so soone as he commeth at, by and by he turneth head back againe [. . .] [A]nd like as by his natural mouing he maketh the Sommer and the Winter; so by this violent moouing he maketh Day and Night. Can such succession of tymes and Seasons be made otherwise than in tyme, or rather be any other thing than tyme?'

12. Michelle O'Callaghan, 'Pastoral', p. 312.
13. Edmund Spenser, *The Shepheardes Calender*, Januarie, Ægloga prima, Argument.
14. Michael Hattaway (ed.), *As You Like It*, p. 96, n. to l. 10.
15. The first occurrence of the word in this sense dates back to 1483. The *OED* gives telling examples, including the following ones: '1525 Ld. Berners tr. J. Froissart *Cronycles* II. clxvi. [clxvii.] 500 The wether was fayre and clere, and the ayre in good temper' and '1604 E. Grimeston tr. J. de Acosta *Nat. & Morall Hist. Indies* i. ix. 33 It is a land of an excellent temper, being in the midst of two extremes.'
16. See for instance Stephen Batman, *Batman vppon Bartholome his booke De proprietatibus rerum, newly corrected, enlarged and amended*, 'De coelo et mundo', Book 8, chap. 45, p. 141.
17. Antony Ellis, *Old Age, Masculinity and Early Modern Drama: Comic Elders on the Italian and Shakespearean Stage*, p. 21.
18. 'Impression', *OED*, noun, 5, 'an atmospheric influence, condition, or phenomenon'.
19. The meaning of 'Hibernia' is notably explained in Raphael Holinshed's *Chronicles*, where its possible connections with the cold climate of the island are dismissed and a mythical Spanish origin is put forward: 'Touching the name *Ibernia*, Historiographers are not yet agréed from whence it is deducted. Some write it *Hybernia* corruptly, & suppose that the estraungers finding it in an odde end of the world, foisty and moisty, tooke it at the first for a very cold countrey, & therof named it *Hybernia*, as to say, the Winterland [. . .] Most credibly it is holden, that the Hispaniardes, the founders of the Irish, has deuotion towardes. Hispayn, called then *Iberia* [. . .] named the land Iberia [. . .] or *Ibernia* [. . .] And fro *Ibernia* procéedeth Iberland or Iuerland, from Iuerlande, by contraction, Ireland, for so much as in corruption of common talke and find that *V.* with his vocabel, is easily lost, and suppressed' (*The firste [laste] volume of the chronicles of England, Scotlande, and Irelande*, sigs. A1v–A2).
20. René Weis, 'Was There a Real Shakespeare?', p. 40.
21. Vitruvius, *The Ten Books on Architecture*, Book 6, chap. 1, p. 77.
22. Philemon Holland (trans.), *The historie of the vvorld*, The Second Booke, chap. LXXVIII, pp. 36–7. Pliny's translator adds that 'in the

middle of the earth [. . .] [t]he fashions and manners of the people are civile and gentle, their sences cleare and lightsome, their wits pregnant and capable of all things within the compasse of Nature' (p. 37).

23. Cornelius Tacitus, *Agricola and Germany*, p. 10.
24. Vegetius, *The foure bookes of Flauius Vegetius Renatus*, chap. 2, sig. Av.
25. Vegetius, *On Military Matters*, 1.2, quoted in Rebecca F. Kennedy, C. Sydnor Roy and Max L. Goldman (eds), *Race and Ethnicity in the Classical World: An Anthology of Primary Sources in Translation*, p. 51.
26. Jean Bodin, *The six bookes of a common-weale*, The Fifth Booke, chap. 1, especially pp. 549–50. For a survey of climate theories, see Lucian Boia, *The Weather in the Imagination*, pp. 38–9. Boia notes that, in his *Anatomy of Melancholy* (1621), Robert Burton devotes a whole section to 'a Digression on Ayr' where, contrary to Bodin, he 'doubts that the congruence of climate and latitude is a convincing explanation'. In fact, he thinks that it is 'the quality of air' which is the cause of human diversity and of 'the explosion' of the human race (p. 39).
27. On this, see A. J. Hoenselaars, *Images of Englishmen and Foreigners in the Drama of Shakespeare and his Contemporaries: A Study of Stage Characters and National Identity in English Renaissance Drama, 1558–1642*, p. 18. See also Lucian Boia's explanations in *The Weather in the Imagination*, p. 34.
28. Mary Floyd-Wilson, *English Ethnicity and Race in Early Modern Drama*, p. 38.
29. Pierre Charron, *Of wisdome*, The First Booke, chap. XLII, p. 164.
30. Ibid.
31. The theme of the 'season's difference' mentioned by Duke Senior is notably developed in *Paradise Lost*, X, 651–707, where Milton explains that one of the consequences of the Fall was the tilting of the earth, i.e. the change of *klima*.
32. Stephen Batman, *The golden booke of the leaden goddes*, fol. 13v. Batman starts his short section devoted to the god of winds by explaining that 'Aeolus was figured standinge in the Mouth of a Caue, in his hand a Tortoyse: Vnder his Feete a payre of Bellowes'.
33. See also the explanations supplied in Pseudo-Aristotle, *Cosmic Order and Divine Power*, p. 31: 'From the dry exhalation, when it is pushed by the cold so that it flows, wind is created, for this is nothing but a lot of air moving together. It is at the same time also called breath [*pneuma*]. Breath is also used in another context for the animating and productive substance found in plants and animals which pervades all things.' For more on *pneuma*, see below, Chapter 4, pp. 127–37. On the 'vitall spirit', see also Philemon Holland (trans.), *The historie of the vvorld*, chap. XXXVIII, p. 18.
34. 'Blow', of course, also means 'blight' here. Cf. Canterbury's words in *Henry V* (probably written around the same period as *As You Like It*): 'when he speaks, / The air, a charter'd libertine, is still' (1.1.47–8).

Tellingly enough, in *As You Like It*, Duke Senior accuses Jaques of having been a sensual 'libertine' (2.7.65).

35. Cf. King James Bible: 'The wind bloweth where it listeth' (John 3:8): the wind is here the metaphor signifying the presence of the Holy Spirit.
36. Rosalind has a more pragmatic vision of the wind. She mentions it when she rebukes Phoebe for her harshness toward Silvius and blames the latter for his obstinacy: 'You, foolish shepherd, wherefore do you follow her / Like foggy South, puffing with wind and rain?' (3.6.49–50). The 'foggy South' evokes the south wind that, in the medieval and early modern periods, was thought to carry pestilence, as is explained in a plague treatise (c. 1364) translated into English by John Jacobus, the royal and papal physician and the Chancellor of Montpellier: 'Also it is good for a patient [. . .] to spar the windows against the south. For the south wind has two causes of putrefaction. The first is that it makes a man, whether whole or sick, feeble in his body. The second cause is as it is written in Aphorisms, chapter 3, the south wind grieves the body and hurts the heart because it opens man's pores and enters into the heart. Wherefore it is good for a whole man in time of pestilence when the wind is in the south to stay within the house all day.' Quoted in Bryon Lee Grigsby, *Pestilence in Medieval and Early Modern English Literature*, p. 107.
37. Note from the author: 'For both experience teacheth & Philosophical reason approoueth, that an Ethyopian may easily in Spaine be smothered with the heat of the countrey though Spaine be more temperate then Ethyopia is.'
38. See Thomas Watson's 1582 sonnet sequence, *The hekatompathia or Passionate centurie of loue*, LVII ('All yee, that greeue to thinke my death so neere'). Preceding the poem are these explanations: 'Here the Authour cheerefully comforting himselfe, rebuketh all those his frendes, or others whatsoeuer, which pitie his estate in Loue: and groundeth his inuention, for the moste part, vpon the old Latine Prouerbe, Consuetudo est altera natura. Which Prouerbe hee confirmeth by two examples; the one, of him, that being borne farre North seldome ketcheth colde; the other of the Negro, which beinge borne vnder a hote climate, is neuer smoothered with ouermuch heate.'
39. Craig Martin, *Renaissance Meteorology: Pomponazzi to Descartes*, p. 59. Martin explains that Luther's own 'decision to enter the Augustinian order allegedly followed his being nearly struck by a bolt of lightning, which, according to his own account, he interpreted as a sign meaning that he should become a friar' (p. 53).
40. For further remarks on the meaning of stoicism at the end of the sixteenth century, see for example Renée Kogel, *Pierre Charron*, p. 53.
41. Lucretius, *De rerum natura*, Book V, p. 207. Unless otherwise stated, I rely on R. C. Trevelyan's translation (1937).
42. Ibid.

43. Ibid. Book IV, p. 135.
44. For the Epicureans, who were materialist philosophers, only atoms (tiny particles always in motion) and the void existed.
45. For a counter-atomist view, see Manilius, *Astronomica*, I, p. 43: 'Who could believe that such massive structures have been created from tiny atoms without the operation of a divine will, and that the universe is the creature of a blind compact?' Counter-atomist ideas in the early modern period are, for instance, developed in Thomas Heywood's 1609 *Troia Britanica*, Argumentum 2, Canto I, sig. B: 'This Universe with all therein contained / Was not at first of Water fashioned, / Nor of the Fire, as others oft have feyned, / Nor of the Ayre, as some have vainly spred, / Nor the foure Elements in order trained. / Nor of Vacuitie and Atoms bred.'
46. Thomas Langley, in *An abridgement of the notable woorke of Polidore Vergile* (1546, fol. 4v), provides an early modern view of atomism: 'Epicurus, one of Democritus' disciples, putteth two causes, atoms or motes, and vacuity or emptiness, of these he saith the four elements come.'
47. See above, Chapter 1, p. 46 and p. 55, n. 76.
48. Neil Rhodes, 'Time', p. 290.
49. Ibid.
50. Barnabe Googe (trans.), *The Zodiacke of Life*, p. 194.
51. Hilary Gatti, *Giordano Bruno and Renaissance Science*, p. 106.
52. Barnabe Googe (trans.), *The Zodiacke of Life*, p. 194.
53. Ibid. p. 210.
54. Ibid. p. 225.
55. This expression first appeared in Thomas Chaloner's 1549 translation of Erasmus's *Praise of Folly*. See Erasmus, *The Praise of Folie*, [trans.] Sir Thomas Chaloner, p. 10, l. 32.
56. Barnabe Googe (trans.), *The Zodiacke of Life*, p. 215.
57. Ibid. p. 222.
58. Ibid. p. 226.
59. See Nancy Lindheim, '*King Lear* as Pastoral Tragedy'. For a climatic analysis of *King Lear*, see below, Chapter 5, pp. 150–75.
60. The parallel has been developed by Frank McCombie, 'Medium and Message in *As You Like It* and *King Lear*', p. 76.
61. See Michael Hattaway's note to 3.4.67, p. 166.
62. Celia to Rosalind as the latter begs to get further news of Orlando: 'Cry "holla" to thy tongue, I prithee: it curvets unseasonably' (3.3.204).
63. Martha Ronk, 'Locating the Visual in *As You Like It*', p. 179.
64. Catherine Bates, 'Love and Courtship', p. 103. Cf. Shakespeare's *The Two Gentlemen of Verona*, in which spring 'now shows all the beauty of the sun, / And by and by a cloud takes all away' (1.3.86–7).
65. Cf. 3.6.26–8, a passage where Touchstone portrays Audrey as a 'foul slut'.

66. See 'foul', *OED*, A.I.1.b: 'Of a disease or a person affected with disease: Loathsome. †the foul disease (also evil): *(a)* epilepsy, *(b)* syphilis, etc.' Syphilis was then 'targeted in relation to sterility' and was thought to be the cause of monstrous births. See Margaret Healy, *Fictions of Disease in Early Modern England: Bodies, Plagues and Politics*, p. 126.
67. It must be noted though that, according to Theophrastus, 'if the winter is dry, the spring will be fair' (*Concerning Weather Signs*, p. 423).
68. Cf. Ovid's *Metamorphoses* (trans. Arthur Golding, The First Book, ll. 296–522, pp. 39–46), which narrate the survival of Deucalion and Pyrrha and the subsequent repopulation of the earth. On the Flood narrative and its origins in Genesis, see Stephen Greenblatt, *The Rise and Fall of Adam and Eve*, pp. 46–7. The author explains how the Genesis storyteller relied on the *Atrahasis*, a Mesopotamian account of the deluge.
69. On this, see Louis B. Wright, 'Elizabethan Sea Drama and its Staging'.
70. Michael Kempe, 'Noah's Flood: The Genesis Story and Natural Disasters in Early Modern Times'.
71. Cf. Benedick in *Much Ado About Nothing*, 2.3.229–30: 'The world must be peopled.'
72. See Sara M. Butler, *The Language of Abuse: Marital Violence in Later Medieval England*, p. 230. See, too, A. C. Cawley (ed.), *The Wakefield Pageants in the Towneley Cycle*, pp. 19–24. In this mystery play, Uxor clearly jeopardises people's lives because, in spite of the patient explanations given by her husband, she refuses to board the ark as the waters rise higher and higher. Instead, she decides to spin wool, forcing everyone to wait for her.
73. Melvin Storm, 'Uxor and Alison: Noah's Wife in the Flood Plays and Chaucer's Wife of Bath', p. 306.
74. In Shakespeare's age, Jaques's propensity for meditation would have been explained by several factors, including the 'temperature' of his brain. According to Pierre Charron, '[t]he temperature which serueth, and is proper to the vnderstanding is drie, whereby it comes to passe that they that are striken in yeeres, doe excell those in their vnderstanding that are yoong, because in the braine as yeeres increase, so moisture decreaseth. So likewise melancholicke men, such as are afflicted with want, and fast much (for heauinesse and fasting are driers) are wise and ingenious' (*Of wisdome*, The First Booke, chap. XIII, p. 48).
75. Alexandra Harris, *Weatherland*, p. 30. Winters, Harris explains, 'were the great measure of Anglo-Saxon life' (p. 31).
76. Ibid. p. 25. See also Lucian Boia, *The Weather in the Imagination*, p. 15. For an early modern reference to Ovid's plight, see for example Backwinter's lines in Thomas Nashe's weather comedy *Summer's Last Will and Testament*, p. 135: 'Ovid could well write of my tyranny, / When he was banish'd to the frozen zone.'
77. Virgil, *Eclogues*, in *Eclogues, Georgics, Aeneid 1–6*, 1.66.

78. Alexandra Harris, *Weatherland*, p. 113. Harris actually refers to *Richard II*.
79. N. to l. 8–17, p. 186.
80. Interestingly, Rosalind-as-Ganymede also refers to these mumming rituals later on in the play, when she pretends to detect an insulting tone in Phoebe's *billet-doux*: 'Why, she defies me / Like Turk to Christian' (4.3.31–2). In a number of St George mumming plays performed at Christmas time (for which the only evidence we have unfortunately dates back to the nineteenth century), the Christian knight fought against a Turkish one 'in the name of "Mahound"' (see Michael Hattaway (ed.), *As You Like It*, p. 188, n. to l. 32). Rosalind's topical reference to Christmas games is part and parcel of a larger network of allusions, which also comprise Amiens's 'Blow, blow, thou winter wind' song (2.7).
81. Bede, *The Complete Works of Venerable Bede*, vol. 6, pp. 178–9.
82. Steve Roud, *The English Year*, p. 469.
83. A slaughtering and feasting ritual strongly reminiscent of this glutton November is present in Arden, where the banquet organised by the lords in 2.7 results from the killing of the deer ordered by Duke Senior in 2.1 ('Come, shall we go and kill us venison?', l. 21). We should also note that the motif of the wounded deer (2.7.26–43) re-enacts the legend of St Hubert, the bishop of Liège known as the 'Apostle of the Ardennes' (c. 657–727). Hubert's feast being celebrated on 3 November, one may better understand why, in Arden, the 'bitter sky' 'freeze[s]' and 'bite[s]' (2.7.186). It is one of these typical November skies that turn the rain into early winter snow. On the medieval figure of St Hubert as an analogue to Jaques, see Robert N. Watson, *Back to Nature: The Green and the Real in the Late Renaissance*, p. 83.
84. *Julius Caesar* is another play teeming with references to butchery. See Richard Wilson, 'A Brute Part: *Julius Caesar* and the Rites of Violence'.
85. In 'Climate', Mike Hulme explains that the 'central European climate was on average perhaps 1 to 1.5 degrees colder during Shakespeare's lifetime than during our own' (p. 32).
86. Mowat, Barbara A. and Paul Werstine (eds), *As You Like It*, p. xli.
87. Several metatheatrical references in the play point to the world of the public stage. For example, Jaques's comment 'O knowledge ill-inhabited, worse than Jove in a thatched house!' (3.4.7) should be read as an allusion not only to the thatched home of Baucis and Philemon in Ovid's *Metamorphoses* but also, in the context of a public performance, to the thatched roof above the actors' heads. See Anne-Marie Miller-Blaise, '"No temple but the wood[en]": Stagecraft and the Staging of the Theatre's Materiality in Shakespeare's *As You Like It*', pp. 202–3.
88. I refer here to John Langshaw Austin's seminal study *How to Do Things with Words*.

89. See above, Introduction, p. 6 and p. 25, n. 21.
90. It may have opened as early as June 1599, but certainly not before. See Steve Sohmer, '12 June 1599: Opening Day at Shakespeare's Globe'.
91. Juliet Dusinberre, 'Pancakes and a Date for *As You Like It*', p. 388. One week later, another play, *Fortunatus*, 'was performed by the Admiral's Men at Richmond Palace on Thursday 27 December'. In the audience sat 'Elizabeth I; ladies'. We know that 'Robert Shaa was later paid £20 for this and one other performance [. . .] dated Sunday 17 February 1600'. See Martin Wiggins, in association with Catherine Richardson, *British Drama 1533–1642: A Catalogue*, vol. 4, *1598–1602*, p. 171.
92. On the correlations between court performances and the festive season, see R. Chris Hassel Jr, *Renaissance Drama and the English Church Year*.
93. On this, see Chris Butler, '"The howling of Irish wolves": *As You Like It* and the Celtic Essex Circle', p. 91.
94. 'As the *dial* hand tells o'er / The same hours it had before, / Still beginning in the ending, / *Circular* account still lending. / So, most almighty *Queen*, we pray, / Like the dial day by day, / You may lead the seasons on, / Making new where old are gone.' This epilogue is reproduced in Hattaway (ed.), *As You Like It*, Appendix I, p. 222.
95. William A. Ringler and Steven W. May, 'An Epilogue Possibly by Shakespeare'. Inserted in Henry Stanford's commonplace book, this epilogue was discovered by Steven W. May in 1972.
96. Shrove Tuesday was 'traditionally situated on the first Tuesday after the first new moon in February'. The 'first possible date for Shrove Tuesday [. . .] coincide[d] almost day for day with the date popularly believed to mark the beginning of spring' (François Laroque, *Shakespeare's Festive World: Elizabethan Seasonal Entertainment and the Professional Stage*, p. 85).
97. For further details, see Juliet Dusinberre who, in 'Pancakes and a Date', notes that Touchstone oddly refers to pancakes in one of his most laborious jokes addressed to Rosalind (1.2.52–3, 1.2.56–63). So, because 'Queen and court were eating pancakes in the Great Hall at Richmond for Shrove Tuesday night' when Shakespeare's company started acting *As You Like It*, Touchstone's rather feeble jest was actually meant to be comic (p. 380).
98. In his 'Seven ages of man' speech, Jaques describes the voice of the 'sixth age' as 'piping' and 'whistling' like the wind (2.7.162–3).
99. On this, see Claude Gaignebet, *À plus hault sens*, vol. 1, pp. 88–9, and François Laroque, '"Motley's the only wear": *As You Like It* ou la bigarrure', pp. 85–6.
100. See below, Chapter 4, p. 132.
101. In this connection, see Thomas Combe's 1593 *Theater of Fine Devices*, Emblem XLIII, 'When one meane faile, then by and by / Another

meane we ought to try', n.p. The engraving represents a galley caught in a sea-storm and the epigram reads: 'When winds do stifly beate against the saile, / Yet Galleys may by the maine force of ore, / So much against the spite of winds preuaile.'

102. Michel de Montaigne, *Montaigne's Essays* (Book 3, chap. IX, 'Of Vanitie'), vol. 2, p. 218.
103. Regarding the episode of the *paroles gelées*, it is worth remembering that Pantagruel and his companions discover the existence of those disembodied voices (resulting from a war fought during a particularly harsh winter) when they approach the confines of the icy seas. The elusive manifestations of language and its commonplaces are one of the main themes in Rabelais's 1552 *Quart livre* (*Le Quart livre* in *Les Cinq Livres de faits et dits de Gargantua et Pantagruel*, chaps. LV and LVI, pp. 1132–40).
104. This refers to 'Fay ce que vouldras', the famous motto of Abbaye de Thélème, in Rabelais's 1532 *Gargantua* (in *Les Cinq Livres de faits et dits de Gargantua et Pantagruel*, chap. LII, pp. 374–6). Another reference to Rabelais appears when Celia tells Rosalind that she 'must borrow [her] Gargantua's mouth first' (3.3.189).
105. On *King Lear*, see below, Chapter 5, pp. 150–75. *Hamlet* is set in a freezing December atmosphere as Francisco, one of the two guards on duty on the ramparts of Elsinore, complains that ''[t]is bitter cold' and that he is 'sick at heart' (1.1.6–7). Later on, Hamlet further acknowledges that '[t]he air bites shrewdly, it is very cold' (1.4.1). On the 'native characteristics of cold Denmark', see Mary Floyd-Wilson, 'English Mettle', p. 142.

Chapter 4

Othello: Shakespeare's *À bout de souffle*

May the winds blow till they have wakened death (2.1.187)

The psychological and spiritual experience of climate in the not-so-green world of the forest of Arden gives way to an elemental and humoral one in the vaster blue world of *Othello*.[1] In a 1601 treatise on passions, the Roman Catholic priest and religious controversialist Thomas Wright (c. 1561–1623), a member of the Essex circle for a brief period of time, describes human emotions in terms of maritime and weather images which are indicative of how climate, humours and the elements combined in the English popular imagination at the turn of the century: 'we may compare the Soul without Passions to a calm Sea; with sweet, pleasant, and crispling streams; but the Passionate, to the raging Gulf swelling with waves, surging by tempests, menacing the stony rocks, and endeavouring to overthrow Mountains', after which he concludes that 'even so Passions make the Soul to swell with pride and pleasure; they threaten wounds, death and destruction by audacious boldness and ire [. . .] and, in sum, never let the Soul be in quietness, but ever either flowing with Pleasure or ebbing with Pain'.[2] In 1596, Wright had warned Essex's secretary, Anthony Bacon, against a possible Spanish invasion, which may explain why his prose imagery here contains possible reminiscences of the battle of the English fleet against the Spanish Armada.

Composed only a few years after the publication of this treatise, Shakespeare's *Othello* also depicts the travails of a passionate soul in a world of elemental passions, full of swelling water and battering winds. Probably written around 1603–4, the play is based on a number of myths, legends and popular romances that perfectly fit the genre of the love tragedy which Shakespeare initiated with *Romeo and Juliet*. Yet it can also lend itself to a symbolic, climatic interpretation. It shows how spring (which began on Valentine's Day in popular calendars), the season of love when the carnivalesque winds of February announce the erotic breezes of Venus, is ruined by the premature advent of autumn under Saturn's influence. This calendrical and astrological scheme is brought to the fore thanks to an overall cosmological pattern that lays bare the links between human emotions and the natural environment. This explains why *Othello* appears as particularly appropriate in a study of climate in Shakespeare, all the more so as the playwright portrays a tragic hero who, shaped as he is by his geographical origins as well as by his cultural environment, proves particularly vulnerable. As he presents the questions, qualms and limits of his protagonist, the playwright revisits the influence of the planets while, in the dramatic economy and imagery of his playtext, he makes rough seas mirror inner conflicts so as to enhance the function of *pneuma*, as argued below.

Right from the start, climatic issues are hinted at when Iago feeds Roderigo's resentment by telling him that Othello dwells 'in a fertile climate' (1.1.70).[3] The lovelorn Roderigo would certainly like to enjoy this 'fertile climate', that is, 'a state of apparent good fortune and happiness'[4] which has so far been refused to him. Iago's image also implies the existence of a sharp contrast between two different climates as it suggests that, if Othello now lives in a prosperous and highly civilised city, he was born in the dry plains of Africa. The 'fertile climate' is the one enjoyed by Venetian women, whose reputation for licentiousness is repeatedly underlined in the play ('I took you for that cunning whore of Venice / That married with Othello,' the Moor tells Desdemona in 4.2.91–2),[5] the better to be pitted against the markedly different atmosphere of male, martial and fortified Cyprus, an island ironically renowned as Venus' birthplace.[6]

Principally set in this Venetian-controlled island, the play owes much to Giraldi Cinthio's *Gli Hecatommithi* (Decade 3, Story 7), a series of 'hundred tales' published in 1565. According to Cinthio, Disdemona and the Moor 'lived in harmony and tranquillity while they were in Venice'.[7] Then the Moor is dispatched to Cyprus and 'he embark[s] upon the ship with his wife and all his train; and having given the order to set sail, he set[s] off on the journey to Cyprus, aided by *a very calm sea*' (my emphasis).[8] He and Disdemona, therefore, travel in the same ship. As to the wicked ensign, he falls in love with Disdemona and, logically enough, he decides to 'prevent the Moor from enjoying her favours', subsequently luring him into believing that his beautiful wife is having an affair with the Corporal. As a result, the Moor 'torment[s] his brain [. . .] and bec[omes] very melancholic'.[9] His melancholy verges on madness as the hero's inner turmoil causes his environment to be altered. But while Cinthio insists on the peaceful meteorological conditions of the Moor's voyage to Cyprus and, in so doing, denies that climate may have had an influence on the mental state of his protagonist, Shakespeare's dramatised version of the story focuses on formidable winds and rough seas, thus suggesting that the weather did contaminate the physiology and emotions of the title character insofar as it contained the seeds of the forthcoming tragedy. Indeed, '[f]rom the beginning of the play Othello is associated with outdoor weather, with openness'[10] for, as Iago puts it, '[t]he Moor is of a free and open nature' (1.3.388). However, this 'open nature' is gradually changed into its contrary and, instead of facing the outside world, the play retreats within closed doors until it eventually moves in a narrow, claustrophobic space.

Planetary conjuctions

Oddly enough, Herbert R. Coursen finds no evidence in *Othello* 'of a set of cosmic coordinates behind or within the world of the play'.[11] I will argue on the contrary that, as was already the case in *Romeo and Juliet*, albeit in a less systematic way, the fate of the eponymous hero is also predetermined here by his name and by planetary influences. That the Moor's destiny should partly reside in Iago's acerbic address as 'his Moorship' (1.1.32) is illustrated in Florio's *Worlde of Wordes*, where the word *moria* is defined as 'an infection, a pestilence, a murrain, a rot or mortalitie that comes among sheepe. Also used for follie and taken from the Greeke.'[12] De facto, the Moor's *moria* plagues his love story after his jealousy has turned to folly. As part of his delusion, Desdemona's handkerchief is associated with the course of the sun after Othello, under Iago's influence, has made it a piece of evidence of her adulterous affair with Cassio:

> A sybil, that had numbered in the world
> The sun to course two hundred compasses,
> In her prophetic fury sewed the work; (3.4.69–71)

While the exact meaning of these lines remains uncertain,[13] the connection between the spotted handkerchief and the sun metaphorically adds cosmological connotations to the textual web of the tragedy. Othello's words corroborate this in the play's final scene:

> It is the cause, it is the cause, my soul –
> Let me not name it to you, you chaste stars:
> It is the cause. (5.2.1–3)

About to strangle his wife in a quasi-orgasmic gasp, the Moor turns to the stars in three jerky lines replete with monosyllables. For him, the planets stand for 'cosmic justice',[14] and superstition prevents him from giving out the 'cause' of his foul deed. Yet the syntax of the first line suggests that Othello actually names the cause without even being aware of it: his diseased 'soul' is the very cause of his act. Moreover, the word 'soul' is associated with the word 'stars' since both appear at the end of two successive lines. In effect, the play shows how stars and planets influence the Moor's destiny – or how Othello, under Iago's pressure, convinces himself that they do. A few lines later, just before he commits suicide, the titular character once more resorts to stellar imagery when he describes Desdemona as an 'ill-starred wench, / Pale as [her] smock' (5.2.271–2).

So, the matrix of the tragedy partly resides in the sky and in its wandering stars whose conjunctions heavily weigh on human fate. Metaphorically speaking, the play is placed under the auspices of the goddess and planet Venus. The love story of Othello and Desdemona begins in Venice (with its possible echoes of the name Venus) before it moves to Cyprus.[15] In spite of their changing places, they seem doomed from the very start: in the Republic of Venice, the couple was already the victim of social and racial prejudices and Othello was openly accused by Brabantio of having resorted to witchcraft in order to seduce Desdemona. In the Serenissima, just as in 'the fair Verona', the planet Venus 'smile[s] not' (*Romeo and Juliet*, 4.1.8) and the lovers' passage to Cyprus will only exacerbate the pre-existing tensions instead of smoothing them over. The astrological impact of Venus here seems partly responsible for the couple's dire destiny, as in Robert Greene's 1585 *Planetomachia*, where the unfortunate and malignant effects brought about by the planet Saturn annihilate the hypothetically favourable influence of Venus.[16] In Shakespeare's play, the beneficent influence of Venice/Venus similarly gradually fades from view as the jealous husband becomes inclined to follow Iago's crooked ways.

Whether or not a simple coincidence, the year 1603, when the play was presumably written, happened to be marked by a solar eclipse with 'a conjunction of Saturn and Jupiter in Sagittarius', a rather ominous phenomenon then supposed to be the cause of a bout of the plague in London.[17] Symbolically present in the tragedy in its implicit opposition to Venus,[18] the planet Saturn was traditionally associated with lead, so that the 'poison' (3.3.327) instilled in the Moor's veins by Iago's words suggests the idea of saturnism, an illness due to slow lead poisoning, which then often afflicted those who worked as printers. Given his proximity to the trade, it is likely that Shakespeare had heard about it.[19]

Besides Saturn, a number of other threatening mythological figures are associated with 'black Othello' (2.3.29). For instance, Iago's description of the general as a 'Barbary horse' (1.1.111) in the obscene account he gives to Brabantio of the love affair between Othello and Desdemona hints at the mythological episode of 'Europa with Jupiter disguised as a bull'.[20] Furthermore, the allusion to an inn reputed to be the Moor's favourite haunt (1.1.154–8) and called the 'Sagittary', with a sign representing a centaur (i.e. a grotesque version of Cupid),[21] concurs with the play's astrological bricolage and helps configure an evil conjunction of the planets. Sagittarius was indeed considered as a 'Descendent' and an 'Autumnall' sign, one not really auspicious for Othello's personal trajectory.[22]

The play as a whole being informed by astrological concerns, it is particularly rich in references to the moon. The most unexpected one crops up in the very first act of the tragedy. After not being engaged in any sort of war for nine months (the time of female pregnancy), the general feels insecure and vulnerable as an idle soldier:

> For, since these arms of mine had seven years' pith,
> Till now some nine moons wasted, they have used
> Their dearest action in the tented field, (1.3.84–6)

As the text substitutes the phrase 'nine moons' for the usual 'nine months', it contributes to blurring the traditional gender borders, thus implicitly introducing a threat to Othello's professed manhood.[23] In act 3, the moon is personified as Diana ('Dian's visage', 3.3.389), a traditional symbol of virginity mentioned by the general in opposition to Desdemona's supposedly stained purity, a passage which looks ahead to the lines when Coriolanus emphasises the chastity of his wife Valeria.[24] However, while the image may seem slightly out of place in a world dominated by patriarchy and military values,[25] it keeps its full strength in *Othello*, all the more so as the cliché is later reworked when the Moor, in deeply insulting terms, portrays his wife as a dark, sullied, contaminated moon. As the play draws to an end, Shakespeare introduces variants which shift from a physiological frame to a cosmological one: in act 4, scene 2, the anthropomorphic moon is once more seen as darkened, this time because of the absence of light possibly generated by the obtrusion of the sun ('the moon winks', 4.2.77). Othello's mention of the chaste winking moon, oddly associated with an unchaste wind, points to a lunar eclipse[26] – maybe inspired by a *real* partial eclipse of the moon which took place on 24 May 1603[27] – a reference which is expanded and intensified as the crisis erupts. In act 5, when he is alone with Desdemona, whom he has just killed, the Moorish general expects no less than 'a huge eclipse / Of sun and moon', wishing that 'th'affrighted globe / Should yawn at alteration' (5.2.102–4).[28] Excessive as it may sound, this adynaton (or hyperbolic ejaculation) is by no means unexpected in such circumstances, given that eclipses, then regarded as bad omens, were thought to foreshadow particularly dark times.[29] As he refers to the brutal and tragic ending of his marriage, Othello's allusion to a simultaneous eclipse of sun (i.e. the Moor himself) and moon (Desdemona) does not simply imply that he is about to enshroud himself in darkness, but it also entails cosmic consequences: the world itself is threatened with collapse as a result of Othello's deed.

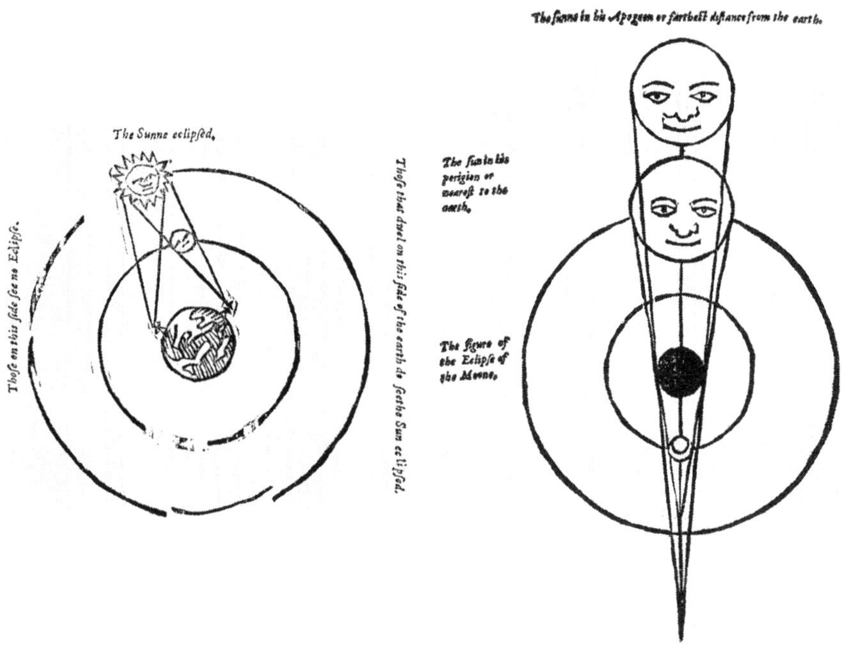

Fig. 7 & 8 Sun and moon eclipses in John Blagrave, *The mathematical ievvel*, 1585. By courtesy of the British Library Board, C.60.o.7, pp. 56–7.

At this moment, the Moor realises that he has triggered a major catastrophe likely to shake the planet as a whole. As Nick Davis explains, the yawning of the earth mentioned by Othello 'can be taken to be an opening of fissures by disturbances of the earth; Pliny's *Natural History* notes a natural coincidentality of earthquakes with eclipse, in an account carrying different philosophical weighting from *Revelations* 6'.[30] Since, in the theory of humours, yawning was regarded as a means of evacuating foul air, this image reinforces the correspondence between Othello's physiology and the cosmos at large. Hippocrates thought that yawning preceded a fever since it allowed the surplus of air to be evacuated through the mouth when bodily temperature increased.[31] Similarly, once Othello's inner temperature has reached its peak, the play is turned into a tragedy of heated passion as much as a cosmological piece in which breath, wind and climate play a central role. The idea that the earth yawns after Desdemona's mouth has been shut forever is a telling trope that highlights the tragic correspondence between the little body of man and nature at large, between micro- and macrocosm.

The Moor suddenly feels constrained by the heavens – both the sky and the roof of the Globe[32] – so that he believes he has no choice but to stage his own tragedy and play the murderer (the 'affrighted globe' designating here the gasping audience in attendance). In this context, the spectacle of his dead beloved wife becomes part and parcel of a dramatic economy and of a cosmological vignette: the presence of Desdemona's dead body seems to point to nothing less than a lunar eclipse prefiguring a tragic denouement.[33]

The idea that the celestial orbs had an influence on human behaviour is entertained by several characters in the play, and by Iago in the first place, since he uses and abuses his interlocutors' credulity to make them blame planetary influences rather than ponder their own actions.[34] In act 2, scene 3, for example, after Montano has been wounded by the drunken Cassio, Othello intervenes and receives the following answer when he asks Iago who started the brawl:

> I do not know. Friends all, but now, even now,
> In quarter and in terms like bride and groom
> Divesting them for bed; and then but now,
> *As if some planet had unwitted men*,
> Swords out, and tilting one at other's breasts
> In opposition bloody. (2.3.170–5, my emphasis)

Besides suggesting that bride and groom may go to bed together and then 'tilt' at each other's breasts, thus anticipating the hoped-for tragic finale he is preparing in secret, Iago evokes early modern astrological beliefs according to which the planets had the power to drive men mad. Othello increasingly sees himself as 'more sinned against than sinning' (*King Lear*, 3.2.59) and mainly as a victim of the skyey influences, then strongly resenting what he interprets as the heaven's cold indifference ('by yon marble heaven', 3.3.460) to his tragic fate which then allows Iago to take the 'elements' (3.3.464) as the main witnesses of the impending tragedy.[35] These 'elements', defined as 'atmospheric agencies or powers' (*OED*, n. 11),[36] become the silent observers of a mock union between the misguided Moor and his Machiavellian manipulator, a grotesque couple about to displace the initially romantic union of Othello and Desdemona. The sky is turned into a desacralised church sealing a deadly, unnatural bond between the two men.

Of moors and humours

In the play, an unfavourable climate causes the links of love to be systematically severed because it deregulates common or natural human behaviour. In order to comfort the lovelorn Roderigo and rally him to his cause, Iago embarks on a prose monologue that offers a perverse variation on the theological commonplace of man's body as a garden ('Our bodies are our gardens to the which our wills are gardeners', 1.3.316–17). This speech primarily testifies to Iago's rhetorical skills. Man must resort to reason to control his passions ('but we have reason to cool our raging motions', 1.3.325–6) and he must take care of his body like a good gardener watering his plants. Besides parodying a well-known biblical image, the passage also presents the human body as a microcosm whose temperature must remain constant. So, when Iago goes on to say that '[t]hese Moors are changeable in their will' (1.3.340–1), he portrays the general as a mutable human being governed by his affects and unable to regulate his own physiological system.

In Venice as in Cyprus, the threat of distemper looms large. In the opening scene, as Iago and Roderigo wake up Brabantio in the middle of the night and summon him to prevent the Moor from eloping with his daughter, the infuriated father refuses to believe such an impossible tale and describes them as two drunkards playing at disturbing his sleep in the Venetian night.[37] Later on, Cassio's drunkenness will represent a similar form of intemperance, a leitmotif in the play. In *A delicate diet, for daintiemouthde droonkardes* (1576), George Gascoigne regards drunkenness as firmly ingrained among the bibulous English and depicts it as a 'monstrous plant, lately crept into the pleasant orchards of England'.[38] As to Thomas Heywood, he observes in a pamphlet entitled *Philocothonista* (1635) that 'the cooler the climates are, the more the inhabitants are addicted to strong and toxic drinks', so that, according to him, the English are 'the most forward to commit this grievous and abominable sin of drunkenness'.[39] As he taps into such clichés, Shakespeare inserts a national stereotype correlated to England's cold climate in an exotic play set in Venice and Cyprus.

In juggling with geo-humoral stereotypes, be they complementary or contradictory, the playwright reconfigures his tragic setting. He once more uses these clichés in act 3, scene 3, when Iago gradually manages to persuade Othello that Desdemona is having an affair with his lieutenant. As he hammers the point home, the ensign insists on Desdemona's defective temperament and he presents her presumed luxuriousness as some sort of unavoidable fact given her geographic origins ('her own clime', 3.3.234). Born in Venice, Desdemona is bound to be sexually voracious,[40] even though the popular saying was then aimed at 'women over the age of 50'.[41] Desdemona, who is of course much younger, is bawdily portrayed by Iago earlier on as 'framed as fruitful / As the free elements' (2.3.326–7) and described as generous and productive, like nature, a sly way of dissociating her from culture and rationality. The ensign then efficiently relies on an eroticised speech to make the foolish Roderigo believe that Desdemona's sexual impulses cannot be satisfied by her ageing husband: 'When the blood is made *dull* with the act of sport, there should be a game to inflame it' (2.1.221–2, my emphasis). The adjective 'dull' here reminds us that onomastics often provided the playwright with a form of shorthand that enabled him to shape the audience's reception. As a matter of fact, the word 'Moor' came from the Latin *morus* and was punningly associated with the Greek *moros* meaning 'dull'. Iago's insinuations thus make the point that Desdemona cannot be satisfied because of her husband's reputed dullness. In such circumstances, adultery with a younger male becomes a viable and credible alternative to satisfy an oversexed young woman.

After Desdemona has begged the Duke to allow her to accompany her husband to Cyprus, Othello's words ('I therefore beg it not / To please the palate of my appetite, / Nor to comply with heat the young affects / In my defunct and proper satisfaction', 1.3.259–62) sneakily equate, in a negative formula ('I [. . .] beg it not'), marital desire with sexual appetite and thus appear in retrospect as mere bragging. Othello pledges to the Duke that he will not lose his martial courage as a result of his marital duties. So, who are we to believe? Othello, who, in the passage just quoted, claims to be an ardent partner, or Iago, who refers to the general as a rather mediocre lover? Shakespeare lets his audience decide, and rather than asserting a definitive truth, he prefers to provide them with various alternatives.

By the same token, any approach of the play in terms of systematic climatic patterns is likely to remain unconvincing. Iago's depiction of Othello is absolutely not coherent, for coherence is not what he cares about: as a true sophist and an arch Machiavel, he resorts to all kinds of arguments to undermine his rival. Contradictory as his successive points of view may be, what matters is that his interlocutor should believe in them. So Iago presents Othello as a quasi-impotent lover in order to please Roderigo while he does not hesitate to portray him elsewhere as a 'lascivious' (1.1.125) or 'lusty Moor' (2.1.286), as in the early modern clichés about blackness and African men.

A masque of blackness

Because the Moor[42] is first and foremost seen as an alien (either as an Arab or as a dark-skinned African), his presence suffices to 'conjur[e] up a complex, confused cluster of images associated with landmarks, geographical characteristics, climate, distinctive faun and flora, and ethnic, linguistic, or cultural diversity'.[43] Just like his Venetian spouse, Othello is regarded as a human being whose behaviour and inclinations are being predetermined by his native climate. He is portrayed by Iago as 'an erring Barbarian' (1.3.348–9), that is, as an inhabitant of the land of Barbary. In his 1526 *Description of Africa* (*Descrittione dell'Africa*), Leo Africanus informs his readers that the name 'Barbar' probably came from the word 'Bar' 'twice repeated' – *Bar* meaning 'desert' in Arabic. The traveller and geographer John Pory reproduces this explanation in a 1600 translation dedicated to Robert Cecil which was probably used by Shakespeare as a source for the play.[44] Like his model, Pory distinguishes the black from the tawny Moors, but Shakespeare blurs the line between the two, and Othello, sometimes depicted as black, sometimes seen as a 'Barbary horse' (1.1.111), corresponds both to the 'white or tawnie Moores' of northern Africa and to the 'Negros or blacke Moores' who inhabited further south.[45] Contrary to the Venetians, his identity in the play remains rather elusive: what matters is that he should be regarded as an outsider, as a stranger '[o]f here and everywhere'

(1.1.136), as Roderigo puts it, and that he is bound to remain a stranger accustomed to the hot temperatures of his native country while unused and unsuited to the damp, cold climate of Venice.

Again, as Pory considers the etymology of the proper noun 'Africa', he explains that Festus derived it from *phrikè*, a Byzantine Greek word signifying 'horror or colde', the letter 'A' in *A*-frica being a privative prefix signifying that 'Africa is a place free from all horror and extremitie of colde, bicause it lieth open to the heauens, and is sandie, drie, and desert'.[46] Pory adds that '[i]n situation & shape this land of Africa is almost an isle, being by a very small and narrowe neckland (passing betweene the Mediterran sea and the gulfe of Arabia, alias the red sea) conioined to Asia'.[47] The 'greater part' of the country was situated between the tropics 'vnder the Torrid or burnt Zone, for which onely cause the ancient writers supposed it to be vnhabitable and desert in so many places'.[48]

In such a context, Othello's behaviour and humours must be the natural consequence of geography and of the hostile climate characterising the burnt zone where he was born. If Richard Knolles's translation of *The six bookes of a common-weale* was not yet available when Shakespeare wrote his play, most of his contemporaries, as made clear earlier in this book,[49] were already conversant with Bodin's views, not least because the French writer relied on a common stock of so-called scientific ideas and offered his readers an educated summary of age-old speculations.[50] As a result, the idea that 'the Southerne nations are weake, little, blacke'[51] and that 'the people of the South are cruell and reuengefull, by reason of melancholic, which dith inflame the passions of the soule with an exceeding violence, the which is not easily suppressed' was certainly not new to Elizabethans and Jacobeans alike.[52] They could all presume that southerners were witty people who were 'inwardly cold', all their heat being absorbed by the hot climate they live in, whereas, by contrast, northerners proved 'inwardly hoat'.[53]

In *Othello*, in line with Bodin's logic, the eponymous hero first appears to be rather calm and cool, able as he is to manipulate his audience with well-chosen words and to resist the prejudices of the Venetians. However, in his revenge plan, Iago wants him hot, so to speak, and he will invest all his dark energy in making and portraying him as such. It is therefore under his ensign's influence that the Moor's emotional reactions begin to change from act 3 onwards.

As Jonathan Gil Harris points out, '[t]here is a long history, dating back beyond Pliny, of attributing dark hues to hot climates' and

'Africans were often thought to be susceptible to an excess of black bile, thanks to the burning-up of their other humours by the heat of the sun'.[54] This is made clear right from the start. When Brabantio addresses Othello as a barbarian stranger ('such a thing as thou'), he is flabbergasted to hear that his own daughter has 'run from her guardage' to his 'sooty bosom' (1.2.71–2). 'Soot' evidently signals the darkness of ash or coal but it also refers to the myth of skin-scorching heat, and the word reappears when the Moor[55] admits that passion has 'collied' or blackened his 'judgement' (2.3.199) with coal.[56]

Such associations were not uncommon in the literature of the period. In *Englands Parnassus* (1600), compiled by Robert Allott, three lines entitled 'Affection' and attributed to William Shakespeare run as follows:

> Affection is a coale that must be coolde,
> Else suffered, it will set the heart on fire,
> The fire hath bounds, but deepe desire hath none.[57]

This finds an ironical echo in Othello's threatening Desdemona that he 'should make very forges of [his] cheeks / That would to cinders burn up modesty' if he spoke her 'deeds' (4.2.74–6). Heated by Vulcan's fire, he is so incandescent in his fury that it is impossible for him to speak of Desdemona's alleged crime without running the risk of catching fire.

These lines once again point to the early modern European climate-based theories of darkness. In this context, the white, immaculate Desdemona, who is like the little 'milk-white' flower of *A Midsummer Night's Dream* before it is hit by Cupid's love-shaft, can only be stained by Othello's ardent passion and end up 'purple with love's wound' (*A Midsummer Night's Dream*, 2.1.167). This is reinforced by the interpretation of her murder as a moment of ultimate *jouissance*. The whole play is teeming with words related to heat. Iago, in act 1, scene 1, alludes to the flare-up of nocturnal fires that sometimes spread 'in populous cities' (1.1.77), an image behind which lies the archetypal destruction of Troy by fire at night. When Michael Cassio, who has been sent to look for him with lights and torches, finds Othello late at night, he tells him that he has been 'hotly called for' (1.2.44) as the island of Cyprus is now in great danger of being seized by the Turks, which he describes as 'a business of some heat' (1.2.40). If the word 'heat' is here associated with a sense of urgency, it is also laden with the geo-humoral stereotypes that characterise

Othello and which he, in turn, has made his. Iago sees him as a dull creature 'rash and very sudden in choler' (2.1.263) and Othello unwittingly endorses his ensign's views. Trying to find out who began the quarrel which forced him to leave his marriage bed, he gradually yields to his choler before realising that his passion, 'having [his] best judgement collied, / Assays to lead the way' (2.3.197–8).

The multiple pun on coal ('coal'/'collier'/'collar'/'choler') in the first scene of *Romeo and Juliet*[58] resurfaces in this mature love tragedy ('collied'), grafted as it is on the myth of Vulcan here embodied by the coal-black Moor who, like the blacksmith god, has his face 'begrimed' (3.3.389) and aims at surprising his wife and his rival in the heat of love-making just as Vulcan had in the case of Venus and Mars. In such a context, the handkerchief may be read as Othello's mock, or grotesque, version of Vulcan's net. Yet, instead of arousing the laughter of the gods gathered on Mount Olympus, it ends up in Desdemona's tragic death. Anger (i.e. his choler) blinds the Moor's judgement and blackens (i.e. 'collies') his behaviour while also mapping, so to speak, his own earthly prison which '[b]urn[s] like the mines of sulphur' (3.3.331). In the throes of agony, waiting for the 'ocular proof' (3.3.362) of Desdemona's adultery, the general calls for hell on earth[59] and asks Iago to rule the sky – to which Iago cynically replies 'O heaven forgive me!' (3.3.375), sharing Othello's line as if he shared his grief, only to add insult to injury. This topsy-turvy world, where the good lover becomes the devil and the wrongdoer rules the sky, leads to a tragic carnival[60] which delineates an infernal topography where 'fair is foul and foul is fair' (*Macbeth*, 1.1.10).

The fact that the general is depicted as hot and dry, that is, as choleric – by his enemies, at least – according to both his bodily humours and his astral determinism subsequently endows his actions with a sense of ineluctability. This seems to deprive him of his own free will, except perhaps at the end when he commits his spectacular suicide trick, and it thus partially accounts for Desdemona's general powerlessness. Incidentally, her perception of her husband is also partly shaped by climate theories that do not necessarily align with those of Iago. Where the ensign insistently depicts Othello as a heated man, she can only see his coldness. To her, he is a southerner at heart, an individual characterised by 'an overly intellectual temperament' (her husband, in the play, appears as more of a wonderful storyteller than a real warrior) which, far from suggesting a hot humour, is in fact indicative of an inward coldness – a disposition which accounts for the Moor's melancholy and which is also the cause of his lustful drives.[61] It is this particular belief that will lead Desdemona to tell Emilia that 'the sun where he was born / Drew all such humours from him' (3.4.30–1).[62] Curiously, in Cinthio's source story, Disdemona says of Othello that he is 'so hot by nature that any little thong moves [him] to anger and revenge'. In other words, Disdemona adopts the Hippocratic view according to which southerners are under the influence of heat and blood. In *Othello*, the playwright takes a radically different stance as the all too naive Desdemona follows Aristotle's precepts and affirms on the contrary that the hot African climate has dried up Othello's jealous humours.[63] Trying to empathise with her husband, she, too, falls victims to climatic stereotypes.

Othello resembles the Moor portrayed in Cintho's source story. Being brought by Iago to an excessive heat, he longs for cool air and cool reason but he unfortunately reaches the expected coldness only at the end when, in front of Desdemona's corpse, he exclaims: 'Cold, cold, my girl? / Even like thy chastity' (5.2.274–5). It is probably no coincidence then that Othello should propose to use 'a sword of Spain', a powerful weapon whose steel had been hardened by being plunged red-hot into ice-cold water (5.2.252). The phrase 'ice-brook's temper', which describes this metallurgical technique in the same line, indirectly refers to the current association between temperature and temperament in the play.[64] For Jonathan Hall, '*Othello* is an exploration of the crisis of the sign within the patriarchy, where the scopic drive seeks stabilization through murder'.[65] This is true, but as Othello's Toledo sword suggests, this scopic drive first seeks stabilisation through icy coldness. The Moor will not 'shed [Desdemona's] blood, / Nor scar that whiter skin of hers *than snow*' (5.2.3–4, my emphasis).

Women and water

If Othello's murder of Desdemona betrays his aspiration to coldness (which Aristotle considered in his *Meteorology* as an active 'determinan[t]', just like heat),[66] the sea-storm which Shakespeare added to Cinthio's storyline draws our attention to the play's wetness (which Aristotle viewed, by contrast, as a passive 'determinabl[e]', together with dryness)[67] and suggests that 'an ontology of torments, waves and turmoil' underpins the play as a whole.[68] Desdemona, in particular, is portrayed as a passive 'weaker vessel' (1 Peter 3:7) after she arrives in Cyprus, and as a wet and cold creature, predetermined by these two humoral features and factors.

Water being the female element par excellence, one realises that, if Othello's spouse seems surprisingly submissive in front of her husband's violence, her strong spirit and personal energy actually permeate a tragedy which presents a gendered climatology. Othello, Roderigo and Brabantio alike try to dominate and repress a female figure alternately presented like a virgin and a whore because they are frightened by her sexual powers. Even if they try to remain in their own male sphere, safely protected from female evil (abandonment, betrayal, cuckoldry), they are contaminated by the feminine elements and fluids (tears, rain, the sea) which, in the play, are let loose in a damaging way.

Roderigo thus hints at his own incontinence, prone as he is to wet dreams (1.3.301) in the absence of the coveted Desdemona.[69] Here, leakiness becomes a masculine trope that questions the traditional clichés about gender difference. Brabantio too is associated with overflowing fluids. Dismayed by the news that his daughter married Othello, he confesses to the Duke that his grief is 'flood-gate', that is, torrential, and that its nature is 'o'er-bearing' (1.3.57). This passage ironically resonates with Othello's desperate tirade on torrential rains later on in the play ('Had it pleased heavens / To try me with affliction, had they rained / All kind of sores and shames on my bare head / [. . .] / I should have found in some place of my soul / A drop of patience', 4.2.47–53). Had he been Job himself facing insufferable torments and weather catastrophes, he would have patiently endured his burden. However, in Othello's world, there is no God to trigger natural cataclysms, only his own heart, traditionally compared to a fountain before being likened to a 'cistern for foul toads' (4.2.57–62). More than a simple humour, melancholy here becomes a microclimate which generates monstrous creatures through some sort of spontaneous generation. In other words, Othello's damp physiology produces 'a sticky cloacal swamp heaving with amphibious life', a life itself representative of 'sexual promiscuity'.[70] The Moor's body is a place of both generation and corruption, just like the earth in the old geocentric astronomy[71] since, for Helkiah Crooke, the human body, which he calls 'this Little worlde', has its own 'Meteorology': '[t]he humor and moistnesse that fals like a Current or streame into the empty space of the throate, the throttle and the chest, resembleth raine and showers' and 'teares do represent the Dew', among other instances.[72] Earlier in the play, Othello regards humidity as a corrosive agent when Brabantio's men draw their swords, ready to attack him in case he resists being arrested. In his answer, the Moor ironically warns them against the dangers of dampness at night: 'Keep up your bright swords, for the dew will rust them' (1.2.59). Moisture later on becomes synonymous with Desdemona's lechery as, to Othello's eyes, her moist, sweating hand betrays her love for Cassio (3.4.34).

Intriguingly, Desdemona is the first who, in the Folio text, alludes to the violence of the storm liable to ruin people's lives. Indeed, in act 1, scene 3, when she confesses the intensity of her feelings for the Moor to the Senate, hence betraying a form of erotic anxiety, the 1622 Quarto and the 1623 Folio texts reveal some strong differences:

> That I did love the Moor to live with him,
> My downright violence and scorn (Q1) [**storm (F, Q2)**] of fortunes
> May trumpet to the world. My heart's subdued
> Even to the very quality (F, Q2) [**utmost pleasure (Q1)**] of my lord.
> (1.3.246–9)[73]

The 'storm of fortunes', in both the Folio and the 1630 Quarto, emphasises the potentially explosive violence of Desdemona, making her more ambiguous than she seems at first sight. Valerie Traub sees in this passage a 'frank admission of erotic desire' as the woman's 'storm of fortunes' (i.e. her sexual desire), paired with the verb 'trumpet', actually 'allows a conflation of consonants, producing "strumpet" – exactly the charge Othello makes' in the last act.[74] In the paintings of the period, particularly in those representing Ovidian stories, landscapes with a large sky full of dark clouds not only suggest an impending storm but also often hint at an imminent rape. Here, the stormy setting depicted by Desdemona foreshadows what is to become a chaotic love relationship with Othello and the increasing possibility of a physical assault finally actualised in Othello's smothering his wife under a pillow, a form of frightening rape followed by death.

The explosive natural energy linked to water and wetness is reiterated in one of Othello's speeches which suggests that, at pains to accept and control his own femininity (symbolised by the ebbing sea), he can only turn it into a form of eruptive violence. In the middle of the play, as his internal turmoil becomes unbearable, Othello compares the intensity of his grief to the power of the Pontic tide in a passage reminiscent of Pliny (3.3.453–60).[75] While the vast sea first served as a metaphor of Othello's intense love (1.2.24–8), it now conveys the irrepressible motion or the incessant flux of his 'bloody thoughts'. These same 'bloody thoughts' are narrowly linked to bloody flows in the play, not least because the naval engagement against the Turks, the cause of Othello's last-minute departure to Cyprus, turns out to be an aborted sea-battle.

The tragedy of *pneuma*[76]

Set at the edge of western civilisation, Cyprus is 'a town of war' (2.3.204) which was also a favourite destination 'for seagoing Jerusalem pilgrims, the last-save-one harbour on the Holy Land itinerary'.[77] This Mediterranean enclave, often presented by critics as a 'colonial outpost',[78] here becomes a place which symbolises sensual passion and represents a new step in Othello's personal (and incomplete) 'pilgrimage' (1.3.153). On yet another level, Cyprus was associated with a number of topical events. Act 2, scene 1 represents an early turning point in the plot which signals the presence of strong gales (a notorious Shakespearean invention) almost immediately after the first soliloquy of the play, in which Iago unveils his plans and his hatred of the Moor.[79] Underscoring the violence of the wind, Montano, the governor of Cyprus, declares that '[a] fuller blast ne'er shook [their] battlements' (2.1.6), a turmoil that foreshadows the fierce domestic storm which is awaiting Othello and Desdemona in Cyprus. The rough and 'ruffian' wind (2.1.7), likely to destroy the strongest foundations (the Duke mentions Cyprus's 'fortitude' in 1.3.221), is metaphorically replicated inside Cyprus's fortress in Iago's repeated efforts to blow away the love-bonds between Othello and Desdemona. In other words, the ensign becomes here the ill wind that blows nobody any good.[80]

Jan Glete makes it clear that, '[a]s the wind was the only available source of propulsion power for major warships, its direction and strength was important for strategy and tactics'.[81] The English fleet actually acquired the necessary skills to face harsh winds and dangerous currents much later than pioneer navigators like the Portuguese and the Spanish, who had been using reliable instruments of pilotage since the mid-fifteenth century. It was not until 1553 that it became 'initiated [. . .] into the secrets of the wind system which controlled the success or failure of voyages to the coast of Guinea' thanks to the help of António Anes Pinteado and Francisco Rodrigues, two Portuguese seamen.[82] By the end of the sixteenth century, the English art of navigation had incorporated a whole gamut of new techniques and England's sailors were now better equipped to avoid destructive winds and to find more favourable ones.

Mastering the forces of nature is a latent obsession in *Othello*, where the roaring gales in the Mediterranean cause the defeat of the Turks in their sea-battle against Venice – a fact duly noted by the reader of the heavily annotated First Folio of the Kodama Memorial Library of Meisei University, Tokyo, who observes in the margins:

'Terrible tempest descried [*sic*] and shipwrak.'[83] Early on in the play, the Turkish threat is presented as a great danger, and it proves all the more terrifying as it is invisible. It thus leaves the spectators free to imagine that the Turks are making for Cyprus 'with a most mighty preparation' (1.3.220). Yet the depiction of their looming presence soon gives way to a less flattering portrait of the Turks, regarded as poor navigators unable to rule the winds from act 2, scene 2 onwards.

In this scene, Othello's herald indeed proclaims 'the mere perdition of the Turkish fleet' (2.2.3), a blunt phrase in which '*m*ere *p*erdition' is aptly substituted for the '*m*ighty *p*reparation' evoked in the first act, and which suggests that love will from then on replace war in Cyprus. However, when in act 3 the word 'perdition' resurfaces in Othello's speech ('perdition catch my soul', 3.3.91), the audience realises that the drowning of the Turkish fleet is nothing but a pyrrhic victory as far as the Moor's personal destiny is concerned. What it leads to in the end is the triumph of chaos and hatred over harmony and marital love. Therefore, the aborted battle against the Turks, with its possible allusions to the Invincible Armada episode of September 1588, when tempestuous winds and extreme weather conditions off the Scottish and Irish coastlines caused many Spanish ships to wreck and led to the victory of the English fleet,[84] drastically changes the course of Othello's life. Interestingly, Shakespeare's subtext implies that the weather, an unpredictable factor which is here supposedly controlled by divine will, deserves more credit than the Venetian fleet for securing the island of Cyprus for the Republic of Venice. The gales are thus presented in a positive light before being implicitly linked to the deterioration of the domestic sphere and to the play's tragic finale.

But if the herald's lines have often been understood as a hint at the Spanish Armada, they also incorporate a side glance at the battle of Lepanto (7 October 1571), which almost exclusively involved oared ships and which saw the victory at sea of the coalition of European Catholic forces against the Turks. The image of the battle remained vivid in the minds of Shakespeare's contemporaries, particulary those who knew of *His Maiesties Lepanto, or Heroicall Song*. James's epic poem on Lepanto was published in 1591, became rapidly famous, and was reprinted in 1603 immediately after his accession to the throne of England, a date which somehow corresponds to the time when Shakespeare was writing *Othello*.[85] During this 'bloudy battle bold',[86] however, the weather, first misty, then breezy, was relatively clement.

A third layer of reference must probably be added if we remember that, before the battle of Lepanto, the Turks had taken Cyprus in August 1571.[87] In his *Generall historie of the Turkes* (1603), Richard Knolles does not only celebrate the 'great and notable battell between the Turks and the Christians, commonly called, the battle of Lepanto',[88] he also reminds his readers of the major 'losse of Cyprus',[89] acknowledging that the conquest of the island had been a crucial victory for the Ottomans.[90] Nightmarish memories of the siege of Venetian-controlled Famagusta in particular seem to have haunted Shakespeare's European contemporaries. It lasted eleven months, and during the winter of 1570–1 the commander of the Ottoman navy, Piali Pasha, began showing signs of weakness as he failed to prevent a Venetian squadron from bringing in supplies. Famagusta's hopes were short-lived, though. Piali Pasha was dismissed and the Cypriot city, desperately lacking relief, was eventually captured in August 1571. The renewed evocation of such an infamous episode in Knolles's *Generall historie* did nothing, of course, to alleviate the fear of the Ottoman savagery.[91] The alert Jacobean playgoers who attended a performance of *Othello* would certainly notice that the first scene of act 2, taking place in a fortified place presumably protected from the elements, was reminiscent of Famagusta.

So, in his evocation of the sea-battle against the Turks, Shakespeare combines memories of the 1588 defeat of the Invincible Armada with the 1571 loss of the last bastion of Christendom in the east and the triumph of the Holy League against the Ottoman fleet in the same year,[92] so as to create a fictional *millefeuille*, rewrite European history, and foreground the unsaid, albeit decisive, element of any maritime battle, namely the weather conditions. He imagines a storm easily interpretable as a form of divine intervention leading to the scattering of the Turkish fleet. At the beginning of act 2, an accelerated rhythm suggests increasingly turbulent weather. The First Gentleman tells Montano, the governor of Cyprus, that the sea is so rough (a 'high-wrought flood', 2.1.2) that he cannot make out any sail in the distance, and the governor himself remarks some minutes later that 'the main and th'aerial blue' (2.1.40) are virtually indistinguishable. The Second Gentleman then observes that '[t]he chidden billow seems to pelt the clouds' and that '[t]he wind-shaked surge with high and monstrous mane / Seems to cast water on the burning Bear / And quench the guards of th'ever-fixèd Pole' (2.1.12–15).[93] At a no less rapid pace, the Third Gentleman eventually announces the good news of the providential tempest ('The desperate tempest hath so banged the Turks / That their designment halts', 2.1.21–2). The audience soon learns that, on his way to Cyprus, Othello has had to face this 'foul and violent tempest' (2.1.35) which temporarily cut him off from his faithful lieutenant Cassio. The same storm, or, as Cassio puts it, 'the great contention of sea and skies' (2.1.92), has also parted Desdemona from her husband. Yet Shakespeare makes the Turks the main victims of the weather while the Venetian camp is obviously spared. So it is a jubilant Moor who reappears to deliver an ecstatic, lyrical speech:

> May the winds blow till they have wakened death,
> And let the labouring barque climb hills of seas
> Olympus-high, and duck again as low
> As hell's from heaven. (2.1.181–4)

The tempest here sounds like an orgasmic natural force which multiplies Othello and Desdemona's sexual desire in erotically loaded lines which take their significance from a long-standing tradition going back to antiquity. From the classical period onwards, scholars regarded *aer*, one of the four elements, as a vital principle, so that what they also called *pneuma* played a crucial function in the regulation of body and mind. In *Othello*, *pneuma*, together with its external counterpart, *ventositas* (i.e. the swelling of the sail), proves a determining factor in the love story of Desdemona and the Moor as the wind was commonly reputed to favour sexual intercourse.[94]

Iago is certainly someone who hopes that his plans will eventually prove successful and who relies on the power of *ventositas* to make his boat 'sai[l] freely, both with wind and stream' (2.3.59). Contrary to the general, he proves able to achieve complete control of the air throughout the play and it may be significant that, at the end, the ensign orders attendants to fetch a chair to carry the wounded Cassio in order to 'bear him out o'th' air' (5.1.103), a reminder that fresh air was then thought non-desirable for injured people.[95] An agent of evil, a past master in Hippocratic medicine and humoral theories, Iago is a proto-climatologist of sorts who perfectly knows how to use the elements to his own advantage and thus easily manipulates Othello. The latter is then progressively portrayed as a sick, frustrated warrior who can only channel his morbid libidinal impulses into a perverted relationship with his young wife.[96] The ensign-bearer therefore gradually becomes the only truly active character in the play.[97]

Following Aristotle and his medieval commentators, early modern thinkers believed that airiness could 'help to extend the genital member but also ejec[t] semen into the female'.[98] Shakespeare's tragedy of desire and jealousy partly relies on the airy element defined as vital heat. Whether lacking or in excess, air shapes the play as a whole. It is no accident that Brabantio feels so oppressed in the opening scene (1.1.142). Brabantio feels so oppressed in the opening scene (1.1.142). While heat and the penury of air stifle the atmosphere in Venice, the action of the wind in Cyprus unleashes human passions and has a strong impact on human behaviour and rhetoric. But as the air is gradually rarified by Iago's insinuations, Othello finds it more and more difficult to breathe under the shocking revelation of his wife's lust and adultery, and his bedchamber finally turns out to be a claustrophobic space.

It is against this background that 'wind instruments' (3.1.6) are given pride of place. In act 3, scene 1, a group of musicians are playing the bagpipes. This short episode, followed by a conversation between a musician and a clown, is laden with scatological innuendoes and half-veiled allusions to the Neapolitan disease (3.1.3–10), that is, to syphilis. Of course, Cassio's carefully planned serenade is spoilt by the clown's obscene remarks. This Bruegel-like scene[99] may be regarded as a grotesque counterpart of the main plot, staging wind instruments that produce a flatulent noise.

Fig. 9 Pieter Bruegel the Elder, *The Peasant Dance*, 1568. By courtesy of the Kunsthistorisches Museum, Vienna.

In other words, this apparently superfluous comic passage literalises the wind trope that permeates the playtext. While there are no bagpipes as such in the main plot, most characters in the tragedy function as wind instruments, or windpipes, issuing forth breath and sounds. Othello, in particular, takes one step further the rough music presented in act 3, scene 1, before turning his wife into a wind instrument. Once in Cyprus, he plays ecstatically upon it until, like a wayward child, he breaks his toy so that no more sound is heard in the end. The brutal obstruction of Desdemona's windpipe signals the cessation of human music and the definitive censoring of feminine voices.

The circulation of air thus has successively comic and tragic connotations in the play. It corresponds to the performance of some dark carnival – carnival being traditionally associated with the feast day of Saint Blaise or Blasius, the patron saint of blasts, who was reputed for his ability to heal throat aches.[100] Here, it is also linked to charivari, a form of popular justice meant to punish cuckolds. In the play, the repeated allusions to these holiday celebrations suggest that part of the action takes place in the month of February, just before the springtime revels of lovers and happy unions, a period which does seem to fit in with the nuptials which Othello means to celebrate and consummate on his arrival in Cyprus. In such a pattern, the Venetian location in act 1 would be linked to the upcoming spring. Yet we have seen that, under the influence of the planet Saturn, the spring festivities are abruptly interrupted in favour of a premature autumnal period preparing for the couple's eternal winter. The play's seasonal cycle is oddly and untimely brought to an end as summertime seems virtually non-existent in the Cypriot enclave where the tragedy unfolds. Recreation and regeneration are replaced by suspicion and melancholy, and the cosmology of fertility which should have characterised the play, had it been a comedy, is suddenly subverted into a cosmology of tormenting blows. The crucial function of the wine, in Iago's plot (the villain's successful scheme largely depends on his ability to make Cassio drunk), also contributes to creating a Dionysian frame coinciding with autumnal celebrations and paving the way for the general debauchery and decadence of the whole Cypriot garrison. Besides, this Dionysiac, autumnal cycle replacing Venice's spring festivities is hinted at in Iago's indirect comparison of Othello with a bear about to hibernate, as the ensign tells the general to 'encave [himself] / And mark the fleers [. . .] / That dwell in every region of his [Cassio's] face' (4.1.77–9). These lines do not simply evoke 'the savage landscape of Othello's travel narrative' of act 1, scene 3,[101] but they also obliquely bring to mind popular accounts of bears starting their winter naps and going back to their caves.[102] The verb 'encave' here indirectly refers to hibernation and reminds the audience that the bear was traditionally regarded as a representative of chthonian forces. As such, the beast is indirectly associated with black Ot*hell*o, a wild man of sorts whose name contains the word 'hell' in its centre, just as the name Des*demon*a ominously points to the underworld (ruled by Dis, or Ha*des*, the god of the inferno), thereby adding dark overtones to the personality of Shakespeare's heroine from act 2 onwards. Moreover, the bear imagery harks back to the Second Gentleman's vivid description of 'the burning Bear' (i.e. the constellation of Ursa Minor, 2.1.14) watered by an exceptionally harsh tempest in act 2, scene 1.[103] It therefore simultaneously refers to the chthonian animal that hibernates in winter and the star helping navigators to find their *bear*ings at sea.

Under the influence of sky and planets, the burning, liberating wind of desire in the Venusian city of Venice[104] gradually becomes a deadly wind of suspicion in Cyprus, where amorous couples are made to suffocate and succumb. More than a simple airy power, this hellish breath becomes an ominous subterranean force when Iago alludes to earthquakes that will soon make Cyprus tremble ('[Othello's] infirmity, / Will shake this island', 2.3.118–19). Prone to epilepsy, the general progressively loses his bearings, causing the whole island to tremble. In his *Natural Questions*, following Democritus' opinion on the subject, Seneca writes that earthquakes are sometimes produced 'by moving air, sometimes by water, sometimes by both'.[105] He assumes for instance that the earth can be 'infiltrated by winds' and that, moving air being 'too subtle to be excluded', its pressure causes earthquakes.[106] This is replicated in the little world of Othello's physiology: confined inside his lungs, the air, for lack of release, makes him quiver and collapse.

Given this analogy, one understands why Iago revels in the blowing out of rumours and fake news ('blown surmises', 3.3.185). His task is facilitated by Othello's 'infirmity' which makes him swollen and shaking.[107] In one of the possible sources of the play, Geoffrey Fenton's *Certain Tragicall Discourse* (1567), the jealous husband, before murdering Desdemona, falls into a fit of frenzy, 'with owlinge, crying, and foaming at the mouth'.[108] In the Venetian part of Shakespeare's play, this particular weakness is never mentioned, while it becomes an issue in windy Cyprus. First referred to in act 4 (4.1.46), it is described by Iago in terms strongly reminiscent of Fenton's narrative: 'The lethargy must have his quiet course; / If not, he foams at the mouth, and by and by / Breaks out to savage madness' (4.1.49–51).

Considered by the ancients as one of the worst illnesses of the soul, epilepsy, whose main cause is attributed to the air, is documented at length in the Hippocratic treatise *On Breaths*.[109] In 1583, Philip Barrough advised patients suffering from epilepsy to 'remaine in an ayer, that is temperate, pure, bright, and cleare' and to 'eschewe grosse and cloudie ayer'.[110] In his *De rerum natura*, Lucretius also describes epilepsy in a particularly vivid and graphic way by repeatedly resorting to powerful storm metaphors:

> [. . .] it often happens that some man,
> Seized suddenly by the violence of disease,
> Falls down before our eyes, as though by a *stroke*
> *Of lightning*: foaming at the mouth he groans,
> Shivers throughout his frame, loses his reason
> [. . .]
> He foams, endeavouring to eject his soul,
> Just as upon the salt sea the waves boil
> Beneath the *mastering vehemence of the winds*.[111]

It is surely no accident that Shakespeare should pay so much attention to rough seas and winds in the first half of the play[112] before turning to Othello's agitated emotional state in the second half, and to his pathetic attempts at 'eject[ing] his soul', to take up Lucretius' phrase. Such frenzy is part and parcel of the 'melancholic symptoms' that characterise the title character.[113] What Robert Burton called 'flatuous' or 'windy melancholy' was for him 'the most grievous and frequent',[114] and this is precisely the very mental illness which Othello suffers from and which affects his 'blown up rhetoric'.[115]

In another influential work, namely *On the Sacred Disease*, Hippocrates 'attribute[s] insensibility to the blocking of *pneuma* in the body'.[116] Totally unable to block the *pneuma* in his body, Othello suffers from an excessive sensibility, which makes him particularly vulnerable to Iago's innuendoes, so much so that he claims he can blow his love away to heaven:

> [. . .] Look here, Iago:
> All my fond love thus do I blow to heaven – 'tis gone! (3.3.445–6)

For Michael Neill, this 'implicit stage direction' means either that Othello is 'puffing his love to heaven' or that he is making an even more violent gesture if 'blow' is understood as 'blast'.[117] Be that as it may, Othello's image is particularly powerful and it suggests that in blowing his love away, the Moor implicitly places himself in the situation of giving up the ghost.

Othello's reference to suffocation ('suffocating streams', 3.3.391) underpins his sexual desire, so that his quest for orgasmic pleasure, now that the coitus with his wife proves impossible, seems only achievable in death, in giving up the last breath. Contrary to this, in Bandello's narrative, before murdering his wife, the jealous husband 'lay down with her, and pleasured her more than usual, so that there was no part of her white, innocent body that he did not kiss, taking that loving pleasure with her that men seek from their ladies'.[118]

Shakespeare expatiates on these erotic connotations thanks to numerous weather images and humoral associations in the play-text. Othello's 'bombast circumstance' is described by his ensign as '[h]orribly *stuffed* with epithets of war' (1.1.13–14, my emphasis) in discourses that are often full of airy metaphors. When Emilia remarks that Othello seems strangely unsettled, he feigns surprise and presents the general as some fearless daredevil whose courage will not be blown away by the cannonballs that whistle in his ears and kill his nearby soldiers:

> Can he be angry? I have seen the cannon
> When it hath *blown* his ranks into the *air*
> And, like the devil, from his very arm
> *Puffed* his own brother – and is he angry? (3.4.130–3, my emphasis)[119]

Once left on her own, Desdemona will reply that '[s]omething sure of state [. . .] Hath puddled his clear spirit' (3.4.136–9).[120] Michael Neill points out the particular meaning of 'spirit' which, in its technical sense, was then thought to be 'concocted out of blood and inspired air',[121] and which was regarded by Galenic medicine 'as the source of all activity in the body'.[122] The spirit was therefore thought to be able to both kindle and channel the passions. As such, Othello's defective 'spirit' leads him to nurse uncontrollable morbid feelings. In other words, his soul is ill because he is unable to breathe properly and turn the surrounding air, present in excess, into a life-giving force. In the play, the epileptic lover is constantly seen sighing, panting and gasping, and his decision to strangle Desdemona in the raptures of passionate grief is probably intimately related to his own physical difficulties. By stopping her breath, he may nourish the vain hope of better controlling his own.

Such associations between wind, breath and the soul appear in other crucial moments of the tragedy. For instance, in act 4, scene 3, as she sings her willow song which provides a lull before the climactic scene to follow, Desdemona suddenly stops and asks Emilia '[h]ark, who is't that knocks?' (4.3.48). 'It's the wind' (4.3.49), her lady-in-waiting replies. Emilia simply answers in order to soothe the anxieties of her mistress. Yet the reference to the wind unmistakably paves the way for the tragedy of breath which is about to take place while, like the legendary swan,[123] Desdemona sings before she dies. The deadly wind re-enters the play at a liminal moment to tell the audience that Desdemona will soon be lacking air and cease to breathe. In *On Breaths*,[124] the wind is called 'breath' when it is inside the body and 'air' when it is outside, and breaths (*physai*) are said to be the main causes of bodily disease.[125] In *Othello*, while Desdemona's breath (i.e. her internal wind) may be the direct cause of her inner grief and melancholy, air (i.e. external wind) here works as a morbid agent that brings about death. In the last scene of the play, the general pathetically calls attention to Desdemona's 'balmy breath' (5.2.16) just before smothering her.[126] The sweet air exhaled from her lungs is an aphrodisiac that almost dissuades him from undertaking what he thinks is an act of justice. Yet, because he associates the eroticism of female breath with his own orgasmic desires, he proceeds in his murderous intentions and stifles her under her pillow. At this very moment, Desdemona is literally *à bout de souffle* and she thus indirectly identifies with the deer collapsing at the end of a ruthless chase in a tragedy that leaves everyone breathless. After the 'mort o'th' deer' (*The Winter's Tale*, 1.2.118), the quarry follows. Once he has realised the senseless cruelty of his deed, Othello imagines himself being blown about in hell by a hoard of devils (5.2.276–9) as he thinks that his 'sail' (5.2.267) on earth has reached shore and may now be unfurled. After he has smothered his good angel, the Moor sees himself as surely damned, and his last lines before kissing Desdemona and falling down on her bed (5.2.358 s.d.) correspond to his last breath, when he publicly and spectacularly cuts his own throat:

> [. . .] in Aleppo once,
> Where a malignant and a turbaned Turk
> Beat a Venetian and traduced the state,
> I took by th' throat the circumcised dog
> And smote him – thus. (5.2.351–5)

Swollen with grief, he bursts like an inflated balloon.[127] In the final 'thus', Othello gives up the ghost and ends his speech on an adverb that lionises his unexpected, spectacular suicide.[128] 'Thus' the Moor ends, dying on his marriage bed, ironically the very same one where, earlier in the play, he had hoped to pluck the 'fruits' (2.3.9) of his furious love for the fair Desdemona.

As to Iago, he eventually finds refuge in silence, refusing to divulge any of his secrets. Contrary to his wife Emilia who, in the end, wants her speech to be 'as liberal as the north' (5.2.226)[129] – i.e., presumably, as liberal as the north wind[130] – he does not yield to such liberality of the wind. Stopping all his sounds in the air, he will control his speech as he has always done: 'Demand me nothing: what you know, you know; / From this time forth I never will speak word' (5.2.301–2).

Elements, cosmos and stage

So, in the play, the elemental havoc of the providential tempest which is first interpreted as fortunate actually foreshadows Iago's insidious and perverse plot against Othello and Desdemona. Deliberately placing itself under the influence of Galenic medicine and, more generally, of a pre-Cartesian system of thought, the tragedy incorporates a variety of humoral beliefs and features a villain who plays a central part in the climatic economy of the play. Iago exploits Othello's apparently native jealousy, partly due to the imbalance of his humoral fluids in a bodily microcosm regulated by a little weather of its own[131] while he ironically is the one who warns the Moor against the dangers of being 'all in spleen, / And nothing of a man' (4.1.84–5).

As the play's language bursts into multiple tiny weather images that proposes a kaleidoscopic view of the prevailing climatic theories, *Othello* gradually turns into a tragedy of the elements heavily dependent on a 'fated sky', whose malignant influence shapes the destinies of the Moor and of his wife. Othello is associated with fire ('Thou art rash as fire,' Emilia tells him in 5.2.134) while Desdemona represents the threatening force of water ('She was false as water', 5.2.134) so that their marriage can be read as the symbolic confrontation of fire and water, a duel whose significance appears when replaced in a broader context.

To the opposition of fire and water, the playwright adds that of light and darkness, turning his work into an 'emotional' – and, I would add, elemental – 'chiaroscuro'.[132] '[B]y this light of heaven, / I know not how I lost him' (4.2.150–1), Desdemona bemoans after she has been violently rejected by Othello. In the following scene, when Emilia suggests the possibility that women may betray their husbands just as men abuse their wives, she swears that she would not 'do such a deed' (4.3.5) 'by this heavenly light' (4.3.60), a phrase immediately taken up in a bawdy sense by her waiting woman, who ironically exclaims: 'Nor I neither, by this heavenly light: / I might do't as well i'th'dark' (4.3.61–2). The epiphora does not simply add a touch of wry humour to Desdemona's naive idealism, but it also calls attention to the materiality of the early modern stage. Indeed, while this 'heavenly light' may refer to celestial and divine powers, it also points to the daylight performance in the Globe – unless it also serves to designate the heavens or the ceiling – or even 'the glittering candlelight' in a court performance.[133] In the tragedy, Shakespeare repeatedly suggests that the stage, sometimes battered by the winds and the rain, and the cosmos, governed by the stars, constantly overlap and cross-fertilise each other to produce new clusters of meaning. If the sky fashions the fates of Othello and Desdemona, drama also sublimates them by putting to the fore a rich network of correspondences that bind together natural phenomena and human behaviour. The strength of performative arts is such, Shakespeare implies, that something as unsubstantial as air may well lie at the core of his text and then re-infuse it with more meaning. The wind in *Othello* is both a paradoxical concrete reality and a metaphorical trope – a double layer which is the prerequisite condition for its dramatic potential – that powerfully conveys the creative and destructive power of desire.

A third layer should finally be added to this as the word 'tempest' then referred to the alchemical phase supposed to remove impurities before reaching 'albedo', the purification process that followed 'nigredo', also presented as a form of alchemical 'eclipse'.[134] In the play, the tempest responsible for the defeat of the Turkish fleet reflects Othello's failure to transform himself, either physically or spiritually, through some sort of alchemical process. The albedo phase is visibly beyond Othello's reach, incapable as he is of whitening himself since, according to the common saying illustrated by one of Whitney's emblems, to 'wash an Ethiop white' is to labour in vain.[135] Having failed to turn his own leaden self into gold, the Moor, as an anti-alchemist, mistakenly regards Desdemona as polluted, thus transforming her golden love into the poisoning lead of suspicion ('Her name, that was as fresh / As Dian's visage, is now begrimed and black / As mine own face', 3.3.388–90). As a result, his lawfully wedded wife is finally turned into 'black weed' (4.2.67).

In Shakespeare's second love tragedy, the contrary and complementary forces of Eros and Thanatos are forcefully released in the outer ambient air so that we, as readers and spectators, may see how weather, stars and the natural elements may alternately stimulate desire and cause the ruin of a loving couple. If Verona's scorching sun leads the lovers to an apotheosis of grief and fondness, while suggesting the possibility of posthumous love, the blowing winds of passion here bring them to 'die upon a kiss' (5.2.358) but deprive us of the promise of a regenerative afterlife.

Notes

1. On 'blue ecocriticism' in Shakespeare's plays, see Steve Mentz, *At the Bottom of Shakespeare's Ocean*, especially chap. 2, pp. 19–32.
2. Thomas Wright, *The Passions of the Mind in General*, pp. 133–4. Quoted by Gail Kern Paster, 'The Tragic Subject and its Passions', p. 154. A recusant priest, Wright wrote the book in prison. It was so successful that it was reprinted several times, and we know that Robert Burton subsequently used it for his *Anatomy of Melancholy*. For more on Wright's career, see Peter Milward, 'Wright, Thomas (*c.*1561–1623)'.
3. As noted by Michael Neill in his edition of the play (p. 201, n. to 1.1.68–9), editors are 'divided' as to whether Iago's lines refer to Brabantio (as the 1622 Quarto suggests) or to Othello (as the Folio's punctuation seems to indicate). I'm following the Folio version here.
4. Michael Neill (ed.), *Othello*, p. 202, n. to 1.1.70–1.
5. See Graham Holderness, 'Moor and Whore of Venice', in *Shakespeare and Venice*, pp. 89–106.
6. See for example Sir John Harington's poem 'Of Cyprus' in Robert Allott (ed.), *Englands Parnassus*, p. 353: 'With filled sayles, in little while, / They came as farre as Cyprus, *Uenus* Ile: / Heere euery place was full of odours sweet, / Of gardens fayre, of spyce of pleasant tast, / The people lustfull, (for dame *Venus* meete).'
7. Giraldi Cinthio, *Gli Hecatommithi Third Decade, Seventh Novella*, in Michael Neill (ed.), *Othello*, Appendix C (Giovanni Battista), p. 434.
8. Ibid. p. 436.
9. Ibid. p. 437.
10. Anthony David Nutall, *A New Mimesis: Shakespeare and the Representation of Reality*, p. 135.
11. Herbert R. Coursen, *Watching Shakespeare on Television*, p. 104.
12. John Florio, 'Moria', in *A Worlde of Wordes*, p. 232. This has been pointed out by Patricia Parker in 'The Novelty of Different Tongues: Polyglot Punning in Shakespeare (and Others)', p. 56.

13. The sybil may be two hundred years old or she may have foretold that the sun still had 'two hundred compasses' of the earth to make before the end of the world.
14. Robert B. Heilman, *Magic in the Web: Action and Language in Othello*, p. 72.
15. See for example Richard Knolles, *The generall historie of the Turkes*, p. 843: 'The people therein generally liued so at ease and pleasure, that thereof the island was dedicated to *Venus*, who was there especially worshipped, and thereof called CYPRIA.'
16. See Nandini Das (ed.), *Robert Greene's Planetomachia (1585)*, pp. 48–9: '[T]he auncient Arabians call [Venus], *astrum genitale* or *prolificum*, for the temperate moisture, and favorable influence that it distilleth uppon all humaine bodies [. . .]: so likewise in the nature of coldness we do appoint *Venus* to be temporate and favourable, and *Saturne* with his melancholy humor to be unfortunate and malignant.'
17. F. P. Wilson, 'Illustrations of Social Life IV: The Plague', p. 125. On eclipses, see Robert Nadal and Arnaud Zucker, 'Les éclipses'. The authors remind us that the Greek mathematician and astronomer Thales of Miletus, who lived in the sixth century BC, was probably the first one who understood eclipses as scientific phenomena, whereas his predecessors ascribed them to a divine origin (p. 477).
18. On 'Saturn-Melancholy', see Frances A. Yates, *The Occult Philosophy in the Elizabethan Age*, p. 60: 'The melancholic was dark in complexion, with black hair and a black face – the *facies nigra* [. . .] induced by the black bile of the melancholy complexion [. . .] He was good at measuring, numbering, counting.' While Othello is generally portrayed as a jealous, paranoiac lover, he should also be seen as an uprooted warrior turned melancholy under Iago's influence and embodying the power of Saturn. Such a transformation is reminiscent of the character of Aaron who, in *Titus Andronicus*, declares to Tamora: 'Madam, though Venus govern your desires, / Saturn is dominator over mine' (2.3.30–1).
19. James Raven, 'Printing and Printedness', p. 226.
20. See Grace Tiffany, *Love's Pilgrimage: The Holy Journey in English Renaissance Literature*, p. 102. Tiffany rightfully remarks that, later on in the play, Iago refers to Desdemona as 'sport for Jove' (2.3.16–17), thereby associating Othello with Jupiter in a rather explicit way.
21. On the Sagittary (or the Archer) as usually figured by a centaur holding a bow and arrow, see Lena Cowen Orlin, 'Desdemona's Disposition', p. 172.
22. Thomas Blundeville, *M. Blundevile his exercises containing sixe treatises*, The First Booke, chap. 24, fol. 158.
23. As Janet Adelman explains, 'the time away from war has been a waste of time', and 'his involvement in the feminised realm of peace and generativity – the real of the "nine moons" bound up with Desdemona's

loving maternal pity – will destroy his masculine identity' (*Suffocating Mothers: Fantasies of Maternal Origin in Shakespeare's Plays, Hamlet to The Tempest*, p. 71).

24. 'The moon of Rome, chaste as the icicle / That's curdied by the forsy from purest snow / And hang on Dian's temple – Dear Valeria!' (5.3.65–7).
25. On this, see Charles and Michelle Martindale, *Shakespeare and the Uses of Antiquity: An Introductory Essay*, p. 162.
26. 'What committed? / Heaven stops the nose at it, and the moon winks; / The bawdy wind that kisses all it meets / Is hushed within thc hollow mine of earth / And will not hear't' (4.2.76–80).
27. On early modern eclipses, see for instance the website of the HM Nautical Almanac Office (HMNAO), which provides a list of eclipses as well as global and local circumstances: <http://astro.ukho.gov.uk/nao/online/> (last accessed 18 May 2018).
28. Incidentally, this image referring to gaping abysses is the exact opposite of Othello's 'hills whose heads touch heaven' (1.3.141) in the first act of the play.
29. Keith Thomas quotes Edward Gresham's 1607 almanac which 'hinted that an eclipse might occasion "much private enmity, malice and secret grudges, death of great beasts, many imprisonments, and the death of some ecclesiastical persons"' (*Religion and the Decline of Magic: Studies in Popular Beliefs in Sixteenth and Seventeenth Century England*, p. 335). See Edward Gresham, *A new almanacke and prognostication for the yere of our Lord God*, sig. CI.
30. Nick Davis, *Early Modern Writing and the Privatization of Experience*, p. 63.
31. Hippocrates, *On Ancient Medicine*, Commentary I.I, p. 127.
32. On this, see above, Introduction, p. 6 and p. 25, n. 21. In the singular form, the 'heaven' was also frequently associated with its homonym, the 'haven'. 'I cannot 'twixt the heaven and the main / Descry a sail' (2.1.3–4), the First Gentleman says in the Folio version of *Othello*, whereas in Q1, the same gentleman speaks of 'the haven and the main'. For Steve Mentz, 'this textual indeterminacy underlines the small representational gap between human havens and supernatural heavens' (*At the Bottom of Shakespeare's Ocean*, p. 20).
33. Luna was also thought to turn men into 'lunatic' creatures. So, when Emilia enters with the news of the 'foul murders' (5.2.108) that have just been committed, Othello remarks: 'It is the very error of the moon, / She comes more nearer the earth than she was wont, / And makes men mad' (5.2.111–13). As he sees it, the moon, straying off its usual course, has caused him to go mad, so that his fury is thus directly imputable to the sky. This idea simply repeats Richard Hooker's observation, in 1593, that disasters occur 'if the Moone should wander from her beaten way' (*Of the lavves of ecclesiasticall politie Eyght bookes*, The First Booke, 1.3, pp. 52–3). See below, Chapter 5, p. 154.

34. Cf. Edmund's views in *King Lear*. See below, Chapter 5, pp. 157–8.
35. 'Witness, you ever-burning lights above, / You elements that clip us round about, / Witness that here Iago doth give up / The execution of his wit, hands, heart, / To wronged Othello's service' (3.3.463–7).
36. See Michael Neill's edition of the play, p. 311, n. to 3.3.464.
37. '[. . .] and now in madness, / Being full of supper, and distempering draughts, / Upon malicious bravery, dost thou come / To start my quiet' (1.1.99–102).
38. George Gascoigne, *A delicate diet, for daintiemouthde droonkardes*, n.p.
39. Thomas Heywood, *Philocothonista, or the Drunkard, Opened, Dissected, and Anatomized*, p. 28.
40. See Thomas Adams's 1616 collection of sermons, *A divine herball*, p. 123: 'These sins being so Nationall, and naturall to the countreys: to ouer-drinke in Germanie; to ouer-eate in England; to wantonize in Italie and Venice; to quarrel in Fraunce, and to be enuious in Spaine; Enuie being euer the bosome-companion of Pride.'
41. Laura J. McGough, *Gender, Sexuality, and Syphilis in Early Modern Venice: The Disease that Came to Stay*, p. 172, n. 108.
42. The ambiguous meaning of the word is stressed in the *OED* ('Moor', 1), where it is primarily defined as 'a native or inhabitant of ancient Mauretania, a region of North Africa corresponding to parts of present-day Morocco and Algeria' but where, in the same entry, one also learns that, 'as late as the 17th cent., the Moors were widely supposed to be mostly black or very dark-skinned'.
43. Geraldo U. de Sousa, *At Home in Shakespeare's Tragedies*, p. 70.
44. John Pory (trans.), *A geographical historie of Africa*, pp. 5–6.
45. Ibid. p. 6.
46. Ibid. p. 1.
47. Ibid.
48. Ibid. p. 2.
49. See above, Chapter 3, p. 84.
50. See for example M. J. Tooley, 'Bodin and the Medieval Theory of Climate', p. 64.
51. Jean Bodin, *The six bookes of a common-weale*, The Fifth Booke, chap. 1, p. 548.
52. Ibid. p. 555.
53. Ibid. p. 551. See below my analysis of *Anthony and Cleopatra*, Chapter 6, p. 180 and p. 182.
54. Jonathan Gil Harris, 'Shakespeare and Race', p. 208.
55. On the 'rich underground network that linked "Ethiopes" or "Moors" with the mulberry', see Patricia Parker, 'The Novelty of Different Tongues', pp. 55–6. Parker usefully reminds us of some of the associations established by early modern glossaries, like the following one:

'*Mora*, a demur, a stay, a delay. Also a Moore-woman. Also a blackberie. Also a Mulberie' (p. 56).

56. In Pory's translation of Leo Africanus's *Geographical historie of Africa*, the origin of the black skin is linked to Cham's lineage. See for example p. 6, 'The originall of the people of Africa': 'For all the Negros or blacke Moores take their descent from *Chus*, the sonne of *Cham*, who was the sonne of *Noë*.'
57. 'Affection', in Robert Allott (ed.), *Englands Parnassus*, p. 7.
58. See above, Chapter 2, p. 73.
59. Cf. Othello's reference to '[d]eath and damnation' in 3.3.398.
60. In Shakespeare's main source, Othello and the ensign eventually 'bea[t] Disdemona dead with a stocking filled with sand' (quoted in Michael Neill (ed.), *Othello*, p. 442). As a result, Cinthio's ending, which seems to turn the bride into a figure reminiscent of the penitential Jack-a-Lent dummy (see François Laroque's *Shakespeare's Festive World: Elizabethan Seasonal Entertainment and the Professional Stage*, pp. 103–4), also combines cruelty with carnivalesque elements.
61. On this, see Jean Bodin, *The six bookes of a common-weale*, The Fifth Booke, chap. 1, p. 557. Bodin writes that '[i]t is the nature of melancholie which abounds most in them of the South, the which being frothie, prouokes to lust'. See also Steven Mullaney, *The Reformation of Emotions in the Age of Shakespeare*, p. 58. Mullaney, however, relies on different early modern views as he attributes the intellectual temperament to someone 'slow to be aroused sexually or emotionally'.
62. Desdemona's view corresponds to the theory developed in Bodin's *Six bookes of a common-weale* and obviously relies on one of those paradoxical climate-based theories that were widespread not just in England but also elsewhere in Europe. In Bartolomé de las Casas's *Apologética historia sumaria*, we learn for example that 'the heat in the tropics was supposed to open the pores, subtracting vital heat from the body', which accounted for the southerners' shyness and cowardice. See Bartolomé de las Casas, *Apologética historia sumaria*, 24, I, 385, quoted in and translated by Nicolás Wey Gómez, 'The Politics of Light: Al-Kindi's Geometrical Optics and the Vindication of the American Tropics in Bartolomé de las Casas's *Apologética historia sumaria* (1527–1561)', p. 40.
63. Once again, both interpretations coexisted at the time. In *English Ethnicity and Race in Early Modern Drama*, Mary Floyd-Wilson explains that Desdemona's 'conclusion that the heat of the African sun would dry and cool the body's humors' turned out to be 'a commonplace in [. . .] early modern writings' such as Stephen Batman's 1582 *Batman vppon Bartholome* (pp. 1–2). And yet, in *Airs, Waters, Places*, Hippocrates maintains the contrary, i.e. that only cold air generated cold (albeit moist) bodies.

64. See *OED*, 'temper': '5. The particular degree of hardness and elasticity or resiliency imparted to steel by tempering' and '†7. The relative condition of a body in respect of warmth or coldness'.
65. Jonathan Hall, *Anxious Pleasures: Shakespearean Comedy and the Nation-State*, p. 180.
66. See Mary Louise Gill, *Aristotle on Substance: The Paradox of Unity*, p. 81.
67. Ibid.
68. Georges Didi-Huberman, *Ninfa profunda. Essai sur le drapé-tourmenté*, p. 46, my translation.
69. See Frankie Rubinstein, *A Dictionary of Shakespeare Sexual Puns and their Significance*, p. 26.
70. Graham Holderness, *Shakespeare and Venice*, p. 99.
71. On the implications of pre-Copernican astronomy regarding the earth, see Arthur O. Lovejoy's 1936 seminal work, *The Great Chain of Being: A Study of the History of an Idea*, p. 103.
72. Helkiah Crooke, *Mikrokosmographia a description of the body of man*, Book 1, chap. 2, p. 8.
73. The whole play is framed by references to the storm. In the last act, when Emilia unveils Iago's villainy, Othello exclaims in a half-choked voice: 'Are there no stones in heaven / But what serves for thunder? Precious villain!' (5.2.233–4). The stones here obviously designate thunderbolts.
74. Valerie Traub, *Desire and Anxiety: Circulations of Sexuality in Shakespearean Drama*, p. 37.
75. On this, see Michael Neill's edition of the play, p. 450, n. to ll. 453–6.
76. On *pneuma*, see above, Chapter 3, p. 86, and below, Chapter 5, p. 158.
77. Steve Sohmer, 'The "Double Time" Crux in *Othello* Solved', p. 224.
78. Virginia Mason Vaughan, *Othello: A Contextual History*, p. 22.
79. This cannot be purely coincidental. As noted in The New Oxford Shakespeare (2016), in 2.1 '[t]he storm has been suggested with sound effects (wind, thunder), which may begin as Iago exits'. Gary Taylor, John Jowett, Terri Bourus and Gabriel Egan (eds), *William Shakespeare: The Complete Works, Modern Critical Edition*, p. 2132, marginal n. to 2.1.0.
80. For instance, the phrase is recorded in John Heywood's 1546 anthology *A dialogue conteinyng the nomber in effect of all the prouerbes in the englishe tongue*, sig. L, The seconde parte, chap. 9: 'An yll wynde, that blowth no man to good, men saie.'
81. Jan Glete, 'Warfare at Sea 1450–1815', p. 32.
82. David Waters, *The Iberian Bases of the English Art of Navigation in the Sixteenth Century*, p. 14.
83. Akihiro Yamada (ed.), *The First Folio of Shakespeare*, p. 266, ss6v,b [TLN 791–854]. Online edition at <http://shakes.meisei-u.ac.jp/ALL.html#45> (last accessed 18 April 2018).

84. The following letter written by the Duke of Medina Sidonia to the King of Spain gives an account of the very harsh meteorological conditions that prevailed: 'I wrote to your Majesty on the 21st ultimo by Don Baltasar de Zuñiga, giving a full account of all that had happened up to that time. Since then we have had, on four separate nights, heavy gales with strong head winds, thick fogs, and rain [. . .] The wind has now veered to the W.N.W., with a more favourable appearance, but the winds on this coast are always more tempestuous than elsewhere; and are so prevalent from the south, that there is no certainty of a continuance of the present fair weather. I pray that God in His mercy will grant us fine weather, so that the Armada may soon enter port [. . .] Pray consider the distress of this Armada after so terrible a voyage, and the urgent need for prompt measures of relief. – On the galleon "San Martin", 3rd September 1588.' See 'Simancas: September 1588, 1–10', in Martin A. S. Hume (ed.), *Calendar of State Papers, Spain (Simancas)*, vol. 4, *1587–1603*, pp. 411–25; British History Online, <http://www.british-history.ac.uk/cal-state-papers/simancas/vol4/pp411-425> (last accessed 30 April 2018).
85. Critics have detected strong echoes of James's poem at the end of Shakespeare's *Othello*. See Alvin B. Kernan, *Shakespeare, the King's Playwright: Theater in the Stuart Court, 1603–1613*, p. 61.
86. Jacobus Rex, *His Maiesties Lepanto* (1603 edn), sig. A4v.
87. The island was to remain Turkish for quite a long time (until 1878). Rhodes, also mentioned in Shakespeare's play (1.3.14), had fallen under Ottoman rule in 1522.
88. Richard Knolles, *The generall historie of the Turkes*, p. 878.
89. Ibid. p. 885.
90. See Jane Hwang Degenhardt, *Islamic Conversion and Christian Resistance on the Early Modern Stage*, p. 68. As pointed out by Degenhardt, even after it 'had been repossessed by the Turks', Cyprus remained a 'site of East-West trade' for the Europeans who 'continued to export coveted Cyprian goods'.
91. In the much-discussed textual crux at the end of the play, Othello compares himself to a 'base Indian' (Q, F2) [**Judean (F1)**] throwing 'a pearl away / Richer than all his tribe' (5.2.346–7). At this point, it may be worth mentioning the gruesome death of Marco-Antonio Bragadin, the Venetian commander of the city of Famagusta, who was captured by a Turkish army and sentenced to death in August 1571. According to Roger Crowley in *Empires of the Sea: The Siege of Malta, the Battle of Lepanto, and the Contest for the Center of the World*, Lala Mustapha Pacha, the commander of the troop, went to great lengths to make sure that the butcher who skinned alive the unfortunate Bragadin was a Jew, thereby adding insult to injury and torture. This dire detail may be at the origin of the F1 reading of 'Judean' instead of 'Indian'. I am here grateful to Charles Watkins for calling my attention to this point.

92. In 1573, George Gascoigne issued his *Devise of a maske for the right honourable Viscount Mountacute*, which runs into one the siege of Famagusta and the triumph of Lepanto. In Gascoigne's text, the presenter of the masque explains that the narrator's father 'being slayne at the last warres against the Turke, and he there taken, he was recovered by the Venetians in their last victorie, and with them sayling toward Venice, they were driven by a tempest upon these coasts' (0.25–8). So, like Gascoigne before him, Shakespeare conflates the Famagusta defeat, on 1 August 1571, with the naval victory of Lepanto in October of the same year. See George Gascoigne, *A Devise of a maske for the right honourable Viscount Mountacute A Hundreth Sundrie Flowres*, pp. 301–12.
93. In the 1622 Quarto version of the text, the Second Gentleman does not actually allude to the 'ever-fixèd Pole' but to an 'euer fired' (i.e. 'ever-burning') one. He says that he 'never did like molestation view / On the enchafèd flood' (2.1.16–17). Even though its figurative sense refers to excitement or irritation, the verb 'enchafe' primarily meant 'to make hot or warm' (*OED*, 1) as it was an English alteration of the French word *eschaufe*.
94. On the behaviour of people living in windy places and on the early modern beliefs attached to them, see Jean Bodin, *The six bookes of a common-weale*, The Fifth Booke, chap. 1, p. 564: 'we see plainely, that those people are more graue and staied, when the ayre is calme and temperat, than those which live in regions beaten with violent winds'.
95. 'Fresh air is ill for the wounded man.' See Maurice Palmer Tilley, *A Dictionary of Proverbs in England in the Sixteenth and Seventeenth Centuries*, A93.
96. In *At the Bottom of Shakespeare's Ocean*, Steve Mentz proposes a different reading. He sees Othello as a failed philosopher (p. 21) and Iago as an 'anti-God' of the sea: since his name, which corresponds to that of 'James' in English, is linked to Spain, i.e. to 'the greatest sea going empire of the early seventeeth century', he 'advertises his connection to the maritime world in a collection of epithets and phrases' and 'scatters sailor-talk liberally' (p. 24).
97. His name may also be derived from the Latin verb *agere*, construed as *ago* in the first person, and hence be understood as 'I perform', 'I act' (or, alternatively, 'I accuse'). See Patricia Parker, 'The Novelty of Different Tongues', p. 43.
98. Philip Lyndon Reynolds, *Food and the Body: Some Peculiar Questions in High Theology*, p. 252. Reynolds actually comments on Albert's account of Aristotle's *De anima*.
99. On this, see François Laroque's analysis in *Shakespeare's Festive World*, p. 49 and pp. 288–9.

100. On Saint Blasius as the patron saint of throat sufferers, see Claude Gaignebet, *À plus hault sens*, vol. 1, pp. 43–6. Among the many carnivalesque allusions that run through the play, one is particularly noteworthy. In one of his asides, Iago comments on Cassio and Desdemona engaging in conversation and cynically exclaims about the lieutenant's behaviour: 'yet again your fingers to your lips? Would they were clyster-pipes for your sake!' (2.1.179–80). Such a scatological depiction of the kiss may then be connected with the 'airy promiscuity' which characterises the play. On this, see Philippa Berry, *Shakespeare's Feminine Endings: Disfiguring Death in the Tragedies*, p. 29.
101. Michael Neill (ed.), *Othello*, p. 330, n. to 4.1.77.
102. According to popular belief, bears emerged from their caves at Candlemas (2 February). On this, see François Laroque's *Shakespeare's Festive World*, p. 48 and pp. 294–6.
103. This transition scene bridging the gap between springtime Venice and autumnal Cyprus also merges water and fire as in the equinoctial gales.
104. See *The Taming of the Shrew*, where Christopher Sly is promised to receive pictures of the goddess Venus behind the reeds '[w]hich seem to move and wanton with her breath / Even as the waving sedges play with wind' (Induction 2, 51–2).
105. Seneca, *Natural Questions*, VI.20, p. 185.
106. Ibid. VI.20, p. 187.
107. It is significant that, in the last scene, Desdemona notices that '[s]ome bloody passion *shakes* [Othello's] very frame' (5.2.45, my emphasis).
108. Geoffrey Bullough (ed.), *Narrative and Dramatic Sources of Shakespeare*, vol. 7, *Major Tragedies*, p. 203.
109. See Charles Segal, 'Lucretius, Epilepsy, and the Hippocratic *On Breaths*'.
110. Philip Barough, *The methode of phisicke*, Liber I, chap. XIV, p. 15.
111. Lucretius, *De rerum natura*, Book III, p. 105, my emphasis.
112. Cf. the famous opening lines of the second book of Lucretius' *De rerum natura*, 'turbantibus aequora ventis', which refer to the winds that disturb the main (Lucretius, *De rerum natura*, trans. W. H. D. Rouse, Liber Secundus, p. 94).
113. David Houston Wood, '"Very Now": Time and the Intersubjective in *Othello*', p. 91.
114. Robert Burton, *The Anatomy of Melancholy*, The First Partition, Memb. V, Subsect. IV, p. 163. While the causes of such melancholy are 'most commonly fear, grief, and some sudden commotion, or perturbation of the mind', they can also be inward: 'Montaltus *cap. 15.* out of Galen recites, heat and obstruction of those mesaraic veins, are an immediate cause, by which means the passage of the chilus to the

liver is detained, stopped, corrupted, and turned into rumbling and wind' (ibid.).

115. François Laroque, *Shakespeare's Festive World*, p. 300.
116. Sophie M. Connell, *Aristotle on Female Animals: A Study of the Generation of Animals*, p. 215.
117. Michael Neill (ed.), *Othello*, p. 309, n. to 3.3.446.
118. Geoffrey Bullough (ed.), *Narrative and Dramatic Sources of Shakespeare*, vol. 7, *Major Tragedies*, p. 204. This passage is played down by Geoffrey Fenton, who simply writes that the Moor 'embraced and kissed her, in such sorte as Judas kissed lour Lorde the same night he betraied him' (ibid.).
119. In act 4, scene 1, as Lodovico wonders if Othello has not turned mad, Iago slyly replies that he 'may not breathe [his] censure / What he might be' (4.1.262–3).
120. In the following scene, the dialogue between Desdemona and her suspicious husband is also a case in point: '*Desdemona* I hope my noble lord esteems me honest. / *Othello* O ay, as summer flies are in the shambles, / That quicken even with blowing' (4.2.65–7).
121. Here Neill quotes Thomas Walkington's 1631 *The Opticke Glasse of Humors*. See Michael Neill (ed.), *Othello*, p. 321, n. to 3.4.139.
122. Ibid. In this regard, see William Bullein's 1595 treatise *The gouernment of health*, p. 16. Describing human anatomy, the physician explains: 'And these animall vertues [the vertues of imagination, fantasie, memorie, &c.], be placed as it were heauenly aboue al the members, communicating their heauenly influences, down vnto the heart, as to a prince, or chiefe ruler within the body, which giueth life to euery part thereof Thou shalt consider, that *the hart was the first that receiued life from the spirites*, and shalbe the last that shall die' (my emphasis).
123. See for instance Virgil's *Aeneid*, XI, 458.
124. See above, p. 134.
125. See Jacques Jouanna, *Hippocrates*, p. 282.
126. Othello later acknowledges that his own hands 'have newly stopped' Desdemona's breath (2.2.200).
127. See Balz Engler, 'Othello's Great Heart'.
128. One should also note the mimetic quality of Othello's speech, whose final voiceless fricatives (5.2.355) reproduce the Moor's last breath.
129. Initially, in Q1, Emilia wanted to be 'as liberal as the ayre'. The reference to liberality is ambiguous here as it also connotes the bawdiness of the wind mentioned earlier on in the play ('The bawdy wind, that kisses all it meets', 4.2.80; cf. the 'wanton wind' in *A Midsummer Night's Dream*, 2.1.129, and the 'strumpet wind' in *The Merchant of Venice*, 2.6.16 and 2.6.19).
130. For another interpretation, see Pamela Allen Brown, '*Othello* Italicized: Xenophobia and the Erosion of Tragedy', p. 156, n. 38. Brown

sees in Emilia's remark a possible 'reference to the greater liberties proverbially enjoyed by Englishwomen'. Incidentally, she quotes a passage from Dekker and Webster's *Westward Ho!* (1607) which connects climatic elements with national stereotypes. In the city comedy, Justiniano tells Mistress Wafer: 'How happy be our Englishwomen that are not troubled with Italian husbands; why your Italians in general are so Sun-burnt with these Dog-daies, that your great Lady there thinkes her husband loves her not if hee bee not jealious: what confirms the liberty of our women in England, then the Italian Proverbe, which saies if there were a bridge over the narrow Seas, all the women in Italy would shew their husbands a Million of light paire of heeles, and flie over into England' (3.3.81–8).

131. As Desdemona notices how much changed her husband seems in the middle of the play, she sadly tells Cassio that 'My lord is not my lord; nor should I know him, / Were he in favour as in humour altered' (3.4.120–1).
132. For more on Shakespeare as the 'master of emotional chiaroscuro', see E. A. J. Honigmann (ed.), *Othello*, p. 77.
133. Gary Taylor, John Jowett, Terri Bourus and Gabriel Egan (eds), *William Shakespeare: The Complete Works, Modern Critical Edition*, p. 2181, marginal n. to 4.3.60.
134. See *Othello*, 5.2.102–4.
135. On this, see Karen Newman, '"And wash the Ethiop white": Femininity and the Monstrous in *Othello*'.

Chapter 5

'The pelting of [a] pitiless storm': Thunder and Lightning in *King Lear*

> How shall your houseless heads and unfed sides,
> [. . .] defend you
> From seasons such as these? (3.4.28–32)

King Lear is yet another tragedy in which the title character yields to folly and, in his blind rage, estranges himself from those who truly love him. Lear thus soon appears as a throneless monarch overpowered by environmental forces in a universe that is neither green nor blue but very black indeed. During the Jacobean era, the working out of a new aesthetics of darkness influenced by northern mannerism probably encouraged Shakespeare to pass from a deconstruction of the traditional pastoral genre in *As You Like It* to the black pastoral of *King Lear*. The wintry landscape of the Forest of Arden has now become a barren heath where the wind blows and where the demented king wanders aimlessly in the Fool's company. The frozen silence of the desert woods is here replaced by the threat and terror of 'hurricanoes', 'cataracts'[1] and roaring thunderclaps.

As an acousmatic play, *King Lear* is replete with rumbling noises and blasting winds as well as with ballads and snatches of songs in which the Fool's lines ('He that has and a little tiny wit, / With heigh-ho, the wind and the rain, / Must make content with his fortunes fit, / Though the rain it raineth every day', 3.2.74–7[2]) remind us of Feste's song at the end of *Twelfth Night* (5.1.385–404).[3] But in the tragedy a fierce coldness, loud thunderclaps and frantically tempestuous weather replace the gentle wind and the falling rain that normally accompany as much as they disturb men and women's daily lives and seasonal routines.

Given Lear's old age, the foregrounding of coldness in the play sounds logical since, in the medical treatises of the time, elderliness was associated with a cold temperament,[4] which provided an explanation, though not a treatment, for the mental confusion of the elderly. Early modern poets agreed with this view, clearly presented in the following poem by Thomas Whythorne, a musician of the 1570s who writes rather despondently in anticipation of his upcoming decay:

> The force of youth is well nigh past,
> Where heat and strength of late took place,
> And now is coming in all haste
> The cold, weak age for to deface
> The show of youth[5]

In *King Lear*, the frailty of old age is viewed in similar terms by the Fool: 'Nay, and thou canst not smile as the wind sits, thou'lt catch cold shortly' (1.4.94–6). Further on in the play, Poor Tom's refrain 'Tom's a-cold' (3.4.56, 3.4.76–7, 3.4.135, 3.4.161, 4.1.52) provides an apt comment on the old man's hostile surroundings while it also reflects the emotional frigidity and detachment of Lear's daughters towards their ageing father – with the notable exception of Cordelia – which is illustrated by Goneril's curt order to Oswald regarding her father: 'And let his knights have colder looks among you; / What grows of it, no matter' (1.3.18–19).

As to the storm, it is announced by Lear's explosion of anger in the play's first scene, when he repudiates his silent daughter before banishing Kent (1.1.155). Worth noting here is the fact that, in sixteenth- and seventeenth-century England, the link between human emotions and the weather partly stemmed from etymological considerations, the Greek word *meteoros* being an adjective applying to a 'doubtful or excited mind'.[6] Lear's inner storm must therefore have been expected to provoke some sort of frightening climatic consequences reflecting the chaotic state of his mind and of his divided kingdom.

'What is the cause of thunder?'

In Chapman's *Bussy D'Ambois*, the deadly fight between Bussy and the valiant warrior Barriso, which is reported onstage by a messenger, leads to a desire for revenge rendered through the image of a brooding thunderstorm comparable to the tense atmosphere which prevails at the onset of Shakespeare's tragedy: 'Sorrow and Fury, like two opposite fumes / Met in the upper region of a cloud' (2.1.110–11). However, in Chapman's play, the storm is a simple comparison, while Lear's fury keeps building up until it gives way to a formidable storm. The old king's many harsh 'words and looks' pave the way for the real 'cataracts and hurricanoes' (3.2.2) that will spout and spurt down from the heavens to wreak havoc and cause apocalyptic conditions on earth.[7] Interestingly, Lear associates the 'sulphurous and thought-executing fires, / Vaunt couriers of oak-cleaving thunderbolts' (3.2.4–5) with women's 'sulphurous pit' (4.5.123), that is, to the vagina, which he presents as being 'all the fiend's' (4.5.122). Such a correlation brands women as devilish creatures and establishes a link between cosmic phenomena at large and female anatomy and sexuality.[8]

While thunder had natural causes in classical meteorology – Aristotle claimed that it arose from a 'collision' between dry and moist exhalations[9] – God's wrath, as we have seen, was often invoked in order to account for the occurrence of natural calamities.[10] Theologians like Calvin did not think otherwise regarding thunder and lightning. The French theologian did acknowledge though that the thunder proceeded from a collision between 'cold and humid vapors' and 'dry and hot exhalations' and he did not question the veracity of such theories. Yet, for him, if meteorological phenomena had natural 'intermediate or secondary causes', they also had a primary cause since God is the origin of all things.[11] So, Lear's question to Edgar disguised as Poor Tom, 'What is the cause of thunder?' (3.4.143), would certainly have been regarded by Calvin as a rather provocative one. To him, the answer was simple: '[t]he Thunder is termed the voyce of God' and therefore '[t]he Thunder and lightning make vs vnexcusable if we do not thereby both know, feare, and magnifie God'. Detecting God's power in nature, Calvin asserted in his sermons that '[t]he Thunder maketh even the Reprobates and Atheistes to knowe and confesse that there is a God'.[12] Obsessed with meteorological phenomena, the French reformer systematically interprets them as divine messages addressed to men. If thunderstorms in particular have attention called to them in his sermons, it is because they are considered the most direct, or tangible, expression of God's presence. God's followers must therefore understand that furious weather conditions reveal His existence and, as a consequence, noisy as it may be, a thunderstorm should not be presented as a source of terror:

> Looke upon a despiser of God, which trampleth all religion under foote: he is carried away with a divelish rage: he shaketh off al difference of good and evill: and yet not withstanding he is inforced to be afraid of the thunder, insomuch that he is as it were out of his wittes and in a traunce [. . .] The thunder can do nothing. And although we see it flie abroad, and shoote from heaven to the earth in a moment: Yet is it still in the hand of God.[13]

Only the 'senseless creatures' who do not believe in God should be afraid of the storm and of its consequences. Such creatures are 'out of [their] wittes', just like Lear, whose 'wits begin to turn' (3.2.67) in the middle of the gale. Unable to see God in the weather-beaten heath which has become his earthly hell, Shakespeare's indignant monarch shares a number of features with Calvin's 'despizer[s] of God'.[14] A suffering old man, he replaces providence with nemesis and keeps calling to punitive 'great gods' (3.2.49) depicted as threatening and revengeful in the spirit of the Old Testament,[15] where God 'commandeth, and raiseth the stormy wind' and 'maketh the storm a calm' (Psalm 107, 25). So, it is this angry God who is desacralised and multiplied by Lear's invocation of the gods.

During the Reformation, cosmographers and natural philosophers paid more and more attention to the mapping of the sky because they had to reconcile theological views with more practical observations. Leonard Digges, who was Calvin's contemporary, for instance, worked out a system that undertook to understand the various manifestations of the weather by putting each of them in connection with the seven days of the week. In it, thunder is given a distinctive significance according to the day on which it occurs:

> Mundayes thundre, the deathe of women: Tuesdayes thundre, plentie of grayne: Wednesdayes thundre, the death of harlottes, and other bloodshede: Thursdayes thundre, plenty of shepe, & corne: Frydayes thundre, the slaughter of a great man, and other horrible murders. Saturdayes thundre, a generall pestilent plague, and great deathe.[16]

In contradistinction to Calvin's sermons, however, Digges's treatise acknowledges the ambivalence of thunder likely to foreshadow 'plentie of grayne' or 'horrible murders' in a short interval of time. And, if we apply Digges's system to *Lear*, the play's thunder must be a 'Frydayes thundre', an ominous foreboding of Cordelia's death.

The idea of God's anger being the cause of harsh weather went hand in hand with the concept of providence[17] which plays such an important role in *King Lear*. To substantiate this, critics often cite the writings of Richard Hooker, a Protestant theologian whose family background 'was clearly Calvinist'.[18] Hooker was particularly interested in the doctrine of divine providence, even though, for him, there was nothing systematic in it. It was 'neither pure Determinism like *Destiny*, nor purely arbitrary, like *Chance*'; it was, rather, 'like virtue in a virtuous human being', that is, 'dependable'.[19] According to Steve Mentz, Lear's rhetoric, enhancing a 'lawless nature' and emphasising 'human insufficiency', owes much to Hooker: 'Hooker's sense of natural "obedience" as the "stay" of universal order clarifies the theological stakes of Lear's storm.'[20] 'Natural obedience' is an important theme in Hooker's *Of the lavves of ecclesiasticall politie*, where the author expatiates on the theory of natural law. He explains in substance that God 'speaks through nature whose voice is his instrument'[21] and that, as the Creator of the world, He logically orders it and governs the elements. Yet 'order is not given once and for all; it has to be sustained and actively kept so as to prevent both the Creation in general, and society in particular, from lapsing back into chaos'.[22] This appears quite clearly in the following passage, worth quoting at length:

> [S]ince the time that God did first proclaime the edicts of his law vpon it, heaven & earth haue hearkened vnto his voice, and their labour hath bene to do his wil: He *made a Law for the raine*. He gaue his *decree unto the Sea, that the waters should not passe his commandement*. Now if nature should intermit her course, and leauve altogether, though it were but for a while, the obseruation of her own lawes [. . .] if the frame of that heavenly arch erected ouer our heads should loosen & dissolue it selfe: if celestiall spheres should forget their wonted motions by irregular volubility, turne themselues any way as it might happen: if the prince of the lightes of heauen which now as a Giant doth runne his vnwearied course, should as it were through a languishing faintness begin to stand & to rest himself: if the Moone should wander from her beaten way, the times and seasons of the yeare blend themselues by disordered and confused mixture, the winds breath out their last gaspe, the couldes yield no rayne, the earth be defeated of heauenly influence [. . .] what would become of man himself, whome these things no do all serue? See we not plainly that obedience of creatures vnto the lawe of nature is the stay of the whole world?[23]

Shakespeare looks at what happens when the 'heavenly arch' meant to protect human beings threatens to 'dissolve itself'. He also complicates Hooker's views by questioning the authority of providence in a tragedy where it desperately fails to manifest itself. *King Lear* offers a deeply pessimistic vision of humankind and, rather than signalling God's wrath, its apocalyptic language and imagery actually point to His absence. The playwright is here poised between a pagan and a nihilistic *Weltanschauung*.

Some of the pre-Christian beliefs found in Pliny the Elder's huge compilation probably influenced part of the general atmosphere and climatic background of the tragedy. Far from seeing nature as a protective entity, Pliny denounces its harshness in his *Natural History*, relentlessly lamenting man's nakedness and vulnerability, thereby paving the way for a 'zoographic critique' of humankind, exposing its 'abject' nature.[24] Laurie Shannon convincingly demonstrates that *King Lear* 'draws extensively on the writings of natural history that had shaped the curricula of the sixteenth century'[25] and that, as a result, Pliny's stark portrayal of man – that is, *nudus in nuda terra*[26] – is repeated in Lear's own anatomy of human beings, regarded in the play as 'naked and depraved'.[27] In Philemon Holland's 1601 translation, Shakespeare could read:

> [O]f all other living creatures, man she hath brought for all naked, and cloathed him with the good and riches of others. To all the rest, given she hath sufficient to clad them everie one according to their kind: as namely, shells, cods, hard hides, prickes, shagge, bristles, haire, downe feathers, quils, skailes, and fleeces of wool. The verie trunkes and stemmes of trees and plants, shee hath defended with barke and rind, yea and the same sometime double, against the injuries both of heat and cold: man alone, poore wretch, she hath laid all naked upon the bare earth[28]

Wandering on the barren heath, unequipped to face nature's calamities, Shakespeare's 'bare-headed' (3.2.60) protagonist endorses a similar view. Depicting himself as '[a] poor, infirm, weak and despised old man' (3.2.20), Lear, after the biblical Job, has become the embodiment of human frailty. The 'contentious storm', Lear says, '[i]nvades us to the skin' (3.4.7) and its very violence arouses in his heart warm feelings of compassion towards the homeless and the destitute:

> Poor naked wretches, whereso'er you are,
> That bide the pelting of this pitiless storm,
> How shall your houseless heads and unfed sides,
> Your looped and windowed raggedness, defend you
> From seasons such as these? (3.4.28–32)

Present in both Pliny's encyclopaedia and in *The Tragedy of King Lear*, those fellow feelings reverberate throughout the text. A few lines later, Lear becomes mad with rage and exclaims as he tears off his clothes: 'Unaccommodated man is no more but such a poor, bare, forked animal as thou art' (3.4.99–100). Alluding to Christian values the better to dismiss them, Lear can only adhere to a pre-Christian set of beliefs in a syncretic vision where archaic views mix with surprisingly penetrating insights.

Astrological charts

In the second book of his *Natural History*, Pliny promotes a natural explanation for the cause of lightning and distinguishes between nightly flashes of lightning and those that occur in full daylight. Doing this, he attributes nighttime *fulmina* to Summanus and daytime *fulmina* to Jupiter. Against all odds, Jupiter turns out to be both a god and a planet in the following lines:

> Most men are ignorant of that secret which, by great attendance upon the heavens, deepe clearkes and principall men of learning have found out: namely, that they be the fires of the three uppermost planets, which falling to the earth, carrie the name of lightnings, but those especially which are seated in the middest, to wit, about *Jupiter*, haply, because participating the excessive cold and moisture from the upper Circle of *Saturn*, and the immoderate heat from *Mars* that is next under, by this meanes he dischargeth the superfluitie: and hereupon it is commonly said, that *Jupiter* shooteth and darteth lightenings. Therefore, like as out of a burning peece of wood a cole of fire flieth forth with a cracke, even so from a starre is spit out as it were and voided forth this coelestial fire, carrying with it presages of future things: so as the heaven sheweth divine operations, even in these parcels and portions which are rejected and cast away as superfluous. And this most commonly happeneth when the aire is troubled, either because the moisture that is gathered, mooveth and stirreth forward that abundance to fall; or els for that it is disquieted with the birth (as it were) proceeding from a great bellied star, and therefore would be discharged of such excrements.[29]

In Shakespeare's tragedy, Lear seems to address Jupiter as a pagan god. This deity, with whom he implicitly identifies, mainly appears as the god of thunder and lightning in his harangues:

> Let shame come when it will, I do not call it.
> I do not bid the thunder-bearer shoot,
> Nor tell tales of thee to high-judging Jove. (2.4.215–17)

Such diatribes, however, make the old man sound pathetic and powerless, for the play as a whole dismisses any sort of supernatural intervention and the only influences acknowledged as such in the text refer to a celestial rather than to a divine origin. In the first scene the old king makes it clear that he believes in the influence of the heavenly orbs ('For by the sacred radiance of the sun, / The mysteries of Hecat, and the night, / *By all the operation of the orbs*, / *From whom we do exist and cease to be*, / Here I disclaim all my paternal care', 1.1.107–11, my emphasis). Lear's 'high-judging Jove' is therefore an indistinct, conglomerate figure encompassing the pagan god of thunder, the biblical Jehovah and the planet Jupiter. In other words, the title character's obsession with apocalyptic climate is what allows the overlapping of biblical, pagan and astronomical layers in the play.

The earl of Gloucester, Lear's alter ego, similarly believes in the impact of the 'fated sky' on human destiny and in the truthfulness of astrology. In act 1, scene 2, he expresses his pessimism regarding the three astrological levels of family, state and cosmos, describing a chaotic situation and a world turned upside down as a consequence of the 'late eclipses of the sun and moon' (1.2.95).[30]

Soon after Gloucester's remark, the grinning Edmund is heard mocking his father's credulousness.

> This is the excellent foppery of the world: that when we are sick in fortune [. . .] we make guilty of our disasters the sun, the moon, and stars, as if we were villains on necessity, fools by heavenly compulsion, knaves, thieves, and treachers by spherical predominance, drunkards, liars, and adulterers by an enforced obedience of planetary influence [. . .] My father compounded with my mother under the Dragon's tail and my nativity was under Ursa Major, so that it follows I am rough and lecherous. Fut! I should have been that I am had the maidenliest star in the firmament twinkled on my bastardizing. (1.2.119–30)[31]

Like the destitute Pierce who, in Nashe's *Pierce Penniless* (1592), already railed against those 'anatomizing [. . .] the sky's entrails in Surgeon's Hall',[32] Edmund pokes fun at his father for giving naive credence to Claudius Ptolemy's writings, in particular when the latter explains that 'the time of conception is as important to know as the time of birth'.[33] The opposition between Gloucester, who is gung-ho about horoscopes, and his bastard son Edmund, who dismisses them as pure nonsense, here somehow reflects the Renaissance controversy over the validity of judicial astrology. In this context, the bastard's view corresponds to the common naturalist or materialistic stance.[34]

In *A Midsummer Night's Dream*, the chaotic weather is perceived as the result of the dissensions between Oberon and Titania, while in *Romeo and Juliet*, the Verona dog days are made indirectly responsible for the tragic events of the play. *King Lear* sits in the middle ground and posits that man and weather are in constant interaction. Gloucester believes that disquieting weather phenomena portend tragic events to come, while Edmund denies that astrology and meteorology may have any impact whatsoever upon human behaviour. As to Lear, he is presented in turn (by others) as a victim of the raging elements and (by himself) as the main force responsible for unleashing the storm and vainly imagining that he can command the elements.

Materialism and 'climatericall' years

When the old king asks the storm to 'drench [. . .] our steeples' and send down 'oak-cleaving thunderbolts' to 'strike flat the thick rotundity o'th'world' (3.2.7), he intends to reappropriate the natural weather forces in order to use them to his own ends. The rain meant to purge the earth is here identified with holy water, the water of baptism.[35] In the play, thunder, wind and water are, as it were, the agents of truth and they serve to open up Lear's so-far blind eyes in order to turn him into a wiser, clear-sighted man. When he breathes his last while desperately holding a mirror in front of Cordelia's mouth to find whether any life and breath is left in her, the scene dramatises the centrality of *pneuma*, the mobile air circulating in the cosmos at large as well as in a body where air is as necessary to life as it is to the fashioning of an ever-shifting self.[36]

If Shakespeare's main source, *The true chronicle history of King Leir*,[37] performed in the 1590s, does not make *pneuma* a crucial concern as no one dies in this play, it nonetheless engages with environmental issues, albeit in a lighter vein: 'My Lord, how do you brook this British ayre?'[38] the jovial Mumford asks the Gallian king in scene 7 – a question left cautiously unanswered by the French sovereign. By and large, climatic phenomena in *King Leir* are infused with religiosity. The play 'highlights a providential order'[39] clearly visible in scene 19, for example, where a '*Messenger or murtherer with two daggers in his hands*'[40] instructed by Ragan to murder Leir and Perillus (i.e. Kent) enters to stab his designated victims. He finds them asleep and, as he prepares to kill them, they suddenly wake up. When the messenger swears 'by hell' that Leir's two daughters hired him 'to this deed', a stage direction interrupts the dialogue and indicates '*Thunder and lightning*'.[41] Perillus warns the murderer against the 'everlasting torments' that he will 'indure, / Even in the hottest hole of grisly hell' and, once again, '*It thunders*'.[42] Frightened, the messenger '*quakes, and lets fall the Dagger*'.[43] Leir then duly thanks the heavens for this miraculous divine intervention and the converted murderer leaves, now decided to serve God's cause.[44]

In *King Lear*, there is no god behind the storm and the thunder acts as a conscience-raiser leading the king to acquire a form of self-knowledge as well as a new understanding of the reality of the world around him. While the sun previously blinded him, the flashes of lightning now begin to enlighten him. So, if Shakespeare provides his audience with a variety of perspectives, he refutes any idea of divine intervention. Thunder is here the voice of an orgasmic nature endowed with human features. The winds blow their cheeks (3.2.1) and are 'eyeless' (3.1.8) in their fury, and the storm is a rumbling belly (3.2.14). Lear, trying to eradicate sin from the earth, actually performs a dreadful exorcism with the help of the storm.[45] It is probably no coincidence that, in Samuel Harsnett's *A Declaration of Egregious Popish Impostures* (1603), one of Shakespeare's main sources, the practice of exorcism is compared to a cunningly staged storm aimed at causing terror:

> Thirdly, it [*the booke of Exorcismes*] served wonderous aptly *ad terrorem et stuporem incutiendum populo:* in steed of thunder and lightning to bring *Jupiter* upon the stage, by these dreadful frightful Exorcismes, thundring, clapping and flashing out the astonishing of Gods names, *Jehovah*, *Tetragrammaton*, *Adonai*, and the rest, to amaze and terrifie the poore people, and to possesse them with an expectation of some huge monster-devil to appeare.[46]

Lear also relies on the terrifying effects of thunder in order to stage the purification process he wishes the whole of humanity went through in his presence. Yet the function of the storm remains ambiguous in the play, just as in Harsnett's treatise, where 'lightning and thunder' are also presented as a sign of people being possessed by the devil. The case of Nicholas Marwood is a telling example:

> *Marwood*, *Westons* patient, beeing pinched with penurie and hunger, did lie but a night or two abroad in the fieldes, and beeing a melancholicke person, was scared with lightning and thunder that happened in the night, and loe, an evident signe that the man was possessed.[47]

In *King Lear*, the storm is the title character's weapon as much as his persecutor, so that Lear appears as both the exorcist and the possessed. But the play goes much further than Harsnett's clinical description which sticks to the enumeration of a number of symptoms since, in the end, nature's 'meteorological violence' is transformed into 'an eschatological horror scene' where the human race is meant to disappear from the face of the earth: 'Crack nature's moulds, all germens spill at once / That makes ingrateful man' (3.2.8–9).[48]

In this particular context, it is worth noting that the word 'germens' harks back to Lucretius' *semina rerum*,[49] that is, combinations of atoms endowed with formative powers from which natural bodies originate,[50] thus suggesting that Lear wants to annihilate at once both the female ('moulds') and the male ('germens') principles of creation. More generally, the weather images of the play owe much to Lucretius' use of meteorological models in order to explain the creation and disintegration of material objects and of living beings. The old king's retort to Cordelia, 'Nothing will come of nothing' (1.1.89), a principle barely rephrased a few scenes later as 'Nothing can be made out of nothing' (1.4.125), also alludes to Lucretius. Such statements obviously echo the atomist philosophy, according to which divine creation *ex nihilo* is presented as utterly impossible: 'nothing can ever be created by divine power out of nothing'.[51] Only the conjunction of atoms can form objects. Ironically, atoms in the play are not just endowed with a generative power but they are also seen as an utterly destructive force. Lear indeed wishes that his evil daughter Regan were 'anatomize[d]' (3.6.33), thereby alluding to her anatomical dissection as well as to her atomisation. The word 'atom' and the verb 'an*atom*ise' are here associated by means of a then current fake etymology, so that 'anatomise' might also be parsed as 'annihilate'.

Like the title character, the characters are tormented and totally unable to experience a state of mind that could in any way be compared to something like the Epicurean ataraxy, a word derived from the Greek *atarachos* ('free from trouble'), a term which actually applied to the weather and to the sea. Most are in fact 'minded like the weather, most unquietly' (3.1.2). Yet in spite of his low morale, Lear addresses Edgar-as-Poor Tom as a 'noble philosopher' (3.4.160) and as a 'good Athenian' (3.4.168).[52] *De rerum natura* also celebrates 'a man of Greece' who '[d]ared to lift up his mortal eyes' against Religion: 'The first was he to stand up and defy her.'[53] According to Lucretius, nothing stopped him, 'neither stories of the gods, nor lightnings, / Nor heaven with muttering menaces'.[54] This bold mortal may well be none other than Epicure himself. In Shakespeare's tragedy, Lear asks the 'good Athenian' (3.4.168) what 'the cause of thunder' (3.4.143) is, trying to ascertain whether he is right in assigning it to supernatural causes. Is it really 'the great Gods / That keep this dreadful pudder o'er our heads' (3.2.49–50), he asks?[55]

The old king echoes here Lucretius' dismissal of superstitious beliefs. In the last book of his poem, Lucretius concludes that all meteorological phenomena, including thunder and lightning, have natural causes. For him, it was 'ignorance of the causes, to assign / All such things to the empire of the Gods, / Acknowledging their power to be supreme'.[56] Like Lucretius' in his own time, Lear's personal cosmology goes against the grain of contemporary ideas. In the seventeenth century, the stability of the firmament, an eminently reassuring fact, was still taken for granted by writers like Henry Church. Nevertheless, as William R. Elton remarks, Lear is convinced that the heavens are 'old' (2.4.180) and that, far from symbolising permanence, they parallel man's decay.[57] By the same token, Gloucester observes that, if the very sea had to endure what the king must now go through, it 'would have buoyed up and quenched the stellèd fires' (3.7.57–8), an image that will incidentally reappear in *The Tempest* ('the sea, mounting to th'welkin's cheek, / Dashes the fire out', 1.2.4–5) a few years later. He also points to the impermanence of the firmament, since the 'stellèd' (i.e. 'fixed') stars are actually liable to be extinguished by the stormy sea.[58] One better understands, then, Lear's last onslaught against the 'heaven's vault' in the play's last act:

> Howl, Howl, Howl! O you are men of stones.
> Had I your tongues and eyes, I'd use them so
> That heaven's vault should crack. (5.3.231–3)

According to Katharine Goodland, 'Lear's allusion to the breaking of heaven's vault'[59] may be a hint at 'the belief that ritual weeping had the power to commune with the supernatural realm', just as in the mystery cycles where women are seen crying before they start cursing Jesus's tormentors and executors.[60] This being said, it would seem that, since the presence of the supernatural is constantly questioned in *King Lear*, Lear's final ejaculation should rather be aligned with the *senectus mundi* imagery in the play. In the dismal and ageing world depicted by Shakespeare, the apocalypse seems near.

The fear that the sky might fall down on men's heads dates back to antiquity. Manilius, the author of a Hellenistic treatise entitled *Astronomica*, had already expressed the same concern as he described the vault of heaven: 'Perchance, [the mortals] wonder, the firmament is seeking to split into separate fragments; with the slackening of the framework cracks are opening and admit new light through a split in the ceiling: what would men not fear might befall them, when they behold the great firmament damaged, and hurt done to heaven strikes their eyes?'[61] At the Globe, the actor playing the title character may have then pointed his finger at the sky in order to convince the audience that the heavenly vault might indeed be crumbling down on them. But more simply, and more probably, he must have pointed at the canopy above him, which stood for the firmament and whose underside was painted with the signs of the zodiac. In a courtly venue such as Whitehall, this would have been a less powerful scene, as the enclosed space provided by the old Banqueting House had no decorated ceiling. However, it could boast of 'at least sixty-three casement windows' and was therefore 'generously fenestrated'.[62] As a result, a courtly Lear could have pointed at one of the hall's numerous windows to show the sky about to collapse.

What is certain is that the weather (as depicted by the characters and by Lear's allies in particular) is so foul in the play that it may be taken to correspond to the reputed dangers of what was then called a 'climactericall' year. In Cotgrave's *Dictionarie*, the word 'climactere' is defined as follows: 'The Climatericall year; everie seventh, or ninth, or the 63 yeare of a mans life; all very dangerous, but the last, most.'[63] The idea of 'climatericall' years came from Greek astrology, which had imported it from old Egyptian and Chaldean traditions.[64] If Lear is definitely older than sixty-three, he is '[f]ourscore and upward, not an hour more, nor less' (4.6.54), which tends to suggest that he is eighty-one and that he belatedly goes through a climacteric year (9 x 9 = 81) which, as a consequence, makes him endure the worst calamities. Kent makes it clear that he has 'years on [his] back forty-eight' (1.4.39), which means that he is now entering his forty-ninth year, that is, yet another climacteric year (7 x 7 = 49).

The performance of thunder

No doubt such disastrous climatic episodes, like those mentioned by Titania in *A Midsummer Night's Dream*, were familiar to Shakespeare's audience. Brian Fagan affirms that '[s]torm activity increased by 85 per cent in the second half of the sixteenth century, mostly during cooler winters', before specifying that '[t]he incidence of severe storms rose by 400 per cent'.[65] Judging from the available testimonies, such frequency did not make people more accustomed to storms and less frightened by them. So, at the Globe, the disquieted spectators could certainly identify with Kent, horrified by '[s]uch sheets of fire, such bursts of horrid thunder, / Such groans of roaring wind and rain' (3.2.46–7), all the more so as a number of appalling thunderstorms had devastated England and northern Europe in the spring and winter of 1606. Terrified by the weather, an eyewitness reported, for instance, that '[t]he nine and twentieth, and thirtieth of March [1606], the wind was so extraordinary great and violent, that it caused great shipwreck; it also caused the sea, and divers rivers to overflow their bounds, and drowned many people, and much cattle'.[66] In 1607, the year when Nathaniel Butter and John Busby registered the play with the Stationers' Company, on 26 November, an anonymous *True report of certaine wonderfull ouerflowings of waters* was published.

Fig. 10 Engraved frontispiece of *1607. A true report of certaine wonderfull ouerflowings of waters*, 1607. By courtesy of the British Library Board, 1103.e.58, front cover.

Making the most of recent meteorological events that had afflicted several parts of England, the report describes an apocalyptic landscape with bridges carried away, houses broken down, men and women drowned. 'God', the author writes,

> [d]oth [. . .] strike our Cattle with diseases: he takes away the liues of our beasts fit for labor: he destroies the Corne-fields, & threatens vs with famine: he vndermines our houses with tempests, to make vs feare a desolation. Read therefore, and reade with trembling these his late dreadful iudgments, mocke not our selues with vaine hopes, but know that if earthly fathers may be drawne away to forget their owne children, our heauenly father may by the vilenes of our souls be drawne to shake off his own people. Listen then how he menaceth, and stand amazed at the wonders of his wrath.[67]

If the readers of Q1 (1608) had the opportunity to peruse *1607. A true report of certaine wonderfull ouerflowings of waters*, they must have drawn a parallel between Lear's perilous condition in the frightening storm on the heath and the devastating weather they were increasingly getting used to in the 'Little Ice Age'. According to Robert Markley, Shakespeare's titular character 'is not wandering through a metaphoric storm that marks his poetic madness and signals the disruption of the natural order, but an all-too recognizable figure who registers the complex connexions between climatic instability and its potential consequences: the loss of agricultural harvests and the fracturing of ideologies of national unity, patriarchal authority, and socioeconomic stability'.[68] All in all, stripping the storm of its metaphoric connotations by making the thunder sound almost real through its rumblings in the wings certainly amounted to making climate almost tangible onstage. Simon C. Estok goes as far as to argue that the tragedy conveys the 'ecophobia' of an audience traumatised by 'bad harvests' and 'cold weather' and that the play proves particularly 'vivid in its foregrounding of environmental unpredictability and in its dramatization of a fear of nature'.[69] Yet, for all its realism in terms of climatic representation, *King Lear* does not depict reality. It basically remains a terrifying dramatic poem, a playtext in which, as Touchstone ironically remarks in *As You Like It*, 'the truest poetry is the most feigning' (3.4.14). Shakespeare's onstage storms first and foremost partake of an artistic process that transforms the landscape into a mindscape open to scrutiny and anatomy.

If we look at the details of the text, we notice that the stage direction '*Storm still*' is repeated six times in the Folio edition. Once again, to a number of scholars, it simply 'evokes [. . .] the severe weather that, if not commonplace, was a far more familiar experience for Shakespeare and his audiences than it [is] for his [twenty-first-century] critics'.[70] The stage directions related to the storm in *King Lear* are present in the Folio but are absent from the 1608 (Q1) and 1619 (Q2) Quartos, a rather unsurprising fact given the overall scarcity of stage directions in the Quartos in general. '*Storm and tempest*' (2.4.273 s.d.) in act 2 and '*Storm still*' (3.1.0 s.d., 3.2.0 s.d., 3.4.59 s.d., 3.4.93 s.d.) in act 3 mark the exterior scenes in the Folio, and it seems to imply that the public performances of the play enhanced both Lear's interior storm and the actual apocalyptic weather sweeping the heath. By contrast, the total absence of such stage directions in the Quartos suggests that Shakespeare initially resorted to the storm as a metaphor of Lear's deranged mind.[71] For Coppélia Kahn, Shakespeare's depiction of the storm as the 'breaking of something enclosed' is a macrocosmic counterpart of Lear's interior microcosm, an outward extension of his own heart 'cracking, letting out the hungry, mother-identified part of him in a flood of tears'.[72] Similarly, the sulphurous thunderclaps in the heavens are replicated in woman's 'sulphurous pit' (4.5.123) as if her allegedly fierce and raging sexuality created the human equivalent of a storm in her bed and belly.

The first recorded performance of *King Lear* was presented at court on 16 December 1606.[73] Performances at Whitehall were generally held in the old Banqueting House,[74] and given the impossibility of achieving spectacular stage effects in such a reduced space (ninety feet long by forty feet wide), Shakespeare may have pragmatically decided to capitalise on 'the penetrative power' of the plays' speeches[75] and to cut down on sensational stage effects in order to focus on the interiority of Lear's storm. Moreover, the powerful stink of fireworks[76] (in his 1573 manual devoted to artillery and explosives, Peter Whithorne affirms for instance that the best saltpeter used for gunpowder 'is made of the dunge of beastes')[77] must have deterred the playwright from any bold experiment in a hall crowded with members of the upper classes. The courtly audience that attended a performance of *King Lear* on St Stephen's Night was in a festive mood and, as they were comfortably seated inside, the spectators were of course safely protected from the whirlwinds and storms of the play. The sense of contrast between images of outside weather conditions and inside comfort would therefore have been particularly vivid and must have marked the performance of the play. Yet, as noted by R. Chris Hassel Jr, performances that took place on St Stephen's were also part and parcel of a liturgical agenda. As such, they were meant to call to mind the biblical values of 'patience in adversity' – an adversity ideally suggested by harsh weather conditions in *King Lear* – and 'these peculiarly Christian virtues of loving one's enemies and forgiving one's persecutors',[78] virtues ideally embodied by Cordelia.

Mortification thus characterised the court performances of the New Year's festivities, and *King Lear* was no exception. Partly because of this distinctive liturgical context and partly because of the material conditions, Shakespeare denounces human folly in a rather sober way. 'O, let me not be mad, not mad, sweet heaven! / Keep me in temper. I would not be mad,' Lear begs the sky before the weather begins to deteriorate (1.5.45–6). These two lines suggest that the title character makes the sky a mirror of his own mind, as is corroborated when the old king evokes '[t]his tempest in [his] mind' (3.4.12). We may imagine that, at this particular moment, a loose piece of canvas was being turned on a wheel producing something that sounded like the howling of strong winds, making the audience aware that this was of course an interior noise, just like the tempest of the play which first and foremost takes place inside Lear's brain.

An experienced playwright by then, Shakespeare knew that his tragedy could be performed in different venues and that he had to write an easily adaptable text. The 'sulphurous pit' (4.5.123) and lightning (3.2.4) he cunningly mentions, for instance, could either be viewed as gripping metaphors by a courtly audience or as strongly realistic hints by the Globe's spectators, forced to endure the stink of pyrotechnics and probably prone to associate it with the odour of hell.[79] As to theatrical acoustics, the playwright was also aware that, if a courtly venue did not allow him to rely on sophisticated weather effects, the technical means available at the Globe, while admittedly more varied, still did not allow him to stage a truly impressive storm.[80] Generally speaking, the drums, cannonballs and fireworks that were used to simulate thunder and lightning (a device already exploited for the performances of *Julius Caesar* at the Globe) created sensational stage effects in a public playhouse, thus giving the audience an idea of what '[t]he wrathful skies' (3.2.43) mentioned by Kent may have sounded like.[81] The audience could easily figure out 'sheets of fire', 'bursts of horrid thunder' and 'groans of roaring wind and rain' (3.2.46–7), impressions that must have been even more sensational on stormy days. Once again, however, one should not overestimate the impact of the Globe's special effects on the spectators. Clearly, the drum rolls in a vast playhouse and the cracking of fireworks in an open space could only give a faint idea of the real thing,[82] so that Shakespeare must have relied on a range of visual and acoustic tricks to set his storm in motion. One can therefore imagine that a public performance of *King Lear* would have involved a swivel, a squib 'to produce the effect of lightning' and a cannonball (rolled around the floor of the 'heavens' or sent down a 'thunder run').[83] All this paraphernalia must have been reduced to a minimum in the small enclosed space of a court performance.[84]

To add grist to this mill, the Folio text makes the Fool voice yet another taunting comment on human callousness and links it to the winter season:

> Winter's not gone yet if the wild geese fly that way.
> Fathers that wear rags
> Do make their children blind,
> But fathers that bear bags
> Shall see their children kind.
> Fortune, that arrant whore,
> Ne'er turns the key to th'poor. (2.4.40–6; in the Folio only)

The upcoming cold conditions heralded by the wild geese heading south here prepare the Globe audience for the 'wintry' stage effects, which would have been superfluous in the case of a private performance where, for lack of spectacular sound machinery, the emphasis was essentially laid on Lear's inner storm. Needless to say, this passage, like other ones in the play, must have taken on fresh resonances in an open-space theatre where it was possible to point at birds in the sky at the time of the performance. This, of course, does not invalidate the hypothesis of those who see in the 'wild geese' a topical reference to Grace Wildgoose, Sir Bryan Annesley's ungrateful daughter. At any rate, different performance conditions triggered different responses from the audience. In act 3, scene 4, Edgar, disguised as Poor Tom, insists on the coldness of the weather and an increasingly mad Lear speaks to himself. In the Folio version, the text reads:

> [*Edgar*:] Bless thy five wits, Tom's a cold. O do, de, do, de, do, de. Bless thee from whirlwinds, star-blasting, and taking. Do Poor Tom some charity, whom the foul fiend vexes. There could I have him, now, and there, and there again, and there.
> *Storm still*
> *Lear*: Has his daughter brought him to this pass? (3.4.55–60)

The Folio and the 1608 Quarto versions present small variants but the text is basically the same. As has already been pointed out, what differs in this case is the presence of the stage direction '*Storm still*'. The 'broad sound' produced in the Globe (as opposed to the 'round sound' produced in private theatres)[85] may have been particularly impressive here and suggested a supernatural presence questioning, if not radically belying, Lear's nihilistic philosophy. By contrast, supernatural effects were not favoured by the intimate configuration of the old Banqueting House. If no sound at all was produced in Whitehall, what the audience kept in mind was the cold mentioned by Poor Tom rather than the storm which they could not hear, being thus forced to concentrate on Lear's rage and internal turmoil.

All in all, the disquieting climate prevalent in *King Lear* echoes some of the contemporary anxieties regarding God's power – is the thunder the voice of God or the voice of folly? – or even the possible impending collapse of civilisation. Here, the playwright seems to adhere to a form of nihilistic world view based on Epicurean philosophy. As a result, in the play, the representation of climate (derived from the Greek *klinein*, 'to slope') is inspired by Lucretius' *clinamen*, a technical as well as epistemological notion referring to the swerve of atoms in the void. For Epicure, the idea of divine intervention had only been meant to allay superstitious anxieties. Shakespeare complicates the issue by staging an old king who, having fallen from grace, seems to espouse Epicurean views but who, at the same time, proves more and more desperate as he realises that there is no god to answer, help or support him. He cannot make any sense of the void and vortex he is caught up in and he proves incapable of transforming the dizzying force of the negative (symbolised by the storm, the wind and the rain devastating the heath) into a propelling strength. Contrary to the poet who has imagination at his service, Lear feels incapable of giving 'airy nothing / A local habitation, and a name' (*A Midsummer Night's Dream*, 5.1.16–17). In his exploration of the multi-layered meanings of the storm, Shakespeare represents the dissection or an*atom*y of a mind gradually losing control and ending in perdition. In his portrayal of Lear's decline in the middle of apocalyptic weather, he shows how much a soul in turmoil may be affected by the perception of its outer environment and how much the course of one's life may be subsequently altered. The imbrication of conscience and climate magnified by the poetic strength of the verse and apocalyptic imagery taking its roots in Lucretius and pagan mythology makes this a powerful tale, told by a mad old king, 'full of sound and fury / Signifying nothing' (5.5.26–7), as Macbeth puts it at the end of his famous absurdist monologue.

Notes

1. As they come after a series of Saxon monosyllables ('Blow, winds, and crack your cheeks! Rage, blow', 3.2.1), the three-syllable 'cataracts' and the four-syllable 'hurricanoes' (3.2.2) used by the infuriated Lear must have had a powerful poetic and dramatic impact on the audience. In their demand for more breath from the actor, these long words mimic the powerful and frightening force of the winds, now raging, now suspended, before blowing even more strongly afterwards. These lines thus poetically evoke a climatic microcosm that uses the sounds and silences of language to produce the equivalent of the actual outward sounds and effects of the storm. Diction, articulation and prosody here combine to create on the linguistic and phonetic levels the forceful impression that the natural elements have indeed been let loose.
2. Unless otherwise indicated, quotes from *King Lear* rely on the Folio version.
3. A similar 'weather-conscious son[g]' (Tiffany Stern, 'Before the Beginning; After the End: When Did Plays Start and Stop?', p. 371) is found in Thomas Dekker's *The Shoemaker's Holiday*, p. 80, 'The Second Three-Man's Song': 'Cold's the wind, and wet's the rain, / Saint Hugh be our good speed. / Ill is the weather that bringeth no gain, / Nor helps good hearts in need.'
4. See Daniel Schäfer, *Old Age and Disease in Early Modern Medicine*, p. 34.
5. Thomas Whythorne, *The Autobiography of Thomas Whythorne*, p. 117. This passage is also quoted and commented upon in Ronald Bedford, Lloyd Davis and Philippa Kelly (eds), *Early Modern Autobiography: Theories, Genres, Practices*, p. 5.
6. *Meteoros* comes from the verb *meteorizo*, 'to rise to a height'. This is explained in '"Minded Like the Weather": The Tragic Body and its Passions' (p. 205), in which Gail Kern Paster relies on Vladimir Jankovic's *Reading the Skies*, p. 15.
7. On the power of words to generate storms, see Philemon Holland (trans.), *The historie of the vvorld*, The Second Booke, chap. LVIII, p. 26: 'It appeareth vpon record in Chronicles, that by certaine sacrifices and prayers, Lightnings may be either compelled or easily entreated to fall upon the earth.'
8. On the analogy with Ariel's imagery in *The Tempest*, see below, Chapter 7, p. 220.
9. Aristotle, *Meteorologica*, II.9, p. 231. See also Pseudo-Aristotle, *Cosmic Order and Divine Power*, pp. 29–31. On early modern 'exhalations' as hot and dry substances thought to be at the origin of storms (vs 'vapours', hot and humid substances thought to cause hail and rain), see Margaret Llasera, *Représentations scientifiques et images*

poétiques en Angleterre au XVIIe siècle. À la recherche de l'invisible, p. 130. Cf. Christopher Marlowe, *Tamburlaine the Great*, Part I, 4.2.43–6: 'As when a fiery exhalation / Wrapt in the bowels of a freezing cloud, / Fighting for passage, makes the welkin crack, / And casts a flash of lightning to the earth.'

10. This is the case, for instance, in *The last terrible tempestious windes and weather*, London, 1613, STC (2nd edn) 25840.
11. Randall C. Zachman, 'Contemplating the Living Image of God in Creation', p. 40.
12. John Calvin, *Sermons of Master Iohn Caluin, vpon the booke of Iob*, trans. Arthur Golding (1574), 'The Table', sig. ij.
13. Ibid. The XXXVII chap. of Job, p. 745.
14. Ibid.
15. On this, see William R. Elton, *King Lear and the Gods*.
16. Leonard Digges, *A prognostication of right good effect fructfully augmented*, sig. B2 r–v.
17. Richard Hooker, *Of the lavves of ecclesiasticall politie Eyght bookes*, The First Booke, 1.3, pp. 52–3.
18. Philip B. Secor, *Richard Hooker: Prophet of Anglicanism*, p. 94. Secor explains that '[h]is tutor at college, who had taught him to follow Calvin's example in all religious matters and whom he supported to the point of being expelled, was the renowned Calvinist, John Rainolds'.
19. W. David Neelands, 'Predestination', p. 192.
20. Steve Mentz, 'Strange Weather in *King Lear*', p. 149, n. 42.
21. W. J. Torrance Kirby, 'Richard Hooker's Theory of Natural Law in the Context of Reformation Theology', p. 690.
22. Pierre Lurbe, 'Theologico-political Issues in Richard Hooker's *Laws of Ecclesiastical Polity* and Thomas Hobbes's *Leviathan*', paragraph 8, <http://lisa.revues.org/4050> (last accessed 2 May 2018).
23. Richard Hooker, *Of the lavves of ecclesiasticall politie Eyght bookes*, The First Booke, 1.3, pp. 52–3. Hooker adds further down: 'But vnto vs there is one only guide of all agents natural, and he both the creator, and the worker of all in all, alone to be blessed, adored and honoured by all for euer' (1.3, p. 55).
24. Laurie Shannon, *The Accommodated Animal: Cosmopolity in Shakespearean Locales*, p. 173.
25. Ibid.
26. i.e. 'naked on the bare earth'. Ibid. p. 20 and p. 172.
27. Ibid. p. 172.
28. Philemon Holland (trans.), *The historie of the vvorld*, The Seventh Booke, 'The Proëme', p. 152. On this passage in particular, see also Laurie Shannon, '"Poore wretch . . . laid all naked on the bare earth": Human Negative Exceptionalism among the Humanists', p. 208. Cf. Shakespeare's *Cymbeline*, when Belarius explains how he was banished: 'Cymbeline loved me, / And when a soldier was the theme my

name / Was not far off. Then was I as a tree / Whose boughs did bend with fruit; but in one night / A storm or robbery, call it what you will, / Shook down my mellow hangings, nay, my leaves, / And left me bare to weather' (3.3.58–64).

29. Philemon Holland (trans.), *The historie of the vvorld*, The Second Booke, chap. XX, p. 14.
30. Up to a point, these eclipses symbolise the political situation of Shakespeare's England and 'figure the eclipse of Tudor England by the glorious sun of the Stuart'. See Willy Maley, 'Critical Review: "Great thing of us forgot"? New British Angles on *King Lear*', p. 164. However, Gloucester's metaphorical allusion does not preclude a topical one as, before the play's first known performance at court in 1606, an impressive sequence of eclipses occurred in England on 3 April, 27 September and 12 October 1605. Partial lunar eclipses had also taken place a few years before, in 1598 and 1601. The sheer number of astrological forecasts mentioning eclipses in the early seventeenth century testifies to the appeal of such meteorological phenomena in early modern England.
31. Edmund's scepticism is interestingly echoed by the reader of the First Folio now held by the Kodama Memorial Library, who comments on Gloucester's beliefs as follows: 'enumeration of the most vnnaturall wickednesse of men folishlie ascriued to the starres'. This annotation seems to confirm the prevailing scepticism about astrology in seventeenth-century England. See Akihiro Yamada (ed.), *The First Folio of Shakespeare*, p. 249, qq3v,b [TLN 419–84]. Online edition at <http://shakes.meisei-u.ac.jp/ALL.html#45> (last accessed 18 April 2018).
32. Thomas Nashe, *Pierce Penniless his Supplication to the Devil* (1592), in *The Unfortunate Traveller and Other Works*, pp. 94–5. Nashe actually alludes to John Harvey's 1583 almanac whose predictions had proved erroneous. See Steane's nn. 211 and 212, p. 94.
33. Harry Rusche, 'Edmund's Conception and Nativity in *King Lear*', p. 161.
34. On this, see also François Laroque, 'The Science of Astrology in Shakespeare's Sonnets, *Romeo and Juliet* and *King Lear*', pp. 40–2.
35. In connection with this, see the Fool's remark: 'O nuncle, court holy water in a dry house is better than this rain-water out o' door' (3.2.10–12). Interestingly, Thomas Becon assured a dying man that baptism was 'a sure token of the fauour of God toward you, and that you are predestinate vnto euerlasting life [. . .] Baptisme is a continuall signe of the fauour of God towarde vs, of the fre remission of sines, of our reconciliation vnto God for Christes sake, and that we be by adoption the sonnes of God.' Quoted in Alec Ryrie, *Being Protestant in Reformation Britain*, p. 323.
36. See Steve Mentz, 'Strange Weather in *King Lear*', p. 142.
37. *The true chronicle history of King Leir* (1605), in *Narrative and Dramatic Sources of Shakespeare*, vol. 7, pp. 337–402.
38. Ibid. scene 7, p. 351.

39. John Dover Wilson (ed.), *King Lear*, p. xxxvi.
40. *The true chronicle history of King Leir*, scene 19, p. 373.
41. Ibid. p. 377.
42. Ibid. p. 379.
43. Ibid.
44. Ibid. p. 380.
45. On Lear as exorcist, see Pierre Iselin, '"Strange things toward": The Dismembered Perspective of *King Lear*'.
46. Quoted in F. W. Brownlow, *Shakespeare, Harsnett, and the Devils of Denham*, p. 287.
47. Ibid. p. 222.
48. Pierre Iselin, '"Strange things toward"', p. 113.
49. Lucretius' *De rerum natura* is already hinted at in Goneril's allusion to the 'epicurism' (1.4.213) of Lear's retinue. The following development is indebted to Jonathan Pollock's 'Of Mites and Motes: Shakespearean Readings of Epicurean Science'.
50. On this, see Antonio Clericuzio, *Elements, Principles and Corpuscles: A Study of Atomism and Chemistry in the Seventeenth Century*, p. 14.
51. See Lucretius, *De rerum natura*, Book I, p. 21: 'But since I have proved above that out of noting / Nothing can be created, and that what / Has been begotten cannot be recalled / To nothing, first-beginnings must need be / Of an immortal body, into which / All things may be dissolved at their last hour.' For further comments on Shakespeare's borrowing here, see Richard Allen Shoaf, *Lucretius and Shakespeare on the Nature of Things*, p. 11.
52. Given Poor Tom's way of life and near nakedness, scholars generally assume that King Lear equates him with the cynical philosophers Diogenes or Crassus, even though Lear is supposed to live before their time.
53. Lucretius, *De rerum natura*, Book I, p. 3.
54. Ibid.
55. Epicure examines the same question in his letter to Pythocles, in which he explains that thunder is not due to divine action but probably caused by the wind rolling in the clouds' hollow parts (*Letter to Pythocles*, in *Diogenes Laertius: Lives of Eminent Philosophers*, vol. 2, par. 100).
56. Lucretius, *De rerum natura*, Book VI, p. 233.
57. William R. Elton, *King Lear and the Gods*, p. 246.
58. Cf. *Othello*: 'The wind-shaked surge with high and monstrous mane / Seems to cast water on the burning Bear / And quench the guards of th'ever-fixèd Pole' (2.1.13–15). See above, Chapter 4, p. 129.
59. In *Romeo and Juliet*, the word 'vault' is used eight times in reference to the Capulet monument, along with Romeo's use of the phrase 'vaulty heaven' (3.5.22). See above, Introduction, p. 23. This confusion of high and low serves as a window to showcase the ambivalence of passion and the death-marked course of the lovers. In *King Lear*, the cracking

of the heaven's vault indirectly describes the whole world as reduced to 'bare ruined choirs, where late the sweet birds sang' (Sonnet 73).

60. Katharine Goodland, 'Inverting the Pietà in Shakespeare's *King Lear*', p. 65.
61. Manilius, *Astronomica*, I, p. 63.
62. R. B. Graves, *Lighting the Shakespearean Stage, 1567–1642*, p. 159.
63. Randle Cotgrave, *A Dictionarie of the French and English Tongues*, 'CLI', n.p.
64. Brendan Dooley (ed.), *A Companion to Astrology in the Renaissance*, p. 430.
65. Brian Fagan, *The Little Ice Age: How Climate Made History, 1300–1850*, p. 91.
66. Stow and Howes quoted in René Weis, *Shakespeare Unbound: Decoding a Hidden Life*, p. 335. In the opening scene of *Coriolanus*, Shakespeare seems to remember the severe winter of 1606–7 that resulted in bad harvests and in the Midland riots which followed the grain dearth. Yet some critics have discarded this parallel, favouring instead a comparison with the Tower Hill riot, which occurred in June 1595. As to the passage where Martius blames the plebeians for being 'no surer, no, / Than is the coal of fire upon the ice, / Or hailstone in the sun' (1.1.170–2), it is often read as an allusion to the great frost of 1607–8.
67. *1607. A true report of certaine wonderfull ouerflowings of waters*, 'Newes out of Summerset Shire', n.p.
68. Robert Markley, 'Summer's Lease: Shakespeare in the Little Ice Age', p. 137.
69. Simon C. Estok, *Ecocriticism and Shakespeare: Reading Ecophobia*, p. 19.
70. Robert Markley, 'Summer's Lease', p. 135.
71. This internal turmoil has a gradual impact on Lear's predicament and contributes to isolating him from the rest of the world. In Q1, as Kent and a Gentleman meet and converse on the heath, the Gentleman describes Lear's battle against the 'to-and-fro conflicting wind and rain', a passage absent from the Folio in which the readers/spectators must wait until act 3, scene 2 in order to discover the climactic as well as climatic scene of the tragedy. This is made clear in René Weis's parallel text edition of *King Lear*, p. 194, 3.1.7–15. Besides, it is worth noticing that the readers of the Folio text, short of seeing the play performed onstage, may also have tended to view the storm as a mental one. This is corroborated by the marginal gloss found in the Kodama Memorial Library Folio: 'The *tempest of the mind* makes ws forget all bodelie troubles' (my emphasis). See Akihiro Yamada (ed.), *The First Folio of Shakespeare*, p. 257, rr3,b [TLN 1798–1861]. Online edition at <http://shakes.meisei-u.ac.jp/ALL.html#45> (last accessed 18 April 2018).
72. Coppélia Kahn, 'The Absent Mother in *King Lear*', p. 46.

73. The title page of the 1608 Quarto edition of the play states that *King Lear* has been 'played before the Kings Maiestie at Whitehall upon S. Stephans night in Christmas Hollidayes'. This is corroborated by the Stationers' Register for 26 November 1606.
74. According to René Weis (ed.) in *King Lear*, A Parallel Text Edition, p. 3, Q1 seems to point to a public performance, 'or stages of it', which took place a few weeks or a few months earlier than the Whitehall one. Yet the Folio text of the play, with its 2,890 lines, seems more playable in a public venue than the Quarto version. The 1608 version, which contains 3,063 lines, must have been a little too long for an open-air playhouse.
75. Bruce R. Smith, *The Acoustic World of Early Modern England: Attending to the O-Factor*, p. 284.
76. On the stinking odour of squibs, see Jonathan Gil Harris, *Untimely Matter in the Time of Shakespeare*, pp. 119–39. Harris notes that, 'like all gunpowder products, the squib combined foul-smelling ingredients – sulfurous brimstone, coal, and saltpeter – that reeked all the more when detonated'.
77. Peter Whithorne, *Certaine vvayes for the ordering of souldiours in battelray*, fol. 24r.
78. R. Chris Hassel Jr, *Renaissance Drama and the English Church Year*, p. 24.
79. See Jonathan Gil Harris, *Untimely Matter in the Time of Shakespeare*, pp. 127–8: 'The association between bad smells and hell had been exploited on the English stage since the Corpus Christi drama. The annual play presented by the Chester Cooks, for example, included a Hell Mouth that emitted smoke and what was evidently a hideous stench.'
80. As performances in public theatres could probably not produce very powerful sounds, they relied on visual effects which their rather rudimentary machinery would have made equally difficult for them to create. Moreover, because the plays were performed in full daylight, night and darkness could not be represented in any really spectacular way. On this, see Gwilym Jones, 'Storm Effects in Shakespeare'.
81. Still in 1638, the Italian architect and stage designer Nicola Sabbattini recommended the use of cannonballs rolled in a tray to reproduce the peals of thunder and the loud noise of the storm. See his influential treatise *Pratiques pour fabriquer scènes et machines de théâtre* (*Pratica di fabricar scene e machine ne' teatri*), chap. 53, pp. 164–5.
82. In *The Acoustic World of Early Modern England*, Bruce Smith specifies that 'the very loudest sound that a sixteenth- or seventeenth-century listener might encounter [. . .] falls within a range of decibel intensities that would nowadays almost rate as normal events', and these sounds were 'thunder, cannon-fire, and bells' (p. 49).
83. See Mary Thomas Crane, 'Optics', p. 265.

84. In 'Reviving the Legacy of Indoor Performance' (p. 238), Bart van Es specifies that spectacular effects were 'technically ill-suited to the smaller indoor stage'. He further affirms that stage tricks 'would have limited appeal for a coterie audience'. Arguably, such remarks could also apply to court performances. Yet van Es also notes that 'Shakespeare's change of dramatic technique [. . .] seems at least in part an independent development' as his plays composed for the Blackfriars do not share the features generally ascribed to indoor plays, often comic or 'politically resonant' ones (ibid.). If we do not know exactly how Shakespeare adapted his plays for the court, the playtexts, however, do give us some clues.
85. Bruce R. Smith, *The Acoustic World of Early Modern England*, p. 213.

Chapter 6

Clime and Slime in *Anthony and Cleopatra*

> [. . .] The higher Nilus swells,
> The more it promises [. . .] (2.7.20–1)

In *Anthony and Cleopatra*, entered in the Stationers' Register on 20 May 1608 and printed for the first time in 1623, Shakespeare dramatically expands the scope of the action in an exotic locale while continuing to dissect the human mind in connection with its geographic environment, the place of the planets and climate in general. In *King Lear*, written around the same time, he examined the meanings of the storm in an atomist universe. Here, he sticks to a similar Lucretian perspective but in a rainless, windless environment as he enhances the material interactions between sun and earth, water and stars, as well as clime and slime, thus foregrounding the importance of the elements and the planets in the building up of his characters' emotions and passions. For Edward Fenton, the translator of Pierre Boaistuau's *Certaine secrete wonders of nature*, Cleopatra, the 'late quéene of *Egipt*', was the embodiment of 'chaste and virtuous love' and she 'made hir self to be deuoured of serpents' after the death of 'hir husband *Anthonie*'.[1] This cameo description and faint praise of Shakespeare's heroine – which, amusingly, makes her stand inside the institution of marriage – cannot of course suffice to give a fair idea of such a complex and highly sophisticated and poetic piece of work, suffused with skyey imagery and a sense of eternity. In the play, the celestial sphere is a much-desired space, as uncertain as it is promising,[2] since the heavens, presented in a climatic perspective, keep opposing temperate Rome against parching Egypt. As in *Romeo and Juliet*, excessive heat here plagues the title characters to such a point that only death, a form of absolute and definitive coldness – 'This case of that huge spirit now is cold' (4.15.89), Cleopatra says as she is gesturing toward Anthony's corpse – will bring them relief and peace of mind. The playwright thus translates meteorological effects into dramatic affects when depicting the oriental locale as the kingdom of death.

In his second Roman play, Shakespeare intertwines cosmic visions with elemental forces and imperial desires. Its cosmology proves so dense that it completely reworks Cleopatra's *Liebestod*. Against such a rich backdrop, the passionate outbursts and erotic drives of the Egyptian queen are correlated with broader geo-humoral concerns and climatic issues. If critics often call attention to the determining presence of the Nile and to the importance of the sublime in the play, few have cogently linked these issues with Shakespeare's specific approach to climate. Yet an exploration of the dichotomy at work between Rome and Egypt in climatic terms, taking Richard Knolles's 1606 translation of Jean Bodin's *Les six livres de la République* as a possible model of influence,[3] does help illustrate and better understand the atmospheric in*temper*ance at the core of Egypt, an exotic land tied up with the Ottoman Empire at the time when Shakespeare was writing his play. Even more importantly perhaps, the traditional readings of the play are being decentred in this chapter. So far, Anthony has repeatedly been regarded as the embodiment of inconstancy ('O, my oblivion is a very Anthony', 1.3.91), but inconstancy should not always be taken as a primarily human characteristic. In the tragedy, I argue, it is first and foremost an environmental feature applying to Egypt and, by extension, to Cleopatra herself more than to her Roman lover.

Two climes, three tempers

For Anthony and Cleopatra, the earth is both a giant Wheel of Fortune[4] and a vast playground offering a condensed representation of the world. Admittedly, the geographical locations of the tragedy cannot be reduced to two places only. The scene being the new and somewhat cosmopolitan Roman Empire, the action takes place in Sicily, Rome, Alexandria, Misenum in the Bay of Naples, and Actium, the locus of the sea-battle off the Grecian coast. Such multiple locations are both indicative of movement and evocative of the threat of political instability. However, the tragedy's main poles are the two locations endowed with strong symbolic and climatic connotations, namely Italy and Egypt, respectively associated with the Tiber and the Nile. These two lands and two rivers are repeatedly compared and contrasted in the play through a series of oppositions grounded in the climate theories of the time. Separated by more than a thousand miles, Rome and Alexandria serve to convey an idea of the immensity of the 'huge sphere' (2.7.14) mentioned by the servants who evoke a Ptolemaic universe made of seven concentric spheres with the earth at their inner centre. This 'huge sphere' is itself separated into climatic zones, as described for instance in José de Acosta's *Naturall and morall historie of the East and West Indies* (1604), where the English translator explains that

> [t]he temperature of the middle region of the world, where the sunne continually runnes his course, is scorched and burnt vp as with a neere fire. Ioyning to the same region, there are two others of eyther side, which (lying betwixt the heat of this burning zone & the cruell cold of the other two extreams,) are very temperate, and can have no communication one with another, by reason of the excessive heate of the heaven.[5]

If we rely on these observations, Egypt is situated in the 'scorched and burnt up' middle region, while Rome lies in one of the temperate zones adjacent to this middle area, right in '*medius mundi locus* (the centre of the world)'.[6] *Anthony and Cleopatra* reveals how Roman and Egyptian self-definitions are tied to Stoic and Epicurean philosophies based on climatic conditions. It does not take long to realise that the interaction between these conditions and the characters' (mis)demeanours follows a characteristic Aristotelian pattern summarised as follows in *Aristotles politiques* (1598):

> But the Meridionall or Southerne people, who are in the second and third climate, through the excessiue heat thereof haue commonly sharper wits, but smaller courage: and they that inhabite the farthest parts of the North, by reason of the excessiue cold and abundance of blood, are very couragious, but of small wit and prudence.[7]

Such views undoubtedly chime with those expressed in Shakespeare's tragedy, where the traditional tripartite division of the world is reflected in the tripartite coalition formed by Anthony, Lepidus and Octavius. In the play, Anthony is first and foremost the triumvir and the representative of Rome, even though, absorbed as he is in his private love affair, he fails to patrol the borders and dallies in Alexandria instead of seeking to maintain the region's loyalty towards Rome.

Unsurprisingly, in Giovanni Botero's 1601 *The worlde, or an historicall description of the most famous kingdomes and common-weales*, Italy is described as 'the most beautifull and goodliest region vnder the sunne, the darling of Nature, the mother of hardie men'.[8] It is situated 'vnder a climate most holsome and temperate, commodious for trafique, and most fertill for corne and herbage'.[9] In *The six bookes of a common-weale*, Jean Bodin is less interested in the climate of Rome itself than in the effects which this climate has upon Rome's inhabitants, and it is precisely this interaction, often more complex than it appears, that Shakespeare explores in turn.

'The people of Rome [. . .] were patient in their losses, constant in their victories, moderat in their passions, hating flatterers and taking delight in graue and seuere men', Bodin writes.[10] For him, Italy is 'in the most temperat situation that can be, betwixt the Pole and the Equator, and in the middest of Asia, Affrike, and Europe'.[11] Italy's temperate clime produces temperate beings and, as a result, for Bodin, the only nations able to command properly are 'the people of the middest', who are the wisest. Southerners are too contemplative, absorbed in 'natural and divine sciences', and northerners, gifted for 'manuall artes', can execute orders but not give them.[12] In his play, Shakespeare is keen to present Rome as a place of common sense and moderation and happy to downplay Plutarch's harsh criticism of the triumvirate – a criticism faithfully reproduced in Thomas North's 1579 translation, *The lives of the noble Grecians and Romanes*. Like Plutarch, North asserts that the triumvirate's members, Octavius Caesar included, had grown 'odious and hatefull to the ROMANES'.[13] Nowhere does this assessment transpire in *Anthony and Cleopatra*, where a good example of 'the people of the middest' can be found in the person of Octavia, who stands in sharp contrast to her hot-tempered second husband. Enobarbus affirms that Caesar's sister and Anthony's second wife 'is of a holy, cold, and still conversation' (2.6.121–2). Caesar also embodies temperance, especially when he declares: 'I had rather fast from all, four days, / Than drink too much in one' (2.7.100–1). The playwright here follows the 1603 enriched edition of North's Plutarch, whose 'Life of Octavius Caesar Augustus' insists on Caesar's soberness and makes it clear that '[h]e was very modest and continent in all the parts of his life [. . .] In his ordinarie diet he banished superfluity of meates [. . .] Also he drunke very little wine.'[14] In short, this means that Caesar is the most apt to command for, as Bodin observes, '[i]f then the inhabitants of the South be wilfull and obstinate [. . .] it is most certaine that the other is changeable, and hauing no constancie' while 'those of the middle regions hold the virtue of the meane, betwixt wilfulness and lightness'.[15] Significantly, Caesar's spirits can never grow too hot; they can simply be warmed by good news ('But it would warm his spirits / To hear from me you had left Anthony', Thidias tells Cleopatra in 3.13.69–70).

If we are to believe Alexas, Anthony would embody this much-praised Roman temperance: 'Like the time o'th'year between th'extremes / Of hot and cold, he was nor sad nor merry' (1.5.51–2). However, the 'well divided disposition' (1.5.53) of the Roman general is reported by one of Cleopatra's followers, who knows full well that his queen refuses all bad news and can only tolerate those which please her delicate ears. Anthony himself certainly wishes he were 'a man of steel' (4.4.33) the better to protect and defend the Roman world but, in actual fact, he does not fit in with this climatic pattern, as Enobarbus makes clear when he says that his general 'use[s] his affection where it is' (2.6.129). In yielding freely to his passions, Anthony behaves exactly like a northerner ill-suited to the clime of Alexandria,[16] a man truly at ease when he crosses mountains and faces the harshness of the freezing air. As Caesar remembers his past feats, he reminds us that, 'like the stag, when snow the pasture sheets, / The barks of trees [Anthony] browsed' (1.4.65–6). Again, when he crossed the Alps, he willingly fed on 'strange flesh / Which some did die to look on' (1.4.67–8).[17] In light of this, his highly inflammable temper can only be made worse by the Egyptian heat that excites his natural irascibility. This is made clear, for example, when his mistress mocks his erratic behaviour and imminent departure for Rome and when, in reaction to her sarcasm, he indulges in a fit of anger and exclaims: 'You'll heat my blood. No more!' (1.3.80). This fits in with Bodin's idea that the northerners' 'inward heat' proves particularly 'vehement'.[18] If 'they of the north[,] hot and moist',[19] suffer from their earthly affections, Cleopatra's lover, who claims the 'dungy earth' (1.1.37) for himself, is no exception in this general climatic approach.

Moreover, in Bodin as in Shakespeare, men are ruled by specific planets, depending on where they come from. For the author of *The six bookes of a common-weale*, Venus is the planet 'proper to the people of the South'[20] and, logically, northerners are under Mars's influence because of their warlike disposition. Those in the middle, fit to rule, depend on Jupiter and Mercury.[21] A similar classification is at work in *Anthony and Cleopatra*, where the playwright endows Caesar with a Jove-like stance and associates Cleopatra with the goddess of love, as in North's Plutarch.[22] With Anthony, she forms a mythical adulterous couple, whose radiating sensuality and eroticism place Mardian on the verge of sexual excitement when the eunuch confesses to the queen that, even though he 'can do nothing' (1.5.15), he is able to imagine '[w]hat Venus did with Mars' (1.5.18). Anthony, at the beginning at least, is therefore unmistakably depicted as the god of war (2.5.118), before being gradually debunked and dismissed because his repeated long stays in Alexandria seriously deteriorate his soldierly abilities as well as his reputation as a Roman general.

The east faintly imagined and idealised in *A Midsummer Night's Dream* thus takes shape in *Anthony and Cleopatra*, where it fails to prove as promising as Titania suggested since Egypt appears almost as desperately muddy as the forest near Athens.[23] Its recurrent flooding, however, is not seen here as a climatic catastrophe but as an agricultural blessing since the mud fertilises rather than destroys the earth in this land of overflowing excess.[24] Inseparable from the Nile river which gives the country its identity, Egypt as a southern region was associated with extreme heat, so that the climatic conditions of Alexandria partly explained the indolence and sluggishness of its inhabitants. Those were typically feminine features according to Johann Weyer's 1563 *De praestigiis daemonum* (*On the Illusions of Demons*), a treatise which established a false etymological connection between *mulier*, the Latin word for 'woman', and *mollities*, i.e. softness.[25] This misogynistic commonplace was sometimes complicated by a no less popular (and no less false) etymology deriving *mulier* from *mollis aer* ('soft air'), present in *Cymbeline* (5.6.448). In the context of Shakespeare's tragedy, such beliefs provide us with a valuable – though problematic – climatic portrait of the Egyptian queen, seen as an airy creature living in a soft environment. Air, in the play, like water and fire, is indeed repeatedly linked to Cleopatra (5.2.88), while Anthony seems to belong to the earth (1.1.37).[26] As to softness, it is not directly associated with Shakespeare's heroine but with the 'o'erflowing Nilus' (1.2.48) which seems to liquefy all human beings – especially ill-adapted foreigners – as well as inanimate matter.[27]

Aristotle, who pays much attention to the case of the Nile in his *Meteorology*, observes that, when it dries up, the land will again be flooded by the next spate:

> For wherever it has encroached on the land because the rivers have pushed it out, it must when it recedes leave behind it dry land: while wherever it has been filled and silted up by rivers and formed dry land, this must again be flooded.[28]

No wonder then that, when she hears of Anthony's second marriage after reluctantly listening to the messenger, the not-so-soft Cleopatra the flooding of Egypt:

> *Messenger*: Should I lie, madam?
> *Cleopatra*: O, I would thou didst,
> So half my Egypt were submerged and made
> A cistern for scaled snakes. Go, get thee thence! (2.5.94–6)

Unable to accept uncomfortable truths, the queen prefers to hear a lie even at the price of the submersion of her homeland. Once again, Shakespeare's heroine is so excessive here that her words plainly appear as what they are, that is, as pure theatrical posturing and verbal exaggeration.

Water is one of the main tropes attached to her as she sees the very idea of destruction in terms of flooding. Thus, Shakespeare subtly adapts Bodin's theory – a theory, as has already been seen, which posits that the people of the south are 'cold and dry' precisely because their 'climat' is hot and 'moist'.[29] In other words, their internal hotness is 'exhaled by the heat and drought of the sunne: whereas the cold doth keepe in the heat in the Northerne regions'.[30] Living in an environment which tends to make her both cold and dry, the queen naturally rejects any hint of internal moisture. As a result, in the first half of the play at least, her chagrin is expressed in terms of torrents of tears and through images of frozen wetness. Metaphoric hail and showers (of riches) are what she promises to the messenger if the news he brings of Anthony are good: 'I'll set thee in a shower of gold, and hail / Rich pearls upon thee' (2.5.46–7). Here, the cajoling Cleopatra plays once again with gender roles as she rewrites the myth of Danaë to stage herself as the king of the gods, Zeus, who appeared in the form of a golden shower to the princess of Argos in order to impregnate her womb.[31] Then, when Thidias begs to kiss her hand, she agrees and remembers the love of Julius Caesar (Octavius Caesar's great-uncle) who, anxious as he was to conquer new provinces for Rome, 'rained kisses' upon her body (3.13.85). Later on, in one of her hyperbolic tirades reminiscent of the biblical plagues of Egypt, rain becomes hail: 'From my cold heart let heaven engender hail' (3.13.160),[32] she declares. So, the dried-up Cleopatra conveys her anger through a meteorological image in which she wishes her 'cold heart' could actually freeze water droplets into ice. Interestingly, her imaginative fantasy corroborates her self-portrayal as a person originally 'cold in blood' (1.5.74), that is, not sexually frigid – even though, in her 'salad days' (1.5.73), Ptolemy XIII's sister was indeed no *femme fatale*[33] – but actually indicative of her humours and basic temperament. Similarly, the reference to the gypsy's 'cold heart' is to be taken less as indicative of her emotional indifference than as a humoral characteristic. We should once again remember that, for someone like Bodin, the 'inhabitants of Affrike'[34] have little heat inside. So, in accordance with this particular logic, Cleopatra contains her natural coldness inside her, which explains why she so often resorts to apocalyptic threats like the 'pelleted storm' (i.e. a hail storm) (3.13.166) which, Lear-like in her fury, she would like to generate. Hail, according to her wishes, would then be melting or 'discandying'[35] on the hot sands of Egypt (3.13.166), thus creating a new deluge that would drown its inhabitants. This naturally ties in with the motif of submersion used earlier on in the tragedy (2.5.94–6), thereby suggesting that the queen's physiological system generates major disturbances entailing both local and global consequences, as Enobarbus makes clear in the tongue-in-cheek portrait he gives of her and of her humours[36] when he engages in a misogynistic 'men's talk' with his general:

> Alack, sir, no – her passions are made of nothing but the finest part of pure love. We cannot call her winds and waters sighs and tears; they are greater storms and tempests than almanacs can report. This cannot be cunning in her; if it be, she makes a shower of rain as well as Jove. (1.2.138–43)[37]

We have just seen that Cleopatra deliberately portrays herself as Zeus and imagines her royal self as a masculine and dominating figure. This self-portrayal is foreshadowed by Enobarbus's cue which, for all its irony, puts her on a par with Jupiter Pluvius, the tutelary god of rain. However, the 'storms and tempests' evoked in this scene are mere verbal explosions which emphasise the queen's irascibility and make up for the remarkable absence of actual winds in the play.

Indeed, no gust of air ever seems to blow in the eastern regions mentioned by Shakespeare,[38] where oars replace sails and where soft breezes give way to the irritating 'breese', or gadfly, of Actium (3.10.14). In this faraway land, life seems paralysed and stagnating, thus causing Pompey to refer to it as 'a Leth'd dullness' (2.1.27), while foreigners, unaccustomed to this hot and dry climate, feel despondent and 'dejected' (4.13.7).[39] The absence of wind symbolically suggests an ageing civilisation. In Book V of his *De rerum natura*, Lucretius explains that the world took its shape out of a storm (*tempestas*),[40] while in Shakespeare's Egypt, the storm has abated, the four elements

(4.11.1–3) have already been assembled, and in spite of the fecundity of Nile's slime, no real birth or rebirth seems possible there.[41] The only wind produced is generated by fans or by the pair of bellows to which the demoted Anthony is compared (1.1.9) in his repeated and senseless attempts to refresh an Egyptian mistress whom he wrongly believes to be hot in temper. Bound as he is '[t]o cool [the] gipsy's lust' (1.1.10), he does not realise that he can only succeed in making her even colder inside.

The more enslaved to Cleopatra he becomes, the weaker he gets. Bodin theorises this situation when he explains that 'the armies that come out from the North [into the south], grow weake and languish': they become 'molten with sweat' and 'languished with heat'.[42] This is exactly what happens in the play, as right from the start Anthony is linked to the wetness of the Nile ('Nay, but this dotage of your general's / O'erflows the measure', 1.1.1–2). Later on, when he has been defeated by Caesar, he is depicted by Enobarbus as a 'leaky' man (3.13.63), while he realises that '[a]uthority melts from [him]' (3.13.90). If in *Cymbeline* thc Roman Iachimo feels that 'the effects of a northern climate "enfeeble" (5.3.4) his southern body when he arrives with Roman troops to Milford Haven',[43] in *Anthony and Cleopatra*, it is the southern climate which debilitates the Roman Anthony, altering his body and his humours. Thus mollified, the great warrior is metamorphosed into a hopeless reveller, now effeminate to the point of blurring and dissolving gender boundaries.

While not totally deprived of lucidity regarding his own temperament, he sometimes describes his Egyptian lover in most cynical terms:

> I found you as a morsel, cold upon
> Dead Caesar's trencher – nay, you were a fragment
> Of Gneius Pompey's – besides what hotter hours,
> Unregistered in vulgar fame, you have
> Luxuriously picked out. For I am sure,
> Though you can guess what temperance should be,
> You know not what it is. (3.13.117–23)

In the context of her hot eastern clime, Cleopatra's intemperance, according to his own male bias, is what seems to guide her actions and promotes a form of triumphant promiscuity. As an embodiment of the southern woman supposedly unable to control her fleshly appetites, she is a seductress and luxuriousness thus runs through her blood. To make things worse for Anthony, she proves sufficiently cunning to stick to power, frail as it may be under Roman rule.

This frailty also reflects the queen's temperament. According to Bodin, the people of the southern region are characterised by their wit and policy, while the people of the north must rely on their force.[44] Southerners, as a result, lack courage. Accordingly, Scarrus accuses Cleopatra of being a coward during the battle at sea between Caesar and Anthony ('The breese' being 'upon her', she, 'like a cow in June, / Hoists sails and flies' (3.10.14–15)),[45] and Cleopatra will later acknowledge that her 'sails' were 'fearful' (3.11.54). Two combined factors may account for her unquestioned cowardice. First, she lives in the east and her representation as a woman lacking courage complies with Bodin's description, according to which 'all auntients have held, That the people of the East are more mild, more courteous, more tractable, and more ingenious, than those of the West, and lesse warlike'.[46] Secondly, Bodin makes this particular characteristic the result of the soft and fertile climates favouring idleness:

> The barrennesse and fruitfulnesse of places doth in some sort chaunge the naturall inclination of the heauens: And therefore Titius Liuius said, That men of a fat and fertill soile, are most commonly effeminat and cowards; whereas contrariwise a barren countrey makes men temperat by necessitie, and by consequence careful, vigilant, and industrious [. . .] So as the barrennesse of the soyle doth not onely make men more temperat, apt to labour, and of a more subtill: but also it makes townes more populous: for an enemie affects not a barren countrey, and the inhabitants liuing in safetie do multiplie, and are forced to traffique or labour.[47]

Eastern lands were thus thought to produce indolent and cowardly creatures.

Now, as easterners, the Egyptians are represented as being able to think, contemplate and invent – activities that, positive as they seem, are opposed to the warlike qualities generally attributed to westerners. The second book of Herodotus' *Histories*, translated into English in 1584, probably influenced the vision that Shakespeare's learned contemporaries had of exotic Egypt. While it describes the Egyptian society in terms of empowered women and dominated men – a paradigm of inversion which Herodotus ascribes to 'the temperature of the ayre, and nature of the riuer'[48] and which evidently transpires in the upside-down world of Shakespeare's Cleopatra – it also sheds light on the mental agility of its people.[49] The Egyptians, the Greek historian explains, 'were alwayes since the first beginning and originall of mankinde'.[50] They presumably invented astrology, improved the art of divination, and were excellent physicians.[51] Bodin hammers the point home when he declares that 'the hidden knowledge of Philosophie, the Mathematikes, and other contemplative sciences, are come out of the South'[52] and sometimes openly relies on Herodotus 'who for good wits and ciuill behauiour commendeth the AEgiptians before all other people of what nation soeuer'.[53] This leads him to make a surprising assumption, given the usual association of easterners with idleness in the geo-humoral writings of the period:

> And after him [i.e. Herodotus] *Caesar* (in his Commentaries of the ciuill warres) gaue them the like priuilege, saying, That the Alexandrians did so artificially counterfeit the Roman engines of warre, as it seemed the Romans were but their apes [. . .] The men themselues were verie wittie and politike: And yet AEgypt is partly vnder the Tropique, whereas the heat is more violent than vnder the Equator[54]

The Egyptians thus stand apart among the eastern peoples as they conveniently cut across all categories: they can be simultaneously contemplative and active, intellectual and warlike. Even though they live in one of the hottest parts of the world, which should make them poor fighters, they can easily overthrow their enemies. For the French writer, the Egyptians are good warriors because of their 'excellent wits'.[55] Herodotus would have added that Egyptian women, in particular, are endowed with a sturdy disposition which makes them stronger than their male counterparts. In the play, Cleopatra's wit is indeed sharp but she uses it to mischievous ends, as Anthony acknowledges when he accuses her of 'hav[ing] been a boggler ever' (3.13.111). When the situation becomes desperate in act 4, his words are even harsher as he portrays her as 'this false soul of Egypt' (4.13.25).[56]

Bodin's praise of the Egyptian civilisation must nonetheless be qualified by his admission that the Egyptians also happen to be exceedingly cruel: 'and in AEgypt at this day they flea them aliue which rob by the high way, then they stuffe the skin full of haie and set it vpon an Asse, by his side that is so fleaed'.[57] So, Shakespeare's multifaceted heroine is perhaps less cowardly than cunning and less frightened than frightening. The discourse on eastern perversity was an early modern commonplace and, in the tragedy, Cleopatra, the 'eastern star' (5.2.307), stands for the Orient as much as for the south in general. While Anthony is blinded by his passion and situates Egypt in a land of pleasure,[58] not a few Jacobean spectators must have associated the Egyptians with the barbarous and infidel Ottomans, whose prolonged contact with climatic extremes (i.e. seasonal rains and arid deserts) was then deemed responsible for their roughness. When the play was written and performed, Alexandria was controlled 'by the Turkish Sultan together with its neighboring Barbary coast'.[59] Egypt being at the time an Ottoman protectorate, Cleopatra must have often been identified with a Barbarian queen,[60] a heroine who could be as captivating as she was threatening.[61]

Water and mud

Cleopatra, Anthony's 'serpent of old Nile' (1.5.25), is a creature born out of the river's mud and slime who eventually collapses all differences between such binaries as wetness and dryness, male and female, or life and death.[62] The tragedy's seemingly gendered approach to climate is thus rapidly undermined and finally denied.

If the Tiber is mentioned in the play, it does not seem to matter much for the Romans, while water is constitutive of Egypt. In his 1595 *Cornucopiae*, Thomas Johnson put forward the paradox inherent to Egypt:

> Among straunge matters, this may seeme to be none of the least, that in Egipt it seldome or neuer raineth: yet by the oueflowing of the riuer Nilus, it is so abundant of all thinges, that it may be compared euen with the chiefest: yet is there not a Uine in all Egipt.[63]
>
> The Eele commeth or is engendred of the earth and mud without anie spawne, neither is there either male or female of them.[64]

Mud, slime and earth in *Anthony and Cleopatra* provide an insight into the very principles of life in the east. As early as in the play's opening scene, a defiant Anthony exclaims: 'Kingdoms are clay. Our dungy earth alike / Feeds beast as man' (1.1.37–8). For an early modern audience, the image of the 'dungy earth'[65] must have contributed to refamiliarising Egypt as an eastern version of rainy, muddy England.[66] As opposed to the 'new heaven' (1.1.17) mentioned a few lines before, 'clay', the stuff which kingdoms are made of, is what the Roman general must be satisfied with.[67] More importantly, Anthony, in his speech, also refers to some sort of generative matrix levelling all living creatures. For Gabriel Egan, this overdetermined presence of an oozy fruitfulness makes love appear superior to mere fecundity since 'the Earth's reproductive principle generates food', while Anthony and Cleopatra's 'coupling is something better because non-reproductive'.[68] The muddy Nile would thus embody the reality principle, that is, the need for reproduction, as opposed to the pleasure principle blindly and selfishly followed by the two lovers in their *amimetobion* contract, that is, in their quest for an 'inimitable life'. Fundamentally linked to the feminine element[69] and to Cleopatra's acting skills and metamorphic being, slime is endowed with the power to sully and smear as much as that of giving shape and birth to all sorts of creatures in the play, as the gullible Lepidus reminds the audience:

> *Lepidus*: You've strange serpents there?
> *Anthony*: Ay, Lepidus.
> *Lepidus*: Your serpent of Egypt is bred now of your mud by the operation of your sun; so is your crocodile. (2.7.24–7)

Serpents, scorpions, crocodiles and the like were then sometimes thought to hatch from the slime of the Nile under the action of the sun. This passage has brought about a lot of comments tracing its influence back to Aristotelian hylomorphism,[70] a theory according to which natural bodies all consist of form and matter and in which slime plays the role of feminine matter (*hyle*) and the sun that of the male shaping power (*morphe*).[71] Ovid's *Metamorphoses* also describe in great detail how new creatures keep emerging from 'the fat and slimy mud' of the Nile under the influence of 'Phoebus' beams'.[72] As far as Shakespeare's play is concerned, it is a drunk Lepidus who, of all characters, bumbles through the theory of spontaneous generation, thus doing little credit to a doctrine which, as a marvellous absurdity, seemed to have mainly been the stuff of dreamers and poets. As such, it was not really taken seriously among classical writers, let alone in the early modern period.[73] Lucretius, for one, expressed serious reservations about it: 'Doubtless it is that stones and logs and earth, / However mingled, yet cannot produce / The vital sense,' he writes in Book II of *De rerum natura*.[74] Anthony (like Shakespeare), rather than trying to demonstrate the validity of this belief, uses it simply to serve his own cause, when he regards himself as a solar force and as the impregnator of Egypt while assimilating Cleopatra to the Nile. In Alexandria, the Roman general thus presents himself as a god-like figure endowed with the power of making things and creatures grow or else rot away.

A 1600 sermon by George Abbott takes up this belief in the powerful agency of the sun to put it in a religious perspective and make it an image of God's potency:

> That the Sunne in the heauen, cannot do more with the creatures, then this with the receiuers. For as the Sunne being one, doth giue light to many, and doth harden the claye, and yet soften the waxe, and maketh the flowers to smell better, and dead carions to sauour worse, and cheareth the springing plants, and cherisheth other growing things, with an influence which cannot be described; so the word of God vttered by one man, doth serue multitudes and great numbers, and fitteth euery one according to his need, as to beat downe him that is proud, and raise vp him that is humble, to threaten where threates are needfull, and to comfort where comfort is expedient, and with a force which cannot be expressed, to frame euery one to that, whereunto he is ordained, the elect to his saluation, the reprobate to damnation.[75]

Shakespeare reworks this idea by removing its Christian overtones and applies it to the Roman world. That Anthony identifies himself with an omnipotent sun from which all life originates is made quite clear when he begs Cleopatra to allow his imminent departure to Rome:

> [. . .] By the fire
> That quickens Nilus' slime, I go from hence
> Thy soldier-servant, making peace or war
> As thou affects. (1.3.68–71)

Anthony swears here by the sun, thus suggesting that he, like the solar fire, is what brings life to an Egyptian queen implicitly associated with the Nile. In *The Faerie Queene*, Spenser also evokes the fecundity of the Nile when he compares the birth of Belphoebe and Amoret to the myriad creatures born from the mud: 'So after Nilus inundation, / Infinite shapes of creatures men do fynd, / Informed in the mud, on which the Sunne hath shynd.'[76] Yet 'this analogy also recalls the repugnance of Error's brood' which the poet links to the ooze of the Nile,[77] a parallel between mud and error often found in the theological writings of the period[78] but in other texts as well. Robert Armin's *A nest of ninnies* (1608), for example, alludes to men 'made sawcie through the mud of their owne minds'.[79] So, for an early modern audience, the muddy ground of the Alexandrian environment must also have symbolised Anthony and Cleopatra's confused, slippery relationship indirectly referred to through the recurring image of the sinking sands.

Added to the presumably weakening effects of a hot, sluggish climate such as Egypt's, this upside-down relationship where the traditional sexual roles are inverted is to be linked with the mythical geography of hell. 'O, my oblivion is a very Anthony' (1.3.91), Cleopatra jeeringly declares as she denounces her lover's departure for Rome. If the Nile, after connoting fecundity in the play, symbolises the nothingness of existence (*nihil*), it also symbolises the river Lethe (2.7.107), the stream of oblivion in the underground: with this parched landscape, where an omnipotent, all-engulfing river flows through the land, Shakespeare does not just make heat a marker of sensuality, but turns it into a feature of hell. An infernal disruption indeed soon becomes general in Alexandria, where the lovers inhabit a riparian area between land and water and where, as a result, they find it difficult to hold on to firm ground and avoid the 'quicksands' (2.7.58) of their unstable union.

The ontological instability which characterises what Plutarch called *amimetobion*, the incomparable lifestyle of the Alexandrian pair in the broader context of the Egyptian world picture, finds its counterpart in the Epicurean philosophy which sees life as a flux in which 'the primal particles of things / Are all in motion'.[80] 'Nature is ever changing and compelling / All that exists to alter', Lucretius asserts in *De rerum natura*.[81] In *Anthony and Cleopatra*, Egypt's mud keeps shaping and unshaping all things just as the shifting clouds create and uncreate their forms in the sky of Egypt. Anthony offers a serious hint when he explains to a sceptical Caesar:

> Thus do they, sir: they take the flow o' the Nile
> By certain scales i'th'pyramid. They know
> By th' height, the lowness, or the mean, if dearth
> Or foison follow. The higher Nilus swells
> The more it promises: as it ebbs, the seedsman
> Upon the slime and ooze scatters his grain,
> And shortly comes to harvest. (2.7.17–23)

Beyond the agricultural information they provide, these erotically loaded lines are certainly suggestive of the intense physical relationships and of the immoderate sexual life of the two title characters. Yet the word 'seedsman' also points to another interpretive path, especially if we remember that Lucretius himself never uses the word *atomus* in his *De rerum natura*, where the atoms are actually called *semina rerum* ('the seeds of things').[82] As a result, 'seeds', 'grains' and 'germens' in Shakespeare are sometimes half-veiled allusions to the atomist doctrine. Read in an Epicurean perspective, Anthony's statement on the swelling Nile reveals the world as both evanescent and placed in a constant de- and re-composition process. The 'grain[s]' scattered by the 'seedsman' refer to the atoms composing the universe of pagan Alexandria, that is, the void/*nihil* of Egypt's atmosphere. Seen in this way, the ooze of the Nile is neither debasing nor destructive, for it simultaneously shapes and dissolves things and beings alike, melting and changing them as swiftly as the clouds in the air. Like the writer's ink, this mud is the material element which allows poetry to come to life. The 'seedsman' throwing his 'grains' in the mud is therefore no mere harvester; he is also an image of the poet playing with words and filling up the void/blank of the page. It must be acknowledged, however, that for all its poetic undertones, the idea of regeneration is also tinged with cynicism in the play. For example, when Anthony learns of the death of his wife Fulvia, Enobarbus casually tells him: 'This grief is crowned with consolation; your old smock brings forth a new petticoat – and indeed the tears live in an onion that should water this sorrow' (1.2.166–9). The idea of change and renewal here clearly predominates at the expense of mourning and compassion.

Reading the clouds

Alexandria connoted mystery, decadence and fine, constant blue skies in early modern England. In *Englands Parnassus*, Egypt is described as follows:

> Of Aegipt.
> The fairest flower that glories Affrica,
> Whose beautie Phebus dare not dash with showres,
> Oer whose climate neuer hung a cloude,
> But smiling Titan lights the Horizon.
> *R. Greene.*[83]

As Robert Greene's short poem makes clear, Egypt's 'climate' was supposedly devoid of any 'cloude'. It is therefore ironical that, in Shakespeare's tragedy, clouds should provide one of the purple patches of the play when the fallen hero proves particularly vulnerable and melancholy.[84] Rather than lending themselves to a univocal climatic interpretation, the clouds obscuring the sky, in this context, may also refer to the *nigredo* phase of the alchemical process.[85] Life is indeed full of impurities for a man who has just been betrayed by his lover.

Repeatedly equated with the sun, Anthony, in whose 'livery / Walked crowns and coronets' (5.2.90–1) and for whom '[r]ealms and islands were / As plates dropped from his pocket' (5.2.91–2), is made to look like an alchemist in Cleopatra's dream, able as he seems to transmute base metal into gold. Yet all he has actually achieved is to turn his own jewel, Cleopatra, into a capricious mistress lacking mettle/metal, a failure which he realises all too late. This painful moment, however, also coincides with a new form of wisdom and self-knowledge when he ponders, in act 4, on life's mutability and on his own insignificance in front of the cosmos at large, thereby relinquishing his former pride. Trying to read the clouds' confused shapes smearing the skies, he identifies in them a multitude of shapes and designs:

> Sometime we see a cloud that's dragonish,
> A vapor sometime like a bear or lion,
> A towered citadel, a pendant rock,
> A forkèd mountain, or blue promontory
> With trees upon't that nod unto the world
> And mock our eyes with air. Thou hast seen these signs –
> They are black vesper's pageants.
> [. . .]
> That which is now a horse, even with a thought
> The rack dislimns, and makes it indistinct
> As water is in water. (4.15.2–11)

Later on, Charmian will ask to 'dissolve' the clouds, so that she 'may say / The gods themselves do weep' (5.2.298–9) after Cleopatra has applied the asp to her breast, thus allowing it to infuse its venom into her blood. As a foreshadowing of heavenly tears, they symbolically bridge the rhetorical gap between the Roman general and the Egyptian queen, each time suggesting their imminent dissolution as well as their ascent in a yet unchartered territory of the universe.

The Roman general's poignant speech takes place when he faces his own sunset. The moment announces the deliquescence of his own identity in the waters of the Nile ('As water is in water') and, by extension, it foreshadows the end of Egypt as an independent country which will soon be dissolved in the great Roman Empire. As he observes the clouds drifting in the sky, Anthony eventually acknowledges that his whole life has been subjected to the flux of time.[86] Furthermore, as he realises 'that his body is like a cloud that "cannot hold its visible shape" [4.15.14], the play registers the dissolution of his masculinity'.[87] By contrast, Cleopatra's vision of the clouds later on in the play is less poetic than Anthony's. She unwittingly takes up his own theme, only to debase it when she imagines herself being paraded in the streets of Rome and expresses her fears that in the citizens' 'thick breath, / Rank of gross diet' (5.2.210–11), she will 'be en*cloud*ed / And forced to drink their vapour' (5.2.211–12, my emphasis). If the clouds open up Anthony's eyes, they blind a Cleopatra obsessed with and offended by bad smells. Yet the word 'vapour' she resorts to traces us back to the origin of clouds, according to Pliny:

> for that certaine it is [. . .] that clouds are engendred by vapours which are gone up on high, or els of the aire gathered into a waterie liquor: that they bee thicke, grosse, and of a bodily consistence, wee guesse and collect by no doubtfull arguments, considering that they overshadow the Sunne, which otherwise may be seene through the water, as they know well, that dive to any depth whatsoever.[88]

Since antiquity, clouds have been regarded as part and parcel of the creative process, and the clouds which Anthony observes can be seen as stains in the shape of bears, lions, citadels, rocks, mountains, promontories or even horses[89] in the heavens: they are part and parcel of nature's *disegno*.[90] If for Erasmus stains or 'clouds' on a wall were 'most similar to nothing' and proved therefore 'too unsubstantial to be expressed by colours',[91] for Leonardo da Vinci such 'nothing' was the stuff of imagination and artistic freedom. Promoting the role of *fantasia*, Leonardo thought that painters had to rely both on their intellectual faculties and on their inventive powers, and that labile stains turned out to be strong natural incentives to artistic creation. In his posthumous *Trattato della pittura*, first published in 1651 in Italian and French, he explains that 'the unauthored and formless stains on a wall or the veinings of marble' are liable to prompt '*varie inventioni*' in the painter's mind.[92] Stains, Alexander Nagel suggests, 'not only encourage the fantasia to make up new formations; once recognized as pregnant with images, they are themselves instances of image formation, externalizations of the image-forming faculty as it were presented to itself'.[93] It was in this spirit that Leonardo asked painters to rehabilitate stains, asking them to focus on the

> power of confused shapes, such as clouds or muddy water [. . .] You should look at walls stained with damp, or stones of uneven colour. If you have to invent backgrounds, you will be able to see in these mountains, ruins, rocks, plains, hills and valleys [. . .] and then battles and figures in violent action, and an infinity of things you will be able to reduce to their complete and proper forms.[94]

While Shakespeare probably never read such words, he must have been acquainted with the cultural consensus over the aesthetic potency of the atmosphere's floating vaporous masses. Moreover, he could have resorted to the early modern version of Pliny's *Natural History* in which Holland explains that '[s]undry colours and diuers shapes are seene in clouds, according as the fire intermingled therin, is either more or lesse'.[95] The translator also remarks that the sky may sometimes be engraved with figures of animals and things and that, 'by the fall of naturall seeds from thence [. . .] there are engendered in the world and the sea especially, an infinite number of strange and monstrous shapes'.[96] All of this suggests that, in act 4, scene 14, the playwright recycles well-known views about clouds and turns them into poetic images of stains smacking of the transcendental rather than of the usual material corruption attached to such ephemeral shapes. Anthony's lines therefore allude to the imaginative power of the clouds' 'confused shapes' which may then be read as an allegory of poetry.

Critics have noted that Andrea Mantegna's *Martyrdom of St Sebastian* (1457–9), now hanging in the Kunsthistorisches Museum in Vienna, provides a pictorial equivalent of this Shakespearean scene.[97] The thick cloud on the left-hand side of the picture suggests a man riding a horse – like Anthony's cloud, which 'is now a horse' before it becomes totally indistinct. For Jacques Darriulat, Mantegna's background cloud could be read as an allegory of painting.[98] It is highly unlikely that Shakespeare knew Mantegna's work as, more than from visual sources, he borrowed from contemporary texts, and horse-shaped clouds appear in the English pamphlets of the period. *The history of strange wonders* (1561), a wonder book originally written by the German theologian Joachim Camerarius,[99] fascinated by celestial apparitions of all sorts, provides us with a case in point:

a

b

Fig. 11a & 11b Andrea Mantegna, *St Sebastian*, 1457–9; detail of the rider in the cloud. By courtesy of the Kunsthistorisches Museum, Vienna.

> Aboute the same tyme, the maner of a longe historye was sene in the cloudes: whiche afterwarde was set furth in a picture, drawen out and published at Belgicke, in thys maner hereafter folowyng.
> There appeared an armed man on horsebacke wyth a speare charged in his hande, and readye for to runne a course. And besydes thys, was sene the lykenesse of the Emperour Charles the fyfth, with a crowne vpon his heade: and neare vnto it a hogges snoute, somwhat hygher there was sene two Lyons rampynge and leaping agaynst thre other Lions: and a lytle beneth them, two great Dragons spuyng out flames of fyre.[100]

Commenting on these strange bestial shapes seen in the sky, Camerarius does not dismiss this 'longe historye sene in the cloudes', but tries on the contrary to interpret its meaning:

> Therefore these thynges shewyng them selues so often tymes, and so wonderfull, verelye we ought to vnderstande and to thynke, that the mouing and workyng of nature is wrasted out of frame, and the state and condicion of men to be tourned out of course, & that the effectes of nature being weake & sicke may no longer continue nor endure[101]

The motif of the horse-shaped cloud would thus foretell impending disaster and, in *Anthony and Cleopatra*, Anthony's visions are soon followed by his and Cleopatra's suicide and by the triumph of Octavius Caesar. This much-neglected prophetic dimension should certainly be taken into account, for if in his plays Shakespeare repeatedly alludes to the forms of clouds in the sky, it is only in this particular piece that he mentions a horse.

Of course, the horse image may also be a subliminal reminiscence of Cleopatra's musings in 1.5.20–1 ('does he walk? Or is he on his horse? / O happy horse, to bear the weight of Anthony')[102] and, in this perspective, it would establish a sort of subliminal link between the lovers from one end of the play to the other, thus circling the circle of their love. But here, this motif seems to incorporate another two meanings, as it can be interpreted both as an ill omen and as a mark of artistic creativity. On the one hand, Anthony's lines echo Demetrius's observation on the metamorphic function of clouds in *A Midsummer Night's Dream* (4.1.186–7), while paving the way for Prospero's remark in *The Tempest* on the 'cloud-capped towers' (4.1.152) that melt and dissolve in the air.[103] On the other hand, such a speech is clearly not just about creation, emphasising as it does the ineluctable dissolution of human beings and of the world itself. In his *De rerum natura*, Lucretius had already suggested that our universe would slowly decline toward disintegration. To him, the clouds, in particular, not only work as simulacra but also serve to illustrate this inevitable process:

> [. . .] we sometimes see clouds quickly gathering
> Aloft together, marring heaven's clear face,
> While with their motion they caress the air:
> For often visages of giants are seen
> To float along, trailing a far-spread shadow;
> Sometimes mighty mountains, or rocks torn off
> From mountains, seem to go before and pass
> Across the sun; then some huge beast appears
> To lead and drag behind it other clouds.[104]

The similarities between Anthony's speech and this passage in Lucretius' poem are noteworthy. If, earlier on in the play, Epicureanism is hinted at in a biased and ironical way as Anthony suspects Lepidus of being 'a very epicure' (2.7.52), that is, a glutton, and is therefore part of a broad network of comic allusions, in act 4, scene 14, it becomes intertwined with the philosophical fabric of the tragedy. Epicure's philosophy is thus being subtly rehabilitated through

Anthony's changing clouds. These clouds, which are strikingly similar to Lucretius' *nubila* (clouds) and are seen as 'spontaneously begotten'[105] forms smearing the vault of heaven, fittingly illustrate the process of dissolution already pointed out by Cleopatra in the third act: 'so / Dissolve my life' (3.13.162–3).

New cosmologies and celestial dislocation

Reading signs in the celestial sphere is not something which is ascribed to Anthony alone. As Janet Adelman explains, both lovers 'participate in the cosmic and natural harmonies signified by divine transexuality' and must, therefore, 'also be seen in that context'.[106] It is this particular context which I now propose to explore. Critics generally agree that the cosmic dimension of the play, replete with skyey images absent in North's Plutarch, associates the lovers with celestial luminaries and leads them to their embracing of death. Yet, the celestial logic underpinning the tragedy is seldom explored per se. According to this poetic and dramatic logic, the two title characters see themselves as king and queen of the firmament. While Anthony is first closely associated with the sun, the Egyptian queen is the incarnation of the moon-goddess, Isis (1.5.70), which entails astronomical consequences. As the subjugated Anthony fails to behave like the sun with which he would like to be identified, he allows himself to be commanded by the moon and, as a result, becomes an 'ebbed man' (1.4.43). Eventually, after the defeat at Actium, he accounts for his failure in terms of cosmic imagery:

> [. . .] Alack, our terrene moon
> Is now eclipsed, and it portends alone
> The fall of Anthony. (3.13.154–6)

Anthony refers here to a lunar eclipse, an event which occurs when the earth passes between the moon and the sun and when, as a result, its shadow obscures the moon. Since antiquity eclipses had been regarded not as natural phenomena but as omens endowed with serious significance. In *Anthony and Cleopatra*, as in *King Lear* and *Othello*, the eclipse foreshadows disaster. The vanquished triumvir thus announces his own demise, which seems all the more inescapable as it is already written in the skies, as it were. Yet Anthony is decided to replace such astral determinism by an act of self will, resolved as he is to become '[a] bridegroom in [his] death, and run into't / As to a lover's bed' (4.15.100–1). 'The next time I do fight / I'll make Death love me' (3.13.192–3), he defiantly tells his despondent followers. At this very moment, as Enobarbus observes, Anthony is ready to 'outstare the lightning' (3.13.195) and defy the elements. As he prepares to fight for the last time, he imagines an elemental contest taking place not just on land and sea (4.11.1–2), but also in the regions of fire and air ('I would they'd fight i'th'fire, or i'th'air', 4.11.3). This histrionic posture and unrealistic aspiration is, of course, aligned with the excesses of his Egyptian existence, a life without measure or limits. As John Gillies convincingly argues, this sensual, exorbitant existence is 'translated into geographic terms' in the play and, I would add, into cosmological ones as well.[107] It is only fitting, then, to see Anthony's afterlife as inscribed in the cosmos, in a Lucretian unbounded space enveloped by the heavens: 'His face was as the heavens, and therein stuck / A sun and moon, which kept their course and lighted / The little O o'th'earth', as Cleopatra puts it when she reports her dream to a flabbergasted Dolabella (5.2.79–81). As his only shroud, the starry night will endow his military defeats with a posthumous magnificence.

Quite logically, once her 'sun' has been eclipsed from her personal horizon, Cleopatra rejects her affiliation with the moon: 'Now the fleeting moon / No planet is of mine' (5.2.239–40), she says to herself before the clown brings her the basket of figs with the asp hidden under them. Dismissing mutability, she aspires to eternity ('I have / Immortal longings in me', 5.2.279–80), so that her end, truly cosmic and erotic, marks her apotheosis:

> [. . .] Husband, I come!
> Now to that name my courage prove my title!
> I am fire and air; my other elements
> I give to baser life. (5.2.286–9)

She now leaves water and earth behind as she identifies with the more noble elements of air and fire in order to join her lover in the sky and become an 'eastern star' (5.2.307). After impersonating the goddess of love in her lifetime, she *becomes* Venus when she reaches the sky to be reunited with her lover. Doing so, she turns her fairly chaotic love affair into a grand mythical story, thus transforming a historical, political and apocalyptic disaster into a personal apotheosis. In the first act, their passionate love relationship was already presented as having nothing earthly or mundane in it:

> [. . .] When you sued staying,
> Then was the time for words – no going then:
> Eternity was in our lips and eyes,
> Bliss in our brows bent; none our parts so poor,
> But was a race of heaven. (1.3.33–7)

Cleopatra then bitterly derided Anthony for the hyperbolic words he used to describe their sacred love, now debased by lies. Yet in the play's last act, she forgets about her initial diatribe and decides to transfigure her mortal body into 'a race of heaven'. And when she turns her political suicide into a triumphant stage act, she at last instils meaning into grand words that had so far been rather devoid of sense.

Curiously, while critics repeatedly associate Cleopatra with the image of the eastern star, Anthony's trajectory is often presented as a downward itinerary.[108] Dan Brayton assumes, for instance, that '[m]ost early modern audience members (and modern readers) would have known the trajectory of Mark Antony's life is between land and water; his demise will take place on Egypt's soggy ground, not Rome's *terra firma*'.[109] It is true that Anthony himself declares in his dying speech that if he once 'lived the greatest prince o'th'world', he now 'basely die[s]' (4.16.56–7). Caesar hints at the 'declining day[s]' (5.1.38) that characterise Anthony's fall and subsequent death. But however stimulating this reading may be, it downplays the place and function of the heavens in the tragedy. We have seen so far that the importance ascribed to clouds in a country where, according to early modern writers and poets, only few or none are ever seen, clearly hints at the key role played by the sky. Even the cold, rational Caesar understands this, for when he remembers Anthony, he removes the word 'fall' from his vocabulary and he associates the existential trajectory of his former friend turned adversary with an upward movement, connecting him with fire ('mine his thoughts did kindle'), especially when he alludes to his and Anthony's 'stars' (5.1.46), turned of late '[u]nreconciliable' (5.1.47).

In doing so, Octavius Caesar does justice to his defunct competitor. Right from the beginning of the play, an overenthusiastic Anthony uttering hypermetrical lines aspires to a heavenly, apocalyptic destiny (1.1.17), pushing back the traditional limits of the sublunary world and echoing the dreams of Renaissance explorers in quest of the New World. The task he assigns himself – that is, 'find[ing] out new heaven' – is grimly proleptic, as his wish does come true earlier than he had imagined, but when he thus voices his dream, his words are words of love unrelated to these ominous associations. If early modern playgoers unmistakably detected apocalyptic tones in his lines and saw a hint of the 'New Jerusalem' (Rev. 21:2)[110] in Anthony's vision – the phrase 'new heaven, new earth' can be found in a myriad of sixteenth- and seventeenth-century sermons and liturgical writings – Shakespeare's Anthony cannot think along such terms, living as he does in a pre-Christian civilisation where he swears by 'Jove that thunders' (3.13.85) as well as by the '[m]oon and stars' (3.13.100).

This discrepancy between the place where he now spends most of his time – that is, slimy Egypt – and the celestial territories he imagines to tread makes him vulnerable and transforms him into a poet instead of the soldier he should remain. More often than not, his speeches give life to splendid dreams of victory and cosmological greatness. Even at the sunset of his life, he dreams of the sounds of the trumpets resounding in the celestial sphere and uniting sky and earth:

> [. . .] Trumpeters,
> With brazen din blast you the city's ear,
> Make mingle with our rattling taborins,
> That heaven and earth may strike their sounds together,
> Applauding our approach! (4.9.35–9)

This vision of a celestial union, be it at the price of dissolution, of course runs counter to the logic of division and fragmentation (during the civil wars) preceding the birth of the Roman Empire. While it echoes the opening lines of the play, when Anthony, as would-be space traveller, already longed for a 'new heaven', it also offers a new utopian cosmography unifying heaven and earth and abolishing human limits. This reconfiguration of space proves obsessive, as the general's cosmological vision suggests in the opening scene: 'Let Rome in Tiber melt, and the wide arch / Of the ranged empire fall! Here is my space!' (1.1.34–5). In a rather similar fashion, Cleopatra's hyperbolic lines echo Anthony's apocalyptic fantasies when she cries out: 'Some innocents scape not the thunderbolt. / Melt Egypt into Nile!' (2.5.78–9). Pointing to the collapse of the existing world picture, with the end of Egyptian independence and the future triumph of the Roman Empire, their parallel phrases are deeply expressive of their common *jouissance* which combines an extraordinary sense of how to enjoy the beauty of the instant with their ultimate death wish.

The lovers thus pervert their passion for life through dreams of inaccessible and impossible grandeur, and they replace it at the end with morbid urges accelerating their own disintegration. Shakespeare conveys this destructive process thanks to the cosmological imagery structuring the tragedy as a whole. As a jealous Anthony gradually loses his bearings and can no longer distinguish his 'eastern star', he finds himself deprived of any benchmark in the sky, and voices his despair in apocalyptic terms:

> [. . .] [Caesar] makes me angry,
> And at this time most easy 'tis to do't,
> When my good stars, that were my former guides,
> Have empty left their orbs, and shot their fires
> Into th'abysm of hell. (3.13.144–8)

The falling stars mentioned here recall John's vision in the Book of Revelation,[111] and the same image is used later in the play to symbolise Anthony's anti-apotheosis, so to speak. Indeed, after the lost battle of Alexandria, the Roman general has become one of the fallen stars he previously described ('The star is fall'n,' the Second Guard says after Anthony has fallen on his sword (4.15.107)). His trajectory in the play is that of a meteor, quick, bright, spectacular, and short-lived.[112] Surprisingly, the erring, falling stars that shoot through the tragedy have attracted few lengthy analyses. It is worth noticing, therefore, that on top of their biblical connotation and of their obvious link with the theme of hubris, 'fallen' or 'falling' stars were then seen as one of the celestial wonders of the time. They presumably belonged to the lowest region of the air, as 'starres of the firmament can not fall, for God hath set them fast for euer, he hath geuen them a commaundement whiche they shal not passe'.[113] Interestingly, in Thomas North's translation of Plutarch's *Lives*, falling stars are mentioned and defined in fairly specific terms:

> For they say, that those which we call falling starres, be no fluxions nor deriuacions of the fire elementorie, which are put out in the ayer [. . .] nor also an inflammation or combustion of any parte of the ayer, which by her to ouermuch quantity doth spread vpwardes: but they are celestiall bodies, which by some slackenes of strength, or falling from the ordinary course of heauen, are throwen and cast downe here beneath, not alwayes in any parte of the earth inhabited, but more ofter abroade in the great Occean sea, which is the cause that we do not see them.[114]

Anthony falls 'from the ordinary course of heaven' in the play, until Cleopatra, by her own spectacular suicide, ultimately turns this fall into an apotheosis.

As noted by Katherine Vance MacMullan, the Egyptian queen 'expands' the image of the falling star as she sees her lover's death as part of a larger phenomenon entailing 'the destruction of the heavenly bodies'.[115] As a result, Cleopatra implores the sun to leave its sphere and leave the earth in total darkness:

> O sun,
> Burn the great sphere thou mov'st in; darkling stand
> The varying shore o'th'world! (4.16.10–12)

The great Egyptian 'whore'[116] here naturally follows Ptolemaic cosmology, a system in which the sun could indeed burn the earth, which she imagines as a vast and desolate shore. Images of flooding now give way to images of burning, a shift which evokes the deregulation of Cleopatra's meteorological/physiological system.[117] Paradoxically, at the beginning of the scene, she longs for airy elevation and cosmic grandeur but still feels attracted by a latent subterranean force. 'Our strength is all gone into heaviness' (4.16.35), she bemoans as her lover dies under her eyes. Yet she quickly makes up her mind: Anthony can only reside in the skies. Immediately after he breathes his last, she describes the world as an insignificant, worthless place: 'And there is nothing left remarkable / Beneath the visiting moon' (4.16.69–70). The sublunary world has been turned upside down and, from then on, only what occurs in the superlunary sphere, Anthony's new playground, is of any interest to her.

Such an emphasis on the celestial forces of the world is accompanied in the play by allusions to cosmic upheavals, cracks and dislocation (i.e. 'displacement' and 'luxation', *OED* 1.a/b), and these images gradually replace the melting metaphors prevailing in the first half of the tragedy. Anthony's tirade on the stars having left their orbs (3.13.144–8) is indicative of the oncoming chaos.[118] The Roman general projects himself in a new, apocalyptic cosmography in which he loses contact with reality. Incidentally, his lines tie in with the servant's piercing remark which, quite early in the play, already foreshadowed Anthony's dis-aster: 'To be called into a huge sphere, and not to be seen to move in't, are the holes where eyes should be, which pitifully disaster the cheeks' (2.7.14–16). If the bleak image of the eyeless face unmistakably evokes a *vanitas* painting, the space/body analogy follows the general trend observed on early modern England maps with their sexualised landscapes and treatment of space in terms of human bodies.[119] Anthony's grotesque body, ready to burst open, already takes on a co(s)mic dimension: dislocation, in this case, is to be understood in both its etymological and its usual senses.

The motif is also hinted at in act 4, when Anthony muses upon the clouds smearing the heavens. Attention must be paid here to his 'rack' that 'dislimns' (4.14.10), 'dislimns' suggesting a dislimbed body, notably because of its proximity with the word 'rack', which designated an instrument of 'torture' (4.15.46). Destruction verges on self-destruction when, in the following scene, Anthony exclaims: 'O, *cleave* my sides! / Heart, once be stronger than thy continent, / *Crack* thy frail case!' (4.15.39–41, my emphasis). 'Crack nature's moulds,' Lear implored (3.2.8) in the eponymous tragedy, begging 'oak-cleaving thunderbolts' (3.2.5) to strike the earth. The same eschatological violence is at work here, but while in *King Lear* it was conveyed through the letting loose of such elements as lightning and thunder, as well as the wind and the rain, it remains an internal phenomenon in *Anthony and Cleopatra*, where it is only betrayed by Anthony's rhetoric: the storm strikes within. Eventually, the theme resurfaces in Caesar's speech when he learns of Anthony's death:

> The breaking of so great a thing should make
> A greater crack. The round world
> Should have shook lions into civil streets,
> And citizens to their dens. The death of Anthony
> Is not a single doom, in the name lay
> A moiety of the world. (5.1.14–19)

After he has passed away, Anthony does not really leave the stage as he continues to live in Cleopatra's imagination:

> His face was as the heavens, and therein stuck
> A sun and moon, which kept their course and lighted
> The little O, the earth. (5.2.78–80)

In such an anthropomorphic cosmology, there are no longer any barriers between physical and celestial bodies. As clear-cut lines vanish, all limits are abolished and earth and sky become interchangeable. Barely interrupted by Dolabella, the queen goes on:

> [. . .] his voice was propertied
> As all the tuned spheres, and that to friends;
> But when he meant to quail and shake the orb,
> He was a rattling thunder. For his bounty,
> There was no winter in't; an autumn it was
> That grew the more by reaping. (5.2.82–7)

In her dream, Anthony is no longer governed by the weather: he has become the weather. Cleopatra's lines take on quasi-mystical inflexions and, in retrospect, they give a new meaning to Anthony's initial challenge to Cleopatra: 'Then must thou need find out new heaven, new earth' (1.1.17). Re-membering her lover, she shapes him anew and gives him fresh life. Gender roles are once again inverted at this very moment but, this time, she takes on the role of the sun giving shape to mud and dust.

Egypt and infinite space

In her stimulating analysis of the play as presenting an epistemological shift in human perception and knowledge, thus making Egypt much more than 'an orientalising stereotype',[120] Mary Thomas Crane contends that 'Egyptians in the play reflect an earlier view that environment shapes subjects while the Romans look forward to a Cartesian mind-body split in which self-contained individuals are separate from and gain mastery over their environment'.[121] Egypt is dominated by natural cycles while the Roman world, thinking space as an ordered, hierarchical territory, obeys a less elemental, more political logic.

Interesting as this may be, I would like, by way of conclusion, to put forward another tentative explanation. Shakespeare, for one, blurs all binaries in the play, and his slimy, saturnalian Alexandria, indulging in spectacles and lavish entertainments, is manifestly closer than austere Rome to the pomp of early modern (and muddy) London.[122] As to Caesar's world, it is of course associated with reason and rule, as well as with political and military strategy. Yet it is so much a world of the past that it can hardly be regarded as representative of the seventeenth-century scientific advances. As Michael Neill underlines, the Romans always look backward to define and justify their political causes: 'Octavius's right to rule, for example, is always implicitly linked to his inheritance of the heroic name of "great Caesar"' and nostalgia seems to be, in the

Roman sphere, 'the determining condition of the heroic'.[123] The 'three-nooked world' (4.6.5) mentioned by Caesar certainly refers to a world made of earth, sea and sky rather than to the triumvirs since, by act 4, the triumvirate has been definitively brought to an end. If yet another argument were needed, the most richly imaginative description in the play, namely that of Cleopatra in her barge (2.2), does not come from an Egyptian, but from a prosaic Roman soldier. This suggests that Egypt cannot once and for all stand for the 'old' science against the 'new', rationalistic science prefiguring Descartes's views and symbolised by Rome. Seducing as such a hypothesis may be, it is easily reversible, particularly if we take Hobbes rather than Descartes as a reference for seventeenth-century natural philosophy in England.

The first, simple assertion that can be made about Shakespeare's Egypt is that, in the play, it highlights the emerging modernity of the Jacobean age: by discarding the traditional attributes of the male hero, Anthony paves the way for a new mode of heroism better attuned to seventeenth- and eighteenth-century ideals. This idea may, however, be pushed much further. In classical literature in general, and for Herodotus in particular, the birthplace of science and civilisation often coincides with Egypt and, as Mary Thomas Crane acknowledges, 'although Egypt was annexed into the Roman Empire after the Battle of Actium in 31 B.C., Alexandria remained a centre of learning to rival Rome and never really adopted Roman customs'.[124] Shakespeare adopts the same logic in his tragedy, making Egypt the land of ambiguities as opposed to Roman certainties, thus paving the way for Hobbes's materialism and ethical relativism. In *Anthony and Cleopatra*, it is Cleopatra's Egypt – much more than the Roman Empire – which the playwright turns into an experimental space, malleable and conveniently spectacular – a place in which Nile, clime and clouds all seem to chime with Epicurean thought and cosmology. Although atomism cannot be said to belong to a 'new' form of scientific knowledge at the time, since it stemmed from Greek antiquity, yet, at the turn of the seventeenth century, it represented a challenging theory, boldly questioning well-established Aristotelian assertions and putting forward the combined notions of flux, movement and void. Seen against this background, rigid Rome seems to fight a rearguard battle. In theory, Caesar is the obvious winner at the end of the play. Ironically, it is not him that the audience will remember but the two adulterous lovers who have turned their earthly defeat into a form of heavenly triumph. The two of them have indeed managed to show that death, far from being the end of all things, is in fact a new beginning, the entrance into a 'new heaven' and 'a new earth' (1.1.17). Caesar stands for a finite universe, enclosed in and limited by neat boundaries, and organised along his political views, as Mary Thomas Crane suggests. By contrast, Anthony and Cleopatra reject the 'Cartesian mind-body split'[125] and promote instead an infinite universe such as the one described in Epicure's *Letter to Herodotus*:

> Moreover, the universe is infinite. For that which is bounded has an extremity, and the extremity is seen against something else: (but the universe is not seen against something else): therefore, since it has no extremity, it has no boundary; and, since it has no boundary, it must be infinite and not bounded.
> Furthermore, the sum of things is infinite both in the multitude of the bodies and the magnitude of the void.[126]

Matter and empty space (i.e. what Enobarbus negatively defines in his soliloquy as 'a gap in Nature' (2.2.225)) broadly define what Shakespeare's Egypt stands for. In this sense, Cleopatra's 'infinite variety' (2.2.243) certainly takes the full measure of the magnitude of the void by making the Nile/*nihil* an extension of her personality.

The main epistemological shift of the play is localised in Egypt and it marks the transition from the hedonistic philosophy of festive, orgiastic Alexandria to an Epicurean one, thus signalling a gradual change from pleasure to ataraxy. It is against this background that the elemental forces sustaining the play are to be understood. 'The elements be kind to thee' (3.2.40), Caesar ironically declares to his sister Octavia, newly wed to Anthony. Yet the elements prove deceptive and the shrewd young Octavius should know better. The fecundity of the earth, so often evoked, sounds like a promise which never materialises (the only 'baby' shown in the play is the asp at Cleopatra's breast (5.2.308)), while the air is so heavy and sluggish there that it debases rather than promotes the warriors' deeds. During most of the play, Anthony is badly affected by his environment and becomes 'leaky [3.13.63], grotesque, constantly melting', while the sneaky, snaky Cleopatra seems at times 'almost interchangeable with the river'.[127]

The tension between building and dissolution in the tragedy is therefore resolved to the detriment of 'monument[s]' (4.16.8): in a play where everything melts, the most impressive of 'citadel[s]' (4.15.4) can only be conjured up in the clouds. Life is defined by its mutability so that, in spite of their predictable interactions with their climatic conditions, the characters evade any form of determinism. As a result, long-lasting views cannot be held, battles cannot be planned, and the course of love cannot and will not run smooth. Such shiftiness is symbolised by the Nile as much as by the clouds in Egypt's sky: they are closely associated for the simple reason that the Nile functions as an inverted sky[128] reflecting the crucial issues dealt with in the tragedy. Yet from act 4 onwards, the Nile becomes less prominent as the sky looms larger and finally appears as an alternative place enabling men to escape from the dungy earth of Egypt. As the space of metamorphosis and of possible re-creation(s), the firmament becomes part and parcel of an erotic and apocalyptic cosmography, of an upside-down, carnivalesque world which makes hell a desirable place and turns defeat and downfall into a paradoxical victory and a grand apotheosis. So, as the only element and force able to transmute raindrops into sunrays and a disappointing, frustrating existence into an eternal, infinite afterlife, the sky ultimately proves the true alchemist of the play. 'Death,' Lucretius writes, does not 'destroy the particles of matter'. It simply 'recombines one element with another'.[129]

Notes

1. Pierre Boaistuau, *Certaine secrete wonders of nature*, p. 67.
2. Anthony's 'Then must thou need find out new heaven, new earth' (1.1.17) sets the tone of the play. Later on, Menas boldly tells Pompey: 'Thou art, if thou dar'st be, the earthly Jove: / Whate'er the ocean pales, or sky inclips, / Is thine, if thou wilt ha't' (2.7.68–70). Pompey, however, refuses Menas's cynical and criminal proposal. He is probably aware that the sky only very rarely satisfies one's secret expectations and, when it does so, as Banquo declares in *Macbeth*, '[t]he instruments of darkness tell us truths / [. . .] to betray's / In deepest consequence' (1.3.122–4).
3. If, as Gary Taylor and Rory Loughnane suggest in *The New Oxford Shakespeare: Authorship Companion* (p. 569), the play was composed in early 1607, Shakespeare may then have had access to the English translation of Bodin's treatise.
4. As he learns of the death of his wife Fulvia, he almost casually declares: 'the present pleasure / By revolution low'ring, does become / The opposite of itself' (1.2.124–6). In act 4, Cleopatra will blame Fortune for Anthony's death, seen as arbitrary: 'let me rail so high / That the false huswife Fortune break her wheel' (4.16.45–6).
5. José de Acosta, *The naturall and morall historie of the East and West Indies*, lib. 1, chap. 10, p. 33.

6. John Gillies, *Shakespeare and the Geography of Difference*, p. 11.
7. *Aristotles politiques*, p. 360. On the identity of the translator, probably John Dickenson, see Gavin Alexander, 'Dickenson, John (c. 1570–1635/6), Author and Government Official'. Alexander explains that in 1598, Adam Islip hired Dickenson 'to translate Louis Leroy's French version of Aristotle's *Politics*', first published in 1568. Alexander contends that 'Dickenson dedicated the work to Sir Robert Sidney, and allowed his own name to be printed only in the presentation copy, now at Shrewsbury School'.
8. Giovanni Botero, *The worlde, or an historicall description of the most famous kingdomes and common-weales*, p. 79.
9. Ibid.
10. Jean Bodin, *The six bookes of a common-weale*, The Fifth Booke, chap. 1, p. 546.
11. Ibid. p. 554.
12. Ibid. p. 561.
13. Plutarch, 'The Life of Marcus Antonius', in *The lives of the noble Grecians and Romanes*, trans. Thomas North, p. 979.
14. Plutarch, 'The Life of Octavius Caesar Augustus', trans. Thomas North, in *Narrative and Dramatic Sources of Shakespeare*, vol. 8, p. 322.
15. Jean Bodin, *The six bookes of a common-weale*, The Fifth Booke, chap. 1, p. 567.
16. See David Bevington (ed.), *Antony and Cleopatra*, p. 15. Bevington observes that Anthony is 'something of a northerner, unprepared to cope with the effeminising, leaky climate of Egypt'. On this, see also Mary Floyd-Wilson, 'Transmigrations: Crossing Regional and Gender Boundaries in *Antony and Cleopatra*'.
17. Shakespeare lifts such details from Plutarch, since in 'The Life of Marcus Antonius', the exploits of Anthony and his army are already emphasised: 'it is reported that euen as they passed the Alpes, they did eate the barcks of trees, and such beasts, as neuer man tasted of their flesh before' (*The lives of the noble Grecians and Romanes*, trans. Thomas North, p. 978).
18. Jean Bodin, *The six bookes of a common-weale*, The Fifth Booke, chap. 1, p. 549.
19. Ibid. p. 567.
20. Ibid. p. 561.
21. Ibid.
22. Enobarbus compares her to the goddess born from the sea (2.2.207) in his famous description of the enticing Egyptian queen as she appeared at Cydnus.
23. For most Elizabethan playgoers, mud must have been synonymous with insalubrity, as the numerous poor of London 'crowded into mean slum tenements of timber eked out with mud and plaster'. See Samuel Schoenbaum, *William Shakespeare: A Documentary Life*, p. 100.

24. On this, see Barbara J. Bono, *Literary Transvaluation: From Vergilian Epic to Shakespearean Tragicomedy*, p. 205.
25. In this, Weyer followed Varro, Lactantius and St Augustine. See Stuart Clark, *Thinking with Demons: The Idea of Witchcraft in Early Modern Europe*, p. 118.
26. See Marjorie Garber, *Shakespeare After All*, p. 726.
27. See Aristotle, *Meteorologica*, IV.6, p. 319. Aristotle writes that '[l]iquefaction takes two forms: the one is condensation into water, the other the melting of a solid'. He also explains that '[c]ompounds of earth and water are solidified both by fire and cold, and are also increased in density by both [. . .] Heat draws out the moisture, and when the moisture evaporates the dry constituents increase in density and pack closer' (p. 320).
28. Aristotle, *Meteorologica*, I.14, p. 109.
29. Jean Bodin, *The six bookes of a common-weale*, The Fifth Booke, chap. 1, respectively p. 567 ('cold and dry') and p. 551. On the presumed coldness of southerners, see my analysis of *Othello* above, Chapter 4, p. 121.
30. Ibid. p. 548.
31. This myth is briefly mentioned in Ovid, *Metamorphoses*, trans. Arthur Golding, The Fourth Book, ll. 749–51, p. 143.
32. Significantly, this line is part of a broader passage in which she imagines her own death in terms of dissolution as she asks the sky to engender 'hail' from her 'heart' and 'the first stone' to '[d]rop in [her]neck', wishing her life were '[d]issolve[d]' (3.13.160–3). On hail, see also above, Chapter 1, p. 47 and p. 55, n. 79.
33. On this, see Gordon Williams, *A Glossary of Shakespeare's Sexual Language*, p. 74.
34. Jean Bodin, *The six bookes of a common-weale*, The Fifth Booke, chap. 1, p. 548.
35. This verb, whose first occurrence is traced back to *Anthony and Cleopatra* by the *OED*, is used twice in the play. The second occurrence is to be found in one of Anthony's tirades: 'The hearts / That spanieled me at heels, to whom I gave / Their wishes, do discandy, melt their sweets / On blossoming Caesar' (4.13.20–3).
36. Enobarbus's comic and climatic version of Cleopatra's histrionic postures is later counterbalanced by his lyrical purple patch directly lifted from Plutarch ('The barge she sat in [. . .]'; see 2.2.198–233), where the queen is said to 'O'er pictur[e] [. . .] Venus' (2.2.216). His tone changes according to his target audience and mirrors the difference between a private conversation and a public narrative. Yet the double, contradictory comparison with Jove (1.2.143), the powerful, oversexed father of the gods, on the one hand, and with Venus (2.2.216), the epitome of feminine wiles and seduction, on the other, is significant as it points to Cleopatra's ambivalence and possible bisexuality. For

a similar ambivalence, see above, Chapter 1, p. 33 and p. 51, n. 13, where Titania is presented as a figure having appropriated the codes of 'patriarchal fatherhood'.

37. Octavia also has her own humoral system. As he sees her crying, having to part from her brother, Anthony says that 'The April's in her eyes; it is love's spring, / And these the showers to bring it on' (3.2.43–4).
38. In the scene depicting the Egyptian queen in her barge on the river Cydnus (then situated in the ancient province of Cilicia now corresponding to southern Turkey), the winds are presented as being 'lovesick' with Cleopatra's sails (2.2.201).
39. A servant mockingly declares of Anthony, Caesar and Lepidus that 'the least wind i'th'world will blow them down' (2.7.1–2).
40. On this, see Mary-Jane Rubenstein, *Worlds without End: The Many Lives of the Multiverse*, p. 49.
41. Children, barely evoked (Caesarion is mentioned once in 3.13.163), are oddly absent from the play, as lovers promote sex over fertility. Anthony proves utterly unable to start a new life in the Orient.
42. Jean Bodin, *The six bookes of a common-weale*, The Fifth Booke, chap. 1, p. 549.
43. Jean E. Feerick, 'A "Nation . . . Now Degenerate": Shakespeare's *Cymbeline*, Nova Britannia, and the Role of Diet and Climate in Reproducing Races', p. 33.
44. Jean Bodin, *The six bookes of a common-weale*, The Fifth Booke, chap. 1, p. 550. See also p. 552 and p. 554.
45. On the Virgilian source (the *Georgics*) and Rabelaisian connotations of this image, see Robert G. Hunter, 'Cleopatra and the "Oestre Junonicque"'.
46. Jean Bodin, *The six bookes of a common-weale*, The Fifth Booke, chap. 1, p. 562.
47. Ibid. p. 565.
48. See the 1584 translation then available in England, namely Herodotus, *The famous hystory of Herodotus*, trans. B. R. [Barnabe Riche?], fol. 78r.
49. Ibid. fol. 78v: 'In this countrey the women followe the trade of merchandize in buying and selling [. . .] whereas contrarywyse the men remayne at home, and play the good huswiues in spinning and weauing and such like duties. In like manner, the men carry their burthens on their heads, the women on their shoulders.' On the carnivalesque Egypt depicted by Herodotus, see John Gillies, *Shakespeare and the Geography of Difference*, p. 120, and Bernhard Klein, 'Antony and Cleopatra', p. 457.
50. Ibid. fol. 73r. The passage is quoted by John Michael Archer in *Old Worlds: Egypt, Southwest Asia, India, and Russia in Early Modern English Writing*, p. 25.

51. Ibid. fol. 90v–91r: 'Many other thinges haue been invented by the Aegyptians [. . .] Likewise in Astrology what fortune is incident to him that is borne one such a day, how hee shall proue in lyfe, by what meanes hee shall miscarry by death: which thinges haue bene vsed, of many that haue laboured in the Arte, and Science of Poëtry [. . .] Phisieke is so studyed and practysed with them that euery disease hath his seuerall phisition, who stryueth to excel in healing that one disease.' See also John Michael Archer, *Old Worlds*, p. 28.
52. Jean Bodin, *The six bookes of a common-weale*, The Fifth Booke, chap. 1, p. 550. The same argument is repeated further on, pp. 559–60.
53. Ibid. pp. 552–3.
54. Ibid. p. 553.
55. Ibid.
56. Plutarch, who downplays Cleopatra's beauty (it 'was not so passing, as vnmatchable of other women, nor yet suche, as vpon present viewe did enamor men with her'), emphasises her cleverness and rhetorical skills: 'so sweete was her company and conuersacion, that a man could not possiblie but be taken'. Plutarch, 'The Life of Marcus Antonius', in *The lives of the noble Grecians and Romanes*, trans. Thomas North, p. 982.
57. Jean Bodin, *The six bookes of a common-weale*, The Fifth Booke, chap. 1, p. 555.
58. 'I'th'East my pleasure lies' (2.3.38), he declares, before assuming later on that '[t]he beds i'th'East are soft' (2.6.51).
59. Patricia Parker, 'Barbers, Infidels, and Renegades: *Antony and Cleopatra*', p. 65.
60. Ibid.
61. On this, see Jonathan Burton, *Traffic and Turning: Islam and English Drama, 1579–1624*, p. 158: 'By the time Shakespeare was most likely writing *Antony and Cleopatra*, the Levant Company charter had been renewed three times and English merchants made regular trips to Alexandria and Cairo [. . .] In his Egyptian sojourns, Antony would have recalled [. . .] the Levant traveller who risked depravity, captivity, and emasculation.'
62. Dan Brayton, 'Shakespeare and Slime: Notes and the Anthropocene', p. 85. In a cogent analysis, Brayton notes that '[t]he fire/slime dichotomy parallels the binaries of empire, gender, and climate that saturate the play: Rome/Egypt, Antony/Cleopatra, west/east, war/peace, civic duty/personal pleasure [. . .] Slime mediates these binaries and, to some extent, collapses them by locating life's origins in an elemental stew [. . .] that pre-exists political boundaries'.
63. Thomas Johnson, *Cornucopiae*, fol. 12r.
64. Ibid. fol. 12v.
65. Cf. *Anthony and Cleopatra*, 5.2.7. The *OED* records that the first use of 'dungy' appears in the play and means '[o]f the nature of dung;

abounding in dung' (*OED*, 1). The word must therefore have caught the audience's attention, even though it had been known as a synonym of 'foul or filthy' or 'vile, defiling' (*OED*, 2) since the fifteenth century. Besides, the word 'dung' in the sense of the '[e]xcrementitious and decayed matter employed to fertilize the soil' (*OED*, 1) had been used since the eleventh century.

66. In the English-like forest of *A Midsummer Night's Dream*, Titania complains that 'the nine men's morris is fill'd up with mud' (2.1.98), while in *The Merry Wives of Windsor* Mistress Ford alludes to a 'muddy ditch close by the Thames' side' (3.3.13).
67. For a similar contrast, see Lorenzo's description of heaven to Jessica in *The Merchant of Venice*: 'Sit, Jessica. Look how the floor of heaven / Is thick inlaid with patines of bright gold: / There's not the smallest orb which thou behold'st / But in his motion like an angel sings, / [. . .] / Such harmony is in immortal souls; / But whilst this *muddy* vesture of decay / Doth grossly close it in, we cannot hear it' (5.1.58–65, my emphasis).
68. Gabriel Egan, *Green Shakespeare: From Ecopolitics to Ecocriticism*, p. 110.
69. By contrast, in *The Faerie Queene*, Spenser sees the slimy Nile as masculine: 'As when old father Nilus gins to swell / [. . .] / His fattie waves do fertile slime outwell' (I.i.21.1–3).
70. See Aristotle, *The Generation of Animals*, Book 3, chap. 11, 762a19–28. Aristotle insists on the fact that *pneuma*, in water, helps to spark life by permeating matter.
71. See Aristotle, *History of Animals*, p. 65: 'So with animals, some spring from parent animals according to their kind, whilst others grow spontaneously and not from kindred stock; and of these instances of spontaneous generation some come from putrefying earth or vegetable matter, as is the case with a number of insects, while others are spontaneously generated in the inside of animals out of the secretions of their several organs.'
72. Ovid, *Metamorphoses*, trans. Arthur Golding, The First Book, l. 498 and l. 499 respectively, p. 45.
73. On this, see Henry Harris, *Things Come to Life: Spontaneous Generation Revisited.*
74. Lucretius, *De rerum natura*, Book II, p. 75.
75. George Abbott, *An exposition vpon the prophet Ionah*, 'The XVI. Lecture', p. 352.
76. Edmund Spenser, *The Faerie Queene*, III.vi.8.7–9.
77. Susan C. Staub, 'While She Was Sleeping: Spenser's "goodly storie" of Chrysogone', p. 26.
78. See for example Thomas Morton of Berwick's 1599 *Treatise of the nature of God*, p. 36: 'men [. . .] in comparison of Angels, are but dolts

and dul-pates, groueling here on earth in the mudde and myre of error and grosse ignorance'.

79. Robert Armin, *A nest of ninnies*, sigs. n.p. and Fr.
80. Lucretius, *De rerum natura*, Book II, p. 54.
81. Ibid. Book V, p. 207.
82. See for instance ibid. Book II, p. 46.
83. Robert Allott (ed.), *Englands Parnassus*, p. 349.
84. Clouds were not of interest only to Shakespeare. Marlowe also pays attention to them in the first part of *Tamburlaine*, for example, when, in act 4, scene 2, the protagonist imagines his army filling up the sky. See Mark Thornton Burnett, *Constructing Monsters in Shakespeare's Drama and Early Modern Culture*, p. 58.
85. On the clouds in connection with alchemy, see Margaret Healy, *Shakespeare, Alchemy and the Creative Imagination: The Sonnets and A Lover's Complaint*, pp. 72–3. In the Sonnets, the references to clouds partake of the 'melancholic symbolism' (p. 73) of the *nigredo*, or dark phase, of the alchemical process. On alchemy in *Anthony and Cleopatra*, see Lyndall Abraham, 'Alchemical Reference in *Antony and Cleopatra*', and Mary Ann Curtis, 'The Joining of Male and Female: An Alchemical of Transmutation in *Antony and Cleopatra*'.
86. On Anthony's clouds, see Yves Peyré, '"Travels in the clouds": Metamorphosis, Doubt, and Reason in the Renaissance'; Sujata Iyengar, 'Shakespeare's Embodied Ontology of Gender, Air and Health', p. 188; and Rhodri Lewis, *Hamlet and the Vision of Darkness*, p. 223.
87. Laura Levine, *Men in Women's Clothing: Anti-theatricality and Effeminization, 1579–1642*, p. 45.
88. Philemon Holland (trans.), *The historie of the vvorld*, chap. XLII, p. 20.
89. Interestingly, the horse image functions here as a gambit preparing the metadramatic intrusion of theatrical illusion in the play. In the Prologue of *Henry V*, horses are already bound up with the audience's imagination (ll. 27–8). In *Anthony and Cleopatra*, Anthony's contemplation of the clouds oddly echoes Agrippa's earlier remark that Caesar, having to part from his sister, 'has a cloud in's face' (3.2.51). It is probably no mere chance that Enobarbus immediately adds a punning remark, according to which '[h]e were the worse for that were he a horse' (3.2.52).
90. Ironically, this speech will be echoed, after Anthony's death, by Cleopatra herself: 'Nature wants stuff / To vie strange forms with fancy; yet t'imagine / An Anthony were nature's piece 'gainst fancy, / Condemning shadows quite' (5.2. 96–9).
91. Erasmus, *Adagia*, 1520, quoted in Raz Chen-Morris, '"The Quality of Nothing": Shakespearean Mirrors and Kepler's Visual Economy of Science', p. 109.

92. Alexander Nagel, 'Structural Indeterminacy in Early-Sixteenth-Century Italian Painting', p. 26.
93. Ibid.
94. Quoted by Ernst Gombrich, *Art and Illusion: A Study in the Psychology of Pictorial Representation*, pp. 154–61.
95. Philemon Holland (trans.), *The historie of the vvorld*, The Second Booke, chap. LXI, p. 29.
96. Ibid. p. 2. On this, see Hubert Damisch, *A Theory of Cloud: Toward a History of Painting*, p. 258.
97. On this, see Yves Peyré, '"Travels in the clouds"', pp. 11–15.
98. See Jacques Darriulat, *Sébastien le Renaissant*, pp. 115–30.
99. Camerarius was a well-know name, if only because in 1538 he published a Latin translation of Aesop, whose fables were 'prescribed in a growing number of school curricula'. Interestingly, 'Shakespeare appears to have studied "Aesop" in that version too'. See Ian Green, *Humanism and Protestantism in Early Modern English Education*, p. 163.
100. Joachim Camerarius, *The history of strange wonders*, sig. D.ir.
101. Ibid. D.iiv.
102. At this juncture we may note that Shakespeare introduces a sly pun on 'horse' and 'whore(s)', thus allowing Cleopatra's last line also to be read as 'O happy *whores*, to bear the weight of Anthony'.
103. On this, see below, Chapter 7, p. 242. The creative power of celestial stains is reaffirmed in *Hamlet*, where the Prince of Denmark, who apparently has a knack for finding shapes in the clouds, plays around with old Polonius to make him believe that he is indeed mad. In order to follow suit, the old councillor plays the Prince's game and accepts to successively see a camel, a weasel and then a whale in the sky (3.2.364–70). This may compare with Chapman's *Bussy d'Ambois*, written around the same time as Shakespeare's masterpiece: 'Sin is a coward Madam, and insults / But on our weakness, in his truest valour: / And so our ignorance tames us, that we let / His shadows fright us: and like empty clouds / In which our faulty apprehensions forge / The forms of dragons, lions, elephants, / When they hold no proportion, the sly charms / Of the witch Policy makes him a little monster / Kept only to shew men for Goddess Money' (3.1.18–26).
104. Lucretius, *De rerum natura*, Book IV, pp. 133–4.
105. Ibid. p. 133.
106. Janet Adelman, *Suffocating Mothers: Fantasies of Maternal Origin in Shakespeare's Plays, Hamlet to The Tempest*, p. 95.
107. John Gillies, *Shakespeare and the Geography of Difference*, p. 116.
108. Dan Brayton, 'Shakespeare and Slime', p. 85: 'Throughout much of the play slime is associated with Cleopatra and Egypt as well as with Antony's downward trajectory.'
109. Ibid. p. 84.

110. See William Whittingham, *The Bible and Holy Scriptures conteyned in the Olde and Newe Testament*: '1 And I sawe a newe heauen, and a newe earth: for the firste heauen, and the first earth were passed away, and there was no more sea. / 2 And I Iohn sawe the holye citie newe Ierusalem come downe from God out of heauen, prepared as a bride trimmed for her housband.' On the New Jerusalem, see also Edmund Spenser, *The Faerie Queene*, I.x.53–7.
111. On falling stars, see Revelation 9:1–2: 'I sawe a starre from heaven unto the earth, and to him was given the keye of the bottomless pit. And he opened the bottomless pit, and there arose the smoke of the pit, as the smoke of the great furnace.'
112. Falling stars, in the early modern era, generally referred to meteorites. See William Fulke's 1563 treatise on meteors, *A goodly gallerye with a most pleasaunt prospect*, fols. 7v–8r. More generally, on Fulke and his conception of 'meteors' (in the very broad meaning then given to the term, i.e. 'any atmospheric or meteorological phenomenon' (*OED*, 2.a)), see Margaret Llasera, *Représentations scientifiques et images poétiques en Angleterre au XVIIe siècle. À la recherche de l'invisible*, pp. 129–31.
113. Ibid. fol. 8r.
114. Plutarch, 'The Life of Lysander', in *The lives of the noble Grecians and Romanes*, trans. Thomas North, p. 486.
115. Katherine Vance MacMullan, 'Death Imagery in *Antony and Cleopatra*', p. 407.
116. See *Anthony and Cleopatra*, 3.6.67, 4.13.13 and 5.2.221.
117. One may here notice that Cleopatra's mercurial temper is reminiscent of Elizabeth I's mood swings. The brief meteorological portrait which Sir John Harington drew of his godmother after her death is rather telling. His letter is reproduced in Gwynne Blakemore Evans's *Elizabethan Jacobean Drama: The Theatre in its Time*, p. 181: 'When she smiled, it was pure sunshine, that everyone did choose to bask in, if they could; but anon came a storm from a sudden gathering of clouds and the thunder fell in wondrous manner on all alike.'
118. A similar dislocation is also at work in *King Lear*: 'O most small fault, / [. . .] / Which like an engine wrenched my frame of nature / From the fixed place' (1.4.258–61). On this, see Muriel Cunin, '*King Lear*: Fabric of the Human Body and Anatomy of the World', p. 97.
119. On this, see Sebastian Münster's *Cosmography* (1544), a book which went through eight editions in Münster's lifetime.
120. Mary Thomas Crane, 'Roman World, Egyptian Earth: Cognitive Difference and Empire in Shakespeare's *Antony and Cleopatra*', p. 12.
121. Ibid. p. 7.
122. On this, see J. R. Mulryne, 'Cleopatra's Barge and Antony's Body: Italian Sources and English Theatre'.
123. Michael Neill (ed.), *Anthony and Cleopatra*, p. 95.

124. Mary Thomas Crane, 'Roman World, Egyptian Earth', p. 12.
125. Ibid. p. 7.
126. See Epicure, *Letter to Herodotus*, 41–2, quoted and translated in Fredericus Antonius Bakker, *Epicurean Meteorology: Sources, Method, Scope and Organization*, p. 181. The theory put forward by Epicure is used extensively by Lucretius in the first book of *De rerum natura*.
127. Susan C. Staub, '"Bred Now of Your Mud": Land, Generation, and Maternity in *Antony and Cleopatra*', p. 78.
128. Cf. Lucretius, *De rerum natura*, Book V, p. 196. Lucretius draws a parallel between the motions of the stars and the streams of the rivers.
129. Ibid. Book II, p. 80.

Chapter 7

The I/Eye of the Storm: Prospero's Tempest

> If as the windes and waters here below
> Do flie and flow,
> My sighs and tears as busie were above;
> Sure they would move
> And much affect thee, as tempestuous times
> Amaze poor mortals, and object their crimes.[1]

The shadow of death by drowning hangs over the very first scene of *The Tempest*: a sensational stage direction, '*A tempestuous noise of thunder and lightning heard*' (1.1.0 s.d.), probably by Ralph Crane, precedes its actual opening.[2] In it, the term 'lightning' is intended as 'an *auditory* effect' rather than as a visual one, since pyrotechnics were avoided in private theatres because of the long-lasting stink they provoked.[3] The distinction we now make between thunder and lightning must therefore have been non-existent for the audience of the Blackfriars, the indoor playhouse for which the play was written. Yet in the 1623 Folio, where it was first published, this difference subsisted, and it provided Shakespeare's early readers with an auditory, visual and sensitive experience right from the start.[4] While the thunder was associated with the horrifying noise of a severe storm, the word 'lightning' not only conjured up the idea of an intense and short-lived light but, for the literati, it could also suggest the blowing of the wind. The pseudo-Aristotelian treatise *On the Cosmos* explains for instance that '[i]n the breaking up of the cloud, the wind, because it is set on fire and made to shine, is called lightning'.[5]

'Blow till thou burst thy wind, if room enough!' (1.1.7–8), the boatswain madly exclaims a few lines later. Shakespeare quickly plunges readers and playgoers into the middle of a frightening tempest, seen from the sea, which threatens the lives of boatswain and captain as well as those of all the onboard passengers. The storm looks like some ultimate ordeal, as was already the case for Claudio, who, in *Measure for Measure*, associates death and afterlife with the 'viewless winds' (3.1.124) with the fear that our spirits may 'bathe in fiery floods' (3.1.122) or 'reside / In thrilling region of thick-ribbèd ice' (3.1.122–3). This ordeal is both an external and an internal one. If Herbert the poet uses the storm to convey psychological turmoil in *The Temple* (1633), Shakespeare the playwright graphically pictures feelings of terror in terms of apocalyptic weather imagery.

In *The Tempest*, the playwright presents us with a character who, like Lear, wants to 'shake all cares and business from [his] age' (*King Lear*, 1.1.39) and who, as a result, finds himself increasingly isolated, loses his dukedom and is left with no other choice but to reconnect himself with nature on a faraway, (almost) uninhabited island.[6] Having shied away from the daily burden of governing his Milan dukedom, Prospero pays a heavy price for his negligence. Antonio, his 'perfidious' (1.2.68) brother, easily ousted him from government before ordering his men to 'hoist' him and his daughter 'To cry to th' sea' and 'to sigh / To th' winds' (1.2.148–50). So, the play's opening scene repeats Prospero's former and traumatic sea passage which had already made him the victim of a 'tempest' (5.1.153). The storm-tossed vessel is now a proper ship rather than the small 'barque' (1.2.144) he had been thrown into, but the gale is equally strong, and the main victim is none other than his brother Antonio, the new Duke of Milan. Incidentally, the play's title, derived from the Latin word *tempestas* ('weather'),[7] may implicitly refer to Machiavelli's *The Prince*, where the tempest evokes some sort of political threat or disturbance.[8] In an even more obvious way, it calls attention to the importance of time (*tempus*) as well as to the breaking out of a violent storm which will remain dangerous all along, at least in the eyes of Trinculo, who, just before his confrontation with the island's monster, Caliban, exclaims:

> Here's neither bush nor shrub to bear off any weather at all, and *another storm* brewing – I hear it sing i' th' wind. Yon same black cloud, yon huge one, looks like a foul bombard that would shed his liquor. If it should thunder *as it did before*, I know not where to hide my head – yon same cloud cannot choose but fall by pailfuls. (2.2.18–24, my emphasis)

In this context, it is only logical that the meteorological conditions of Prospero's island should arouse critical attention. Because of what the magus tells his daughter Miranda in the second scene of the play, suggesting that there was neither a real shipwreck nor any actual drowning, the harsh weather conditions plaguing the crew of Alonso's ship have often been 'exposed as magically derived rather than natural'.[9] However, they are still regarded as a fairly marginal issue, if we are to judge by the relatively small number of studies entirely focused on Prospero's storm. The closing chapter of this book is an attempt to fill this gap.

Travel and travail: Harrying hurricanoes

The Tempest is often related to a number of topical issues, such as James I's politics, dynastic alliances, the mapping of the New World, deforestation (one of Prospero's tasks apparently consisting in cutting trees),[10] the movement of enclosures, as well as . . . the weather. Bad weather and enclosure practices did contribute to price escalation and sometimes to starvation among the poorer subjects of the realm, which explains why so many riots took place between 1590 and 1610. In the play, the repeated allusions to the disastrous effects of foul weather and to the forceful maintenance of enclosed spaces (Ceres, in particular, calls attention to 'this short-grassed green', 4.1.83, and, even more strikingly, Iris mentions 'pionèd and twilled brims', 4.1.64) must have struck a chord in the audience of the indoor Blackfriars theatre, among them many landowners likely to endorse enclosures.[11] The first performance we know of took place on Hallowmas night, 1611, a particularly wet year according to contemporary testimonies. Part of the literature of the period echoes these rather terrible weather conditions. In *The commons complaint* (1611), for instance, Arthur Standish offers advice on agricultural practices, claiming that he had 'conference with many of the best Commonwealths-men for [his] better understanding', and that he intends to offer his readers 'perfect knowledge how to Plant, preserue, and maintaine the blessings of God as well in this our destroying age'.[12] So, with its opening scene full of howling winds, roaring seas and apocalyptic threats, *The Tempest* must have met the concerns of all and sundry in such difficult times.

As the play's title suggests, however, Shakespeare emphasises the pouring rain less than the impressive strength of gale-force winds that often turned sea travels into sea travails for sailors and passengers alike. From the time of Columbus onward, there have been many accounts of bad storms, but it was the Genoan explorer who, in a 1494 letter to Queen Isabella of Spain, provided the first account of a hurricane:

> The tempest arose and worried me so that I knew not where to turn; eyes never behold the seas so high, angry and covered by foam. We were forced to keep out in this bloody ocean, seething like a pot of hot fire. Never did the sky look more terrible; for one whole day and night it blazed like a furnace. The flashes came with such fury and frightfulness that we all thought the ships would be blasted. All this time the water never ceased to fall from the sky [. . .] The people were so worn out, that they longed for death to end their terrible suffering.[13]

The image of the 'pot of hot fire' resonates in *The Tempest* at the moment when Ariel proudly tells his master how he produced terror among the crew: he 'flamed amazement' (1.2.198) by making fire and lightning assail the king's vessel. Prospero's 'brave spirit' (1.2.206) presents the storm as hell on earth, and 'the fire and cracks / Of sulphurous roaring' (1.2.203–4) which he describes hark back to the 'sulphurous and thought-executing fires' (3.2.4) as well as to the 'sulphurous pit' (4.5.123) mentioned in *King Lear*, another weather-related play.[14]

This means that, rather than being derived from Columbus's letter, Ariel's fire imagery first and foremost relies on a Shakespearean intertext. Nonetheless, the motif of the fiery ocean had initially been exploited by early modern travellers before being recycled in the travel narratives that gathered, rewrote and sometimes forged the explorers' testimonies.[15] The comparison with hell, in particular, was so fruitful that it was captured in a number of maps charting the New World and displaying personified winds associated with devilish creatures. For instance, from 1544 onwards, a world map drawn in 1540 by the Dutch cartographer Gemma Frisius was inserted in Petrus Apianus's *Cosmographia*. On this heart-shaped map, the winds appear in the shape of bearded old men in the top left-hand corner and as curly-haired children blowing out little suns in the right-hand corner, while, in the bottom parts, they are frightening skulls with dishevelled hair and sharp teeth. These hot, plague-bearing winds are actually presented as fire-breathing devils.[16]

Columbus's sea voyages were regularly vexed by such devilish winds. In the course of his second voyage to the New World, in September 1495, the explorer was battered by a tropical storm in Isabella harbour. Six ships were reported sunk, but Columbus's

Fig. 12 The 1540 map of Gemma Frisius, 'Charta cosmographica, cum ventorum propria natura et operatione', in Petrus Apianus's *Cosmographia*, Paris, 1551. By courtesy of Rare Books, Special Collections, J. Willard Marriott Library, The University of Utah, GT3200 1540 A65.

Niña remained intact. His experience proved very useful two years later, during his fourth voyage (1502–4), as he experienced another tropical cyclone. Indeed, in June 1502, 'Columbus and his men deduced the approach of a hurricane from the direction of high clouds and an ominous southeasterly swell'[17] and, as a result, they sought shelter at the harbour of Santo Domingo. Columbus's ships were not destroyed by the high winds because he was able to detect the signs of the approaching storm, but the governor of Hispaniola (which now corresponds to the Dominican Republic and Haiti), who had ignored the warnings of the weather, lost most of his ships in the straits between Hispaniola and Puerto Rico.

Pamphlets relating Columbus's eventful sea voyages were soon incorporated into literary writings. In 1516, Pietro Martire d'Anghiera, who knew Columbus personally, collected the accounts of his crew and published his *De orbe novo*, a book devoted to the discovery of America and translated into English in 1555 as *The decades of the newe worlde or west India*. In 1552, Bartholomé de las Casas, who had accompanied Columbus to the Indies and had been with him to Hispaniola in 1502, published his *Brevísima relación de la destruición de la Indias* (*The Devastation of the Indies, A Summary Account*). The book was easily accessible in Shakespeare's England, having been translated as *The Spanish colonie* (1583) by 'M.M.S.'. The narrative pays no lasting attention to the gales sweeping the West Indies. Yet in just a few words, its author makes it clear that the hurricane which struck Hispaniola was the manifestation of divine wrath after God had been infuriated by the brutal treatment of the Indians in the Spanish settlement.[18] Even more interestingly for us, another famous account, *La historia del mondo nuovo* (*The History of the New World*), written by Girolamo Benzoni, was published in 1565 and translated into Latin in 1578. This compelling narrative, which shed light on the Spanish cruelty to the Indian natives, was particularly well received in Protestant countries. In the 1590s, Theodore de Bry, the Dutch print-seller and engraver, produced a series of engravings to illustrate Benzoni's text.[19] One of them notably depicts the English settlers in the throes of a violent storm but lucky enough to have disembarked while their ship, drawn in the background, is about to sink.[20] In fact, this engraving represents a *Huracán*, a word used by the West Indian natives to designate the Indian god of evil,[21] and which the Spaniards subsequently brought back to Europe. Impressed by the raging storms they had had to face in the course of their four voyages to the New World, they borrowed the Indian term to portray them in an accurate way.[22]

The word 'hurricane', or rather 'hurricano', crops up twice in the Shakespearean canon. In *King Lear*, the maddened protagonist begs the storm to strike the earth and exclaims 'Rage, blow, / You cataracts and hurricanoes' (3.2.1–2),[23] while in *Troilus and Cressida*, as he realises the deceptive nature of his beloved Cressida, who has easily yielded to Diomed's advances, Troilus compares his imminent sword strike to a 'hurricano' or a 'dreadful spout' (5.2.174–9). These occurrences designate 'a waterspout' (*OED*, 2) rather than violently blowing winds, while the word 'hurricane', used earlier in the sixteenth century in its variant forms of 'furacane' or 'furicano',[24] pointed to a 'violent wind-storm of the West Indies' (*OED*, 1.a). As a result, if a hurricane is indeed described in the first scene of *The Tempest*, the word itself does not appear in the text because Shakespeare associated it with the rain rather than with the wind.

Because of its very ambiguity, the word 'tempest' presented an undeniable advantage. Early modern lexicographers gave it a rather broad definition encompassing wind, storm and foul weather, as Cotgrave's 1611 entry makes clear: 'A tempest, storme, bluster, boisterous weather.'[25] More often than not, it referred to 'a violent storm of wind' (*OED*, 1.a) and it therefore frequently appeared in the travel narratives telling of the discovery of the New World. In a 1604 translation of José de Acosta's 1590 *The naturall and morall historie of the East and West Indies*, the discovery of new lands was explicitly related to the advent of tempests.[26] In *The Tempest*, the banned Prospero becomes an unwitting discoverer of sorts who settles in the middle of nowhere – the island is not, and will never be, named, as if he had somehow failed to fully appropriate this foreign space – and he thus charts new territory, the place where the wind has taken him, before reproducing a similar pattern later on when he brings about a terrible storm which causes his enemies to shipwreck. The word 'tempest' was also metaphorically understood as 'a violent commotion or disturbance' (*OED*, 2.a), a figurative sense found in *Titus Andronicus* when Queen Tamora tells the Emperor 'cheer the heart / That dies in tempest of thy angry frown' (1.1.454–5). Here, Alonso's ship is described by Miranda as a 'brave vessel' (1.2.6), the word 'vessel' being then likely to be understood in the biblical sense of 'receptacle of the soul' (*OED*, 3.†b.), a meaning making the initial wreck of the ship serve as a metaphor for the punishment of a corrupt humanity.

Still, the playwright must have been aware of the threat represented by the *real* cyclones that struck the fleet of many an explorer in quest of the New World. If Setebos, the god worshipped by Caliban's mother, unmistakably evokes Patagonia,[27] Bermuda is often quoted as the possible location of Shakespeare's allegorical island.[28] Ariel reminds Prospero that he called him 'at midnight to fetch dew / From the still-vexed Bermudas' (1.2.228–9), a site reputed to be incessantly plagued by storms ('still-vexed')[29] and which was thus associated with the torments of hell ('the first man that leapt, cried "Hell is empty, / And all the devils are here"', 1.2.214–15).[30]

It is probably no coincidence that, in 1609, a hurricane was responsible for the first permanent English settlement in Bermuda. The play's main source is indeed the account of the *Sea Venture*'s wreck on Bermuda's shores.[31] In *The Tempest*, Steve Mentz observes, 'the spirit-storm', namely Ariel, 'drives the play as the trade winds drove the *Sea-Venture* across the Atlantic to Bermuda's reef'.[32] The ship was on its way to Jamestown, Virginia, when a huge storm blew it on to one of the uninhabited islands of the Bermuda archipelago. In a private letter, William Strachey gave a strikingly detailed account of the expedition.[33] A manuscript version of this narrative was first circulated in England in 1610, before being put to print fifteen years later in the travel narratives compiled by Samuel Purchas. If critics are uncertain as to how Shakespeare may have been acquainted with Strachey's manuscript letter, most agree on the fact that the playwright borrowed several ideas and phrases from a few 1610 pamphlets using Strachey's narrative, including Sylvester Jourdain's *A Discovery of the Barmudas, Otherwise Called the Ile of Divels* and the Council of Virginia's *True Declaration of the Estate of the Colonie of Virginia*.

Purchas his pilgrimes (1625) gives us an idea of the excitement initially caused by Strachey's narrative. In it, readers are made to relive the adventures of a ship's crew facing '[a] most dreadfull Tempest' in the Bermuda islands[34] and to realise that the violence of the gale was such that even experienced mariners were frightened to death by it:

> When on S. *Iames* his day, Iuly 24. being Monday (preparing for no lesse all the blacke night before) the cloudes gathering thicke vpon vs, and the windes singing, and whistling most vnusually, which made vs to cast off our Pinnace towing the same vntill then asterne, a dreadfull storme and hideous began to blow from out the North-east, which [. . .] at length did beate all light from heauen; which like an hell of darkenesse turned blacke vpon vs, so much the more fuller of horror, as in such cases horror and feare vse to ouerrunne the troubled, and ouermastered sences of all, which (taken vp with amazement) the eares lay so sensible to the terrible cries, and murmurs of the windes, and distraction of our Company, as who was most armed, and best prepared, was not a little shaken.[35]

While these words provide a graphic rendering of the atmosphere described in the Bermuda pamphlets, they surpass them in their expressiveness and dramatic potential. The heavens look black, the clouds are huge and the storm, as noisy as it is terrifying, turns the sea into a howling hell. Strachey describes a maritime inferno when he explains that 'Windes and Seas were as mad, as fury and rage could make them', before adding that 'there was not a moment in which the sodaine splitting, or instant ouer-setting of the Shippe was not expected'.[36] Among the many verbal parallels between the 'Reportory' and *The Tempest*, one finds the long description of the 'Sea-fire' or St Elmo's fire which 'might have strucken amazement',[37] echoed in Ariel's claim to have 'flamed amazement' (1.2.198) as he boarded the vessel. Ariel makes it clear that, as a result, the ship was 'all afire with [him]' (1.2.212). In other words, the elusive spirit of the air personifies a natural force at the beginning of the play when he literally transmutes himself into the furious energy of St Elmo's fire.[38] Characterised by the presence of balls of fire glowing with blue flames in the rigging, this phenomenon is duly recorded in Hill's 1574 *Contemplation of*

Fig. 13 A ship caught in a tempest and struck by St Elmo's fire in Thomas Hill's *Contemplation of mysteries*, 1574. By courtesy of the British Library Board, C.56.b.15, fol. 22r.

mysteries, in which an engraving represents a ship hit by 'two burning Candels, named Castor and Pollux'[39] which are said to 'appeare after the kinde of a fyre, cleaving to the top of the maste'.[40]

According to Alden T. and Virginia Mason Vaughan, Shakespeare probably had direct access to Strachey's account, either by word of mouth or in manuscript form,[41] while Gabriel Egan posits that the playwright may even have had contact with 'those involved in the Virginia project'[42] which entailed a number of scientific weather experiments before 1610. Indeed, a thermoscope was discovered among the artefacts dug out in the Jacobean settlement of Jamestown.[43] Prospero's attempts at mastering the climate may thus reproduce some of the Jamestown experiments. 'The play is utterly ambiguous about the kind of control over the physical world that Prospero's knowledge gives him,' Egan observes, 'and by probing this question (what *is* his "art"?) we begin to perceive the ecological significance of *The Tempest*.'[44]

Important as it certainly is, this 'ecological significance' was not systematically one of the concerns of the travel narratives that circulated in Jacobean England. As J. M. Nosworthy notes, Shakespeare surely 'did not need an actual shipwreck and its concomitant pamphlets to tell him that vessels sometimes come to grief in squally weather', and the fact is that, in *Pericles* and *The Winter's Tale*, 'storm and shipwreck find a place without any topical prompting'.[45] The playwright undoubtedly found part of his inspiration not only in Ovid's *Metamorphoses*[46] but in the vast reservoir of storms, sea travels and magical practices in Virgil's *Aeneid*, as well as in various biblical and theological sources more discreetly woven into the texture of the play but nonetheless present. Staging a proto-scientific experimentation with climate actually allowed Shakespeare to engage in a debate opposing those who ascribed foul weather to natural causes to those who explained and justified it by resorting to some godly intervention. Aristotle's *Meteorology* argued that meteorological phenomena were due to natural rather than divine causes, while Luther openly mocked such beliefs in the powers of nature when he wrote in 1522: 'So, the blind man's guide Aristotle has written a whole book on the heavenly signs. He ascribes them all to nature and makes that they are no signs. These our scholars follow; so, one fool makes a world full of fools.'[47] Following Luther's precepts, the representatives of the Church of England tended to think that none other than God himself raised the winds. For instance, George Abbott asserted in *An exposition vpon the prophet Ionah* (1600):

> Who so walketh by the land, or passeth by the sea, if winds promote his businesse, or hinder his purpose, and disquiet him in his enterprise, let him assigne it to his prouidence who ruleth all with power, who sent that tempest here to Ionas; for from him they do all come.[48]

Demons, in Christian theology, are moreover presented as airy creatures moving around in 'thin air' (4.1.146) as well as able to manipulate this particular element. Ariel's role in the play brings to mind Hebrews 1:4, where God 'maketh his angels spirits, and his ministers a flame of fire'. While Shakespeare interprets and dramatises this for the Blackfriars, his own spirit, Ariel, embroiled in a complicated master-servant relationship, is an inadequate angel who is just as pagan as he may be Christian. His multiple and dizzying appearances lead the audience to wonder about the source of his power: is it pure magic or should we see in Ariel the embodiment of natural forces abated or increased by human will?[49] All sorts of possibilities are explored but none is definitely advanced and the playwright visibly refuses to provide any definitive answer. *The Tempest* thus enhances the complex relationships between nature, science, magic and the Scriptures, and lets the audience have the last word.

If the spectators were presented with a number of critical issues during the performance, they could fortunately rely on many familiar images. Laurie Shannon reminds us that '[t]he story of Noah's ark', for instance, 'taxed the early modern imagination',[50] and it was indeed frequently quoted in the sermons and the literature of the period. With its insistence on flooding (making the old Gonzalo wish that he could 'die a dry death', 1.1.67–8), the opening scene proposes a spectacular variation on the biblical myth of Noah's ark.[51] The initial tempest which so terrifies the passengers on board can therefore be interpreted as signalling the presence of an angry God. No wonder then that, for his spectral show within act 4, scene 1, Shakespeare has Iris, the messenger of the gods and the goddess of the rainbow,[52] move between heaven and earth.[53] With her 'saffron wings' (4.1.78), the goddess is directly imported from Virgil's *Aeneid*,[54] but her function in the play is also pregnant with Christian meaning. The arrival of the 'Nymphe of the Ayre', as she is described by Pierre Le Loyer in his *Treatise of Specters* (1603),[55] puts an end to the island's stormy weather and announces the happy union of Ferdinand and Miranda. In the Bible (Genesis 9:14), a similar process is at stake, since the apparition of a rainbow in the sky signifies the end of the flood and reminds Noah and his sons of the covenant that has been established between them.[56]

Temperate climate and temperance

Myths, whether biblical or classical, are thus intertwined in Shakespeare's meteorological romance,[57] and the tempest conjured up at the opening of the play provides spectators and readers alike with several interpretative tracks to follow. In his 1622 *Historia ventorum* (*History of Winds*), Francis Bacon differentiates clement breezes from harsh winds. Interestingly, he contends that the west wind belongs to the first category and is, as such, a characteristic of the Golden Age:

1. The south wind with us is rainy, the north wind clear; the former collects and nurtures clouds, the latter breaks and dissipates them. Poets therefore in their descriptions of the deluge represent the north wind as at that time imprisoned, and the south wind let loose with full powers.
2. The west wind is reputed by us as the wind of the Golden Age, the companion of perpetual spring, and the nurse of flowers.
3. The school of Paracelsus, seeking a place for its three principles even in the temple of Juno, that is the air, established three winds. For the east they found no place.
   Tincturis liquidum qui mercurialibus Austrum,
   Divitis et Zephyri rorantes sulphure venas,
   Et Boream tristi rigidum sale.
4. In Britain the east wind is considered injurious, insomuch that there is a proverb,
   'When the wind is in the east,
   'Tis neither good for man nor beast.'[58]

It is hard to decide what kind of wind Shakespeare actually describes. We may be tempted to think that he focuses on the north wind, as Ariel must 'tread the ooze of the salt deep', 'run upon the sharp wind of the north' and 'do [Prospero] business in the veins o'th'earth / when it is baked with frost' (1.2.252–6). However, if frost is mentioned from time to time, the climate of the island seems rather mild if we are to believe Gonzalo's observation on the 'lush and lusty' (2.1.53) grass. In *A contemplation of mysteries*, Thomas Hill writes that '[t]hunder with an Easterly or Westerly winde, declare raine, or a mightie tempest to follow'.[59] The storm being accompanied by a momentous tempest in Shakespeare's play, one may conclude that the wind blowing on the island is either easterly or westerly, and Bacon's previous remarks, read in connection with Gonzalo's 'commonwealth speech', tend to suggest that a strong western wind keeps blowing on Prospero's island.[60] This is corroborated by Theophrastus' *Concerning Weather Signs*, where the Greek writer declares that, '[w]hen the west wind is accompanied by lightning from the north, it indicates either storm or rain'.[61]

The combination of rain and mild temperatures brought by the west wind probably accounts for the verdant landscape of the island. The 'lush and lusty grass' mentioned by Gonzalo is presented as a 'short-grassed green' by Ceres (4.1.83), while Iris makes it clear that the masque takes place '[h]ere on this grass-plot' (4.1.73) before she invites the 'temperate nymphs' to 'celebrate / A contract of true love' (4.1.132–3). The adjective 'temperate', here linked to chastity, appears as something of a surprise in a betrothal ceremony from which Venus is curiously absent and where love and fertility should come in the foreground. At this very moment, Ferdinand notices that Prospero has just remembered the imminence of Caliban's 'foul conspiracy' (4.1.139) and that he is 'in some passion / That works him strongly' (4.1.144–5), to which Miranda replies: 'Never till this day / Saw I him touched with anger, so distempered' (4.1.145–6). Prospero's sudden fit of anger breaks the spell of the revels he had himself ordered, just as the initial tempest had created discordance and commotion on the island. Now, 'the name Prospero signifies both material prosperity and, in its classical and Renaissance sense, good weather [. . .] Prospero, then, is God on earth, an absolute and controlling force in human and meteorological terms.'[62] Prospero, who appears as a regulator of time with its cognate words of 'season' (*tempestas*), 'temperance' and 'temper', is thus also to be seen as a master of disruption and bad weather.

Temperance is a trope narrowly linked to weather issues, one that Shakespeare had already used and explored in *The Winter's Tale*, an earlier romance where the word 'climate' appears on three occasions. Delphos, which in those days corresponded to the island of Delos in the Greek Cyclades, rather than to the actual city of Delphi in continental Greece, is a place associated with temperance and fertility by the Sicilian emissaries of the King, as Cleomenes exclaims that 'The climate's delicate, the air most sweet; / Fertile the isle' (3.1.1–2).[63] The surprisingly 'temperate' climate of Delos, seemingly modelled on Virgil's island in the *Aeneid* (Book 3),[64] is as misleading as that of Prospero's island. Both islands actually serve as counterparts to the greater English island, where the weather may be rough but where the inhabitants' temper appears more balanced.

In *The Tempest*, moreover, temperance is related both to an impossible fertility (Ceres's masque is aborted) and to a fragile chastity (Miranda's virginity is threatened by Caliban's luxuriousness and by Ferdinand's flirtatious mood). In act 2, scene 1, temperance is indeed described by Antonio as a 'delicate wench' (2.1.45), an image which reads like an antiphrasis since Gonzalo had first compared the leaky vessel to an 'unstanched wench' (1.1.48). The irony is reinforced in the next scene when Prospero tells Miranda of how his brother Antonio usurped his Milan dukedom. In doing so, he calls her 'wench' (1.2.139), presumably 'an endearing form of address' (*OED*, 1.c). For the audience, however, the term has become rather negative at this point so that, instead of being presented as a temperate and innocent young maid, Prospero's daughter may immediately be envisaged as a young lady ready to yield to her passionate love for Ferdinand.

Furthermore, the word 'temperance' had a specific relevance regarding climatic issues as it had come to designate 'moderate temperature; freedom from the extremes of heat and cold; mildness of weather or climate; temperateness' (*OED*, II.†4). It also corresponded to one of the four cardinal virtues and it was broadly understood as 'self-restraint and moderation in action of any kind' (*OED*, I.1.a). As such, it was highly praised in early modern Europe as the readers of the *Nicomachean Ethics*, a text then available in Latin,[65] were aware that Aristotle promoted temperance and associated it with serenity, arguing that the temperate man finds 'no pleasure contrary to reason'.[66] For the Greek philosopher, temperance required no particular effort or the need to fight against one's desires. In Book I of his *De officiis*,[67] Cicero also dealt with the Greek *sophrosyne*, regarding it as a capacity for restraint as well as 'an art of timing'.[68]

In *The Tempest*, temperance seems to apply exclusively to women and weather, a context where the word functions each time as an ironical qualifier associated with presumably changeable entities or states. Its many occurrences in the play turn it into a leitmotif which lays bare the social microcosm of the island. While temperance becomes a mantra for virtually everyone yearning for serenity and restraint, excess would in fact best qualify the actions of its characters. Similarly, the climate on the island is continually defined as idyllic while it is plagued by a succession of violent storms. John Gillies has aptly summarised the function of 'temperance and fruitfulness' in the play, showing that it is primarily a geographic and climatic one likely to subsume other dimensions, be they moral or political.[69] He has also shown that the first – necessarily propagandist – reports from Virginia, for instance, insisted on the temperate climate and, as a result, on the fertility of a virgin soil ready to be tilled and abandon its riches to the European newcomers, no matter what the reality was like. A tongue-in-cheek Shakespeare reproduces this propagandist discourse in the play. No wonder then that, in an act of self-persuasion, Gonzalo, one of the shipwrecked lords, trying to cheer himself up, declares that the island 'must needs be of subtle, tender, and delicate temperance' (2.1.43–4) after he has nearly drowned in the recent tempest. The good old councillor prefers to try and delude himself by idealising the island rather than describe the sad reality of his predicament. Gonzalo's 'commonwealth speech' (2.1.145–66), strongly influenced by Montaigne's 'Of Cannibals', thus portrays a prelapsarian world in which the sun is always shining and where nature by itself produces everything man needs '[w]ithout sweat or endeavour' (2.1.158).

Yet this kind of temperance is sheer wishful thinking as one quickly notices that the shipwrecked victims are debilitated in a climate reported as constantly mild. Supposing that the weather on the island is truly comforting (except when Prospero raises storms), one may then wonder why the characters feel so weak once the illusory tempest has quieted down. This apparent contradiction actually ties in with a particular trend of geo-humoralism illustrated by Hippocrates, who thought that, if a 'rugged, high, and watered'[70] region produced endurant individuals, a more welcoming environment (i.e. a well-watered one with hot summers and cold winters) logically resulted in weaklings, not used to hard labour and seen as 'fleshy, ill-articulated, moist, lazy, and generally cowardly in character'.[71] In *The Tempest*, Shakespeare clearly relies on this tradition associating temperate climate with dullness and a lack of endurance.[72] This is why, in act 2, scene 1, Alonso and most of his court fall asleep instead of mustering their energy to search for the lost Ferdinand. Similarly, Caliban reveals later on in the play that Prospero regularly takes a nap at noon ('Within this half hour will he be asleep', 3.2.111), thereby making himself vulnerable and exposed to a plot against his life.

On top of that, Alonso and his court are stranded so far from home that they have difficulties in adapting themselves to their new environment. William Vaughan explained in his 1612 health treatise that the best air one could find was '[t]hat which is a mans natiue soyle, and Countries ayre is best'. Hammering the point home, he added that '[t]his by the Philosophers is approued in this principle: Euery mans naturall place preserueth him, which is placed in it'.[73]

Shakespeare follows the same logic: his newcomers are totally unprepared to endure those rather unusual conditions and, as rational beings, they attribute their 'drowsiness' to the climate rather than to Prospero's magic tricks.

The audience is of course aware that Ariel is the one who first puts the mariners to sleep (1.2.230–1) before doing so with the rest of the crew (2.1.198 s.d.) except for Sebastian and Antonio, since their aggressiveness and constant Machiavellian plotting keeps them awake most of the time.[74] We have seen that Ariel, in the play, embodies the elements of air and fire and that he strikes his victims as a stroke of lightning would. He is the one in charge of making people fall asleep all at once and, according to the superstitions of the time, a flash of lightning was believed to be capable of shutting the eyes of anyone awake and opening those of anyone asleep. This theory is notably developed in *A contemplation of mysteries*, in which Thomas Hill writes that

> [t]he lightning striking a person sléeping, doth open the eyes of the sléeper: but of [the] person then waking or awake, it shutteth the eyes. The cause of which is, in that the lightning causeth the sléeper to awake, but the person awake, it astonisheth in such maner, that he shall sooner dye, then that he can open the eyes againe for the tyme, as the auncient report.[75]

Yet when Sebastian exclaims '[w]hat a strange drowsiness possesses them!', Antonio understandably fails to appreciate that the sleepiness of his confederates may have been the direct consequence of a flash of lightning, especially as Ariel does not appear as such then. He therefore replies unhesitatingly: 'It is the quality o'th' climate' (2.1.198). The climate of his native homeland may not be better, but at least it suits his temper and, above all, it contributes to fashioning his identity. Alonso and his court have become misfits under the sun of Prospero's new dukedom.

Purgation and exhalations

In *King John*, the sun is said to 'pla[y] the alchemist' (3.1.4) as it turns '[t]he meagre cloddy earth to glittering gold' (3.1.6). In *The Tempest*, by contrast, wind and sea create a sort of 'heavenly alchemy' (Sonnet 33, l. 4) that reverberates throughout the island,[76] while the title itself refers to 'the alchemical term for the boiling of the alembic to remove impurities and transform the base metal into purest gold'.[77] This purgation process partakes of Prospero's aims as the storm is basically designed as a long and uneasy transforming and improving process to transmute the ship's 'base' passengers.[78] Shakespeare, who must have known something of the current alchemical treatises, was certainly aware that the elemental force of the wind was needed for the production of the Philosopher's Stone. The earliest extant alchemical treatise, the *Emerald Table* (*Tabula Smaragdina*), which dates from the fourth century AD, describes 'the Philosopher's Stone, the goal of every alchemical process, as the child of the Sun and the Moon, borne by the Wind and nourished by the Earth'.[79] In *The Tempest*, the wind brings reconciliation – albeit an imperfect one as no one properly repents – and it allows Prospero to recover the dukedom he had lost.

Ariel makes this clear in his 'Full fathom five' song (1.2.397–405), in which he observes that Ferdinand's father 'doth suffer a sea-change / Into something rich and strange' (1.2.401–2): thanks to the experience of the tempest, human creatures are transmuted and become much better creatures afterwards. 'Those are pearls that were his eyes' (1.2.399), Ariel sings, referring to the transformation process brought about by the storm. Paracelsian alchemy also implied the apparition of iridescent colours at the moment of the regeneration or resurrection process. In the masque scene, Iris naturally embodies such iridescence. One should not forget either that Shakespeare's play shortly followed Ben Jonson's *Alchemist*, a biting satire of Paracelsian alchemy in which the charlatan Subtle compares alchemists to carriers of tempests (1.1.60–2).[80] In his critical version, Shakespeare ironically takes Subtle at face value and reaffirms the power of illusion, when Jonson dismisses it, by making it consubstantial to the understanding of reality.[81] So, as soon as Prospero exposes his enemies to the elements in the opening scene, the storm works like an alchemical formula aiming for the regeneration of humankind.

Elsewhere in his plays, Shakespeare uses the wind to cleanse or purge the world from its sins.[82] That is the case in *King Lear*, which also gives pride of place to the connection between the wind and men's passions, as tempests were then often taken as external manifestations of an inside turmoil. The French bishop Nicolas Coeffeteau declared, for instance, that 'as there were foure chiefe winds which excite diuers stormes, be it at land or sea; so there are foure principall *Passions* which trouble our *Soules*, and which stir vp diuers tempests by their irregular motions'.[83] This, of course, remains true in *The Tempest*, a play which showcases the passions of a father estranged from power and bitterly resenting his predicament. Because the human body and the natural environment were thought to interact, bad weather necessarily impinged on our humours.[84] A shared line between Gonzalo, Sebastian and Antonio hammers the point home:

> *Gonzalo*: It is foul weather in us all, good sir,
> When you are cloudy.
> *Sebastian*: Foul weather?
> *Antonio*: Very foul. (2.1.139–40)

Gonzalo's remark to Alonso is reminiscent of the Gentleman's reply to Kent in *King Lear*, when the latter asks who is there: 'One minded like the weather, most unquietly' (3.1.2). Yet whereas Shakespeare relied on Epicurean references in *Lear*, the playwright here seems conversant with the Aristotelian view of the weather. It is certainly no coincidence that he should pay attention to sounds, if one is to believe Caliban's 'Be not afeard. The isle is full of noises, / Sounds and sweet airs, that give delight and hurt not' (3.2.133–4), lines that may echo Aristotle's *Meteorology*, as the Greek philosopher asserted that the

> [w]ind is also the cause of noises beneath the earth, among them the noises that precede earthquakes, though they have also been known to occur without an earthquake following. For as the air when struck gives out all sorts of noises, so also it does when it is itself the striker; the effect is the same in either case.[85]

In the same treatise, Aristotle explains further down that '[w]hen the wind is too fine to communicate any impulse to the earth [. . .] it gives out all sorts of noises, so that sometimes the earth seems to bellow as they say it does in fairy stories'.[86] He also states that winds come up from the earth before observing that hot and dry exhalations are the matter of winds,[87] a theory widely endorsed by Shakespeare's contemporaries. The radical Puritan William Fulke, for example, fully adhered to the exhalations theory in order to account for the development of thunder:

> A great thonder, is when the *Exhalation* is muche in quantitie, and verye hote and drye in qualitie, the cloude also very thycke and stronge, that easely wyll not geue place to the wynde, to escape out.
>
> Wherfore if the *Exhalation* do vehemently shake the cloude, though it doe not at the first disperse it, it maketh a longe and

> fearefull romblyng against the sydes of the cloude, vntill at the last being made stronger by swyfter motion, it dissolueth the cloude, and hath lybertie to passe out into the open ayer. The cloude resolued, droppeth downe, and then followeth a showre of rayne.[88]

Similarly, a few years later, Thomas Hill asserted that '[t]he matter or subtaunce of the windes, is a hote and dry exalation, grosse and earthly, forced downearde through the coldnesse of the middle region of the ayre'.[89]

As in *Othello*, the air circulates vigorously in *The Tempest* and it is being expulsed both by the earth (if one is to believe Aristotle) and by human bodies. At the beginning, Prospero reminds Ariel of his initial desperate situation: '[Sycorax] died / And left thee there, where thou didst *vent* thy groans / As fast as mill-wheels strike' (1.2.279–81, my emphasis). The verb 'vent' is echoed later on in the grotesque scene where Stephano and Trinculo meet Caliban. Trinculo is at the origin of this encounter as he is seeking shelter 'for fear of the storm' (2.2.106) and notices a gabardine – belonging to Caliban – likely to protect him from thunderstrokes. And when his friend Stephano finds the jester half-hidden under the gabardine in the company of a very strange creature, he sardonically asks: 'How cams't thou to be the siege of this mooncalf? Can he *vent* Trinculos?' (2.2.101–2, my emphasis). Here, the expulsion mechanism all at one reproduces the female process of bringing forth a baby, the act of farting, and the tempest phenomenon itself.[90]

As to the 'brave monster' (2.2.183) of this 'brave new world' (5.1.183), a devil-like, carnivalesque figure, he is characterised by his venting of farts and of foul words. We learn early on that he has been taught how to curse (1.2.363), that is, to mouth abuse at his new master. It is his way of venting his grief. According to Rebecca Totaro, cursing first 'occurs within the body after the subject has experienced some form of marginalization'.[91] The curse is comparable to a disease that must be expelled and, in the process, it is turned into an exhalation polluting the air. In his 1541 *Image of gouernance*, Thomas Elyot describes this phenomenon as he praises the usefulness of sane places of exercise in Rome at the time of Alexander:

> To the sayde baynes and places of exercyse were assigned a numbre of kepers and ministers, to whome were gyven sufficient salaries. And those places were alway kepte so nete and dylectable, that to the eyen or nose, was neuer any thynge vnpleasaunt or noyfull: by the

> whiche clennesse the citie was meruaylouselye preserued from sundrye syckenesses, whiche vndoutedly do grow of corrupt exhalations, ventyng out of mens bodyes. And for that cause to auoyde occasyon of pestylence and other horryble dyseases, this Emperour prohybyted by speciall lawes, ingurgitations, bankettes, late suppers & longe.[92]

If the word 'exhalation' does not occur in *The Tempest*,[93] Caliban's words can nonetheless be assimilated to 'corrupt exhalations' that pollute the air of the island. As Totaro puts it, '[m]eteorologically speaking, the symptoms of the curse were animated by the same materials and processes that created and put in motion comets, earthquakes, and volcanoes – each hot, dry bodily projectiles that Aristotle calls "exhalations".'[94]

Now, if Caliban's curses correspond to the windy exhalations of the island atmosphere, Antonio and Sebastian's sardonic interruptions of Gonzalo's 'commonwealth speech' similarly function as parasitical verbal storms. They 'form part of a pattern of linguistic disruption' which sees Alonso 'repeatedly interrupt[ing] the efforts at consolation' or Prospero 'persistently interrupt[ing] the conversation between Ferdinand and Miranda'.[95] By constantly chiming in with remarks that prevent the smooth flow of conversation, the two villains also unwittingly mimic the severe weather which plagues the island.

A shifting storm

In *A contemplation of mysteries*, Thomas Hill presents the consequences of thunder for different months in the calendar:

> *Hermes*, an obseruer of celestiall matters wryteth: that the thunders of Ianuarie, to promise mightie windes, and the aboundaunce of corne. The thundring in February, threatneth the death of manye, and especially of the riche. The thundring in the Moneth of March, portendeth mightie windes, firtylitie, and populare battels. The thunders of Aprill, promise a good and ioyous yere. The thundering in May, thretneth famine and scarcitie. The thunders of Iune, signifie a fruitfull yéelde, and the plentie of victuals. The thunders of Iuly, promise a plentifull increase of victuals, but the scarcitie of fruites. The thundring in August, portendeth a prosperous estate of the common welth, but many diseases and sicknesses. The thunders of September, signifieth firtilitie, but the death of the Nobilitie.[96]

In *The Tempest*, however, things are far from clear. Ceres excludes winter from the cycle of seasons and wishes that Ferdinand and Miranda lived in a 'brave new world' (5.1.183) of eternal spring.[97] Shakespeare bridges the gap between temperate and rough weather and allows opposites to coexist or coincide on the island in the name of a complex interaction between the darker and lighter sides in nature and mankind. '[T]his thing of darkness I / Acknowledge mine' (5.1.275–6), Prospero finally says of Caliban after he has forgiven his enemies and abjured his magic.

Such interaction is rendered by the recurring storm, which associates light and darkness and carries positive as well as negative connotations in the play. In 1999, Leslie Thomson argued that the stage direction '*thunder and lightning*' had a fixed meaning in early modern plays and functioned as a signal for the production of stage effects from the tiring house aimed at establishing a supernatural atmosphere.[98] Since then, this reading has been nuanced, and rightly so, for *The Tempest* shows, if anything, that the significance of the storm changes from scene to scene. Shakespeare adopts here a number of different – and sometimes antagonistic – perspectives, which he finally reconciles in the space of his tragicomedy.

In the first scene, the storm looks natural enough, which is not the case in *Macbeth*, for instance, because the concomitant presence of the weird sisters points to its supernatural origins. However, such naturalness is almost immediately questioned, as in act 1, scene 2 we learn that Prospero has been 'raising this sea-storm' (1.2.177). Like the witches in *Macbeth*, he has 'untie[d] the winds and let them fight' (*Macbeth*, 4.1.68). In Jacobean England, the belief that some individuals could harness supernatural powers in order to effect changes in the natural world was still held, and scholars like John Dee and Cornelis Drebbel were thought capable of controlling the weather.[99]

Prospero seemingly interacts with the natural world as a man of science able to control the elements, as Miranda suggests: 'If by your art, my dearest father, you have / Put the wild waters in this roar, allay them' (1.2.1–2). Yet it is hard to decide whether he, as an early modern magus, devotes most of his time to the pursuit of wisdom or if he, as a necromancer dabbling in black arts, is mainly interested in power and manipulation.

Compared to a 'sorcerer' (3.2.41) by Caliban, he must share a number of features with the dead witch Sycorax, once known for her 'sorceries terrible' (1.2.264). In early modern Europe, witches were often held responsible 'for the high frequency of damaging climate anomalies, especially in winter', and some of them had been burnt

Fig. 14 A witch causing a storm. Woodcut in Olaus Magnus's *Historia de gentibus septentrionalibus*, Rome: Johannes Maria de Viottis, 1555, p. 117. By courtesy of Cornell University Library, Division of Rare and Manuscript Collections.

for changing the climate.[100] In Scotland, the most famous case of early modern weather-working witchcraft remains that of the North Berwick witches who, in 1589, presumably raised several storms in order to drown King James VI and Anna of Denmark on their way home. In his *Symbolaeographie* (1594), William West of the Inner Temple writes that '[witches] shake the air with lightnings and thunder, to cause hail and tempests, to remove green corn or trees to an other place, to be carried of her familiar which hath taken upon him the deceitful shape of a goat, swine, or calf etc. into some mountain'.[101] West, however, makes a sharp distinction between witches, poor and old women, and 'soothsaying wizards', that is, men consorting with the upper circles who 'divine and foretell things to come and raise up evil spirits by certain superstitious and conceived forms of words'.[102]

Oddly enough, Prospero occupies an in-between position, partly because he has taken up Sycorax's prerogatives, and partly because, in his books, he has learnt how to raise spirits, either good or evil. The wizard's portrait gradually darkens from scene to scene, notably through the character of Caliban, the magus's 'id', or darker self, as it were. In act 2, scene 2, seemingly contributing to the deforestation of the island under the orders of his new master, the so-called monster enters 'with a burden of wood' (2.2.0 s.d.) and 'a noise of thunder' (2.2.3 s.d.). Afterwards, he is heard cursing Prospero, and a little later in the same scene, Trinculo hears the clap of thunder and exclaims: 'Alas, the storm is come again!' (2.2.36). As Gwilym Jones cogently puts it, the play posits early on the 'supernatural origin' of the thunder, then 'consolidates' this suggestion 'by directing the sound effects to be produced again', and, as a result, Shakespeare creates a 'tension between natural and supernatural', causing the natural to be 'subsumed by, and subject to the work of, the supernatural'.[103]

Plays-within: From iridescence to dissolution

The tension increases during the interrupted banquet scene of act 3, scene 3, when the storm once again echoes the spectacular tempest of the beginning but is endowed with yet another dimension, a spiritual one this time, so that critics have defined this passage as 'a commonplace theophany which revels in its theatrical tricks'.[104] While it is true that '[t]hunder and lightning' (3.3.52 s.d.) are concomitant with a supernatural apparition, since Ariel is said to enter 'like a harpy' (3.3.52 s.d.), there is no truly divine manifestation here, as the playwright makes it clear that the harpy is none other than Ariel in disguise. In fact, the harpy theme is principally climatic, at least when Ariel first appears; at any rate, it is not the result of a simple coincidence.

Much has already been said about the Virgilian dimension of this particular episode and of the play in general. While this has widened our understanding of *The Tempest*, it has contributed to downplay other sources, namely the Greek ones, resorted to by the playwright. In *King Lear* and in *Anthony and Cleopatra*, Greek atomism probably shaped some of the philosophical underpinnings of the plays due to its interests in nothingness and the concept of *clinamen* as etymologically related to the word 'climate'. I now suggest that, in *The Tempest*, we should reconsider the part played by Homer's *Odyssey* (20.61–82), a poem which, like *The Tempest*, centres on a 'romance wanderer'[105] and abounds in references to the states of the sky and to weather-related issues,[106] and where harpies[107] (Ἀρπυῖαι) and stormy winds appear to be interchangeable.[108] For Homer, harpies indeed personified whirlwinds or hurricanes carrying people away,[109] while for Hesiod they are said to be the sisters of Aello and Iris. One should therefore see in the harpy-like Ariel the embodiment of the 'mutinous winds' (5.1.42) alluded to by Prospero at the close.

The religious intertext of this set piece then overlaps with the meteorological sequence, but only when Ariel begins to speak.

Indeed, the disguised spirit accuses Alonso, Sebastian and Antonio of being 'three men of sin' (3.3.53) and he explains why they cannot fight against the retributive 'elements' of the island:

> [. . .] I and my fellows
> Are ministers of Fate – the elements
> Of whom your swords are tempered may as well
> Would the loud winds, or with bemocked-at stabs
> Kill the still-closing waters, as diminish
> One dowl that's in my plume. My fellow ministers
> Are like invulnerable. (3.3.60–6)

Because Shakespeare's contemporaries believed that the world was made of the four elements,[110] Ariel insists that fire and earth (i.e. 'the elements / Of whom [the] swords are made'), associated with the ship's passengers, cannot vanquish the elements of wind and water, at the origin of storms and tempests. Winds and waters thus act as the natural agents of justice and they relentlessly strike the guilty insofar as the storm is characterised by its punitive nature and can henceforth be interpreted as a sign revealing the presence of a divine power to an amazed audience of shipwrecked survivors. The vocabulary used ('sin' and 'Destiny' l. 53, 'good' l. 70, 'requite' l. 71, 'innocent' l. 72, 'Ling'ring perdition' l. 77, 'wraths' l. 79) points to a religious presence behind Ariel's stormy show. After Ariel has 'vanish[ed] in thunder' (3.3.82 s.d.), followed by the other shapes, the stunned Alonso acknowledges that 'the thunder' actually 'pronounced / The name of Prosper' and 'did bass [his] trespass' (3.3.97–9). So, each time Alonso's conscience begins to nag him, the bass voice of thunder enhances his guilt, while the music of the storm must be contrasted with the 'heavenly music' (5.1.52) invoked by Prospero when he abjures his magic at the end of the play. In 'The Storm', Herbert asserts that 'tempestuous times / Amaze poore mortals, and object their crimes' (ll. 5–6). The poet seems to suggest that it is less the storm itself that provokes guilt than it is in fact men's sense of guilt which triggers the storm when he writes that 'A throbbing conscience spurred by remorse / Hath a strange force' (ll. 9–10). This somewhat corresponds to what happens in act 3, scene 3 of *The Tempest*, in which it is impossible to decide whether Alonso is made to realise his sins because of Ariel's new storm, or if the latter is a mere illusion prompted by Alonso's pang of conscience. On the one hand, the tempest of the beginning seems real as Alonso and his court find themselves stranded on an island in the middle of nowhere. On the other hand, Prospero keeps resorting to illusion and deception, so much so that, when the boatswain narrates how he has finally succeeded in finding Alonso in the play's last act, his speech is saturated with references to 'sleep' ('We were dead of sleep', 5.1.230) and 'dream' ('Even in a dream, were we divided from them', 5.1.239).

It has often been argued that the tempest is a mere illusion created by a god-like Prospero,[111] so that, as in Herbert's 'The Bag', the storms simply illustrate 'the triumph of his art' ('The Bag', l. 5).[112] Actually, Shakespeare's *tour de force* consists in the way he reconciles shadow and substance, science and magic. Whether actual storms do or do not occur in *The Tempest* matters little; what the play is really about is the problematic nature of illusion. The playwright thus makes the weather issue come in handy as a way of challenging the reality/fiction dichotomy. This is also emphasised by the interrupted masque, which presents yet another level of reality and which provides a commentary on the climatic forces at work in the play.

In the masque, the main characters are Ceres, the goddess of wheat and harvest, and Iris, the rainbow goddess, whose appearance may have led some members of the audience to entertain reasonable hopes of oncoming good weather on Prospero's island.[113] In *The Fountaine of Ancient Fiction* (1599), Richard Linch translates Vincenzo Cartari and explains that Iris is the daughter of Thaumante, who has the power to command 'the changes and alterations of the aire, making it sometimes faire, sometimes tempestuous, rainie, and cloudie' and who, as such, can send down 'haile, snow, thunder, and lightning'.[114] This explanation justifies the arrival of Juno in the masque (4.1.102 s.d.), which can also be seen as a wink at Ben Jonson's *Hymenaei*, performed in 1606 for another (real) marriage, that of the Earl of Essex and Frances Howard.[115] Indeed, if Iris's mother was responsible for the alterations of the weather, Juno was able to predict these changes. Stephen Batman, who depicts the latter as 'sitting under a Raynbow, with a Crowne of white Lyllies on her head, a Scepter in her right hand, and in her lefte hande a payre of Sheares, on ech side a Peacoke',[116] gives a purely climatic explanation of such a portrait:

> By *Iuno* is mente the Ayre, the Raynbowe betokeneth Rayne, whereof it is a token. Her Crowne of Lyllies, signifieth the cause to bee in the Ayre, that ye Earth doth yelde foorth sweete Flowers: her Scepter is a representation of the power that is geeuen to the Ayre, aswel for preseruation, as encreasing of natural thinges. Her Sheares in her left hand sheweth Hotenesse and Moystnesse of the Ayre, into which two qualities it is sundered and parted. The two Peacockes are added,

> because theyr cry doth prognosticate chaunge of Weather. The Lyon is a figure of the Sunne, and the Lawne, of thynne Uapours: for the force and vertue of the Sunne is greatest, when hee entreth into *Leo*: hereby aduertising, that the heate of the Sunne draweth thyn vapors into the Ayre, where beeing congealed, by the extreeme colde of the middle Region, they are agayne by the scortchinge heate of the Sunne resolued into droppes.[117]

While the peacocks are absent from the pageant, the arrival of Juno truly coincides with a change of weather since, from this time onwards, no other storm ever occurs in the play.

Oddly enough, the masque does not include Venus and Cupid, as Ceres refuses the society of the goddess of love and her blind son on the grounds that they were both involved in the abduction of her daughter. Fortunately for her, the goddess of love is said to be busy finding her way through the clouds and will not be present (4.1.91–4). Of course, the haunting threat of Proserpina's rape (which accounts for Ceres's anger) echoes Caliban's own attempt against young Miranda (1.2.347–8) and it aims at showing the audience that festive pageantry does not preclude darker issues to come. Yet the masque is characterised by an ideal and idealised landscape with no winter in it (4.1.114–15), as in Virgil's fourth *Eclogue*.[118] It is ironic, then, that during the masque, old Gonzalo should be seen as the visual embodiment of winter in Ariel's description ('His tears runs down his beard like winter's drops', 5.1.16). The stereotyped image of old age was commonly associated with winter and Ariel explicitly affiliates the old man with this particular season. For the first time in the play, however, we feel no desire to draw a satirical portrait of the councillor, whose naive and irritating enthusiasm has now given way to despair. Ceres's blessing and promise of a perpetual spring are therefore belied by the predicament of the old man and the necessity of restoring a complete seasonal cycle is then insistently put to the fore.

The pageant itself may be seen as the overall illusion created by strangely shaped clouds, that is, by a meteorological phenomenon remarkable for its instability and its poetic nature. Indeed, as Prospero pronounces his disillusioned speech on revelry and actors as 'spirits' being 'melted into [. . .] thin air' (4.1.148–50), he alludes to 'cloud-capped towers', 'palaces' and 'the great globe itself' about to 'dissolve' and '[l]eave not a rack behind' (4.1.152–6). While in *A Midsummer Night's Dream* 'far-off mountains' were 'turned into clouds' (4.1.186–7), things work the other way round as the clouds morph into towers and palaces. Of course, just as in *Anthony and Cleopatra*, what is described here is a typical Lucretian phenomenon: the transformed clouds we see in the sky are *simulacra* produced in the atmosphere, distorted by the air, causing our senses to err.[119] These clouds are nothing but travelling idols gliding though the air, and we 'must admit / That bodies are sent forth with power to strike / The eyes and provoke vision'.[120] Shakespeare here rehashes a well-known *topos*, but he subtly adapts it by rehabilitating the power of the dream. What Lucretius saw as false ideas is being redefined by the playwright as representations (in the theatrical sense of the word) that help man to live through the various ordeals of human existence.

The flexibility of the verse in the above-quoted passage perfectly conveys the elusive quality of the air. The actor playing Prospero is indeed forced to mimic the airy evanescence that is the very focus of the speech by taking his breath before 'into thin air' (4.1.146). This will then enable him to 'laun[ch] into elaborated simile'.[121] The masque, Prospero says, has left 'not a rack behind'. A 'rack' here evokes some sort of mist driven by the wind (*OED*)[122] and it also suggests 'the cloud effects of masque scenery'.[123] This scenery was crucial to Shakespeare's courtly audience, accustomed as it was to a spectacular environment. Clouds, in the dramatic economy of the Jacobean masque, were particularly important since they served to transport deities and supernatural creatures while introducing a form of continuity with the maskers' vaporous costumes. Besides, they were also used to hide an all too visible machinery and to create a varied and contrasted background that gave spectators the sense of a dense, smoky mass shifting in the heavens.[124]

As it happens, Ben Jonson's *Hymenaei*, mentioned above in connection with Juno, and *The Haddington Masque*, performed in 1608, both introduce their dancers in a 'great globe' similar to the one mentioned in Prospero's 'revels speech' (4.1.153).[125] In *Hymenaei*, it consisted of a huge rotating microcosm containing eight men representing the Humours and the Affections. In *The Haddington Masque*, a revolving celestial sphere contained the twelve masquers placed in a band representing the zodiac. Moreover, the 'cloud-capped towers' (4.1.152) evoked in Prospero's lines may also be picked up from *Hymenaei*, as the upper part of the scene was built with clouds carrying the female masquers to the floor. Jonson himself indicated that they were 'made artificially to swell and ride like the rack'.[126]

Prospero's great 'Our revels now are ended' speech is thus partly indebted to Ben Jonson, and during the performance of the play at court, it was probably taken as an ironical tribute. For when

Prospero evokes the material racks of the pageant, he simultaneously refers to his art as an illusion[127] – something never really acknowledged by Jonson, who had much more serious views of his own plays. Shakespeare, however, does not simply lift his ideas from Jonson's masque, as the comparison of clouds to fantasies and the mention of 'racks' are already present in his *Anthony and Cleopatra* when, shortly before his death, Anthony associates deception with 'dragonish' clouds and argues that men see substance where there is actually none (*Anthony and Cleopatra*, 4.14.2–11).[128] In *The Tempest* as in *Anthony and Cleopatra*, everything melts and dissolves, and life is defined by its mutability.[129] How could one better characterise the fundamental instability of human existence, whose fortunes and misfortunes are just as hard to predict as the weather?

Such an idea of life as a *trompe l'oeil* must of course be related not only to 'the transformation scenes in court masques'[130] but also to the specific venue chosen for the performance of *The Tempest*. Written for the Blackfriars (and probably for the court), *The Tempest* capitalises on increased visual possibilities. The storm provides Shakespeare with one of his most spectacular openings, and as such it clearly had to be recreated in the course of the performance. In this context, the amazement of the first audiences perhaps unwittingly mimicked that of the onstage characters. We have already noted in Chapter 5 that thunder was elaborately staged in amphitheatres: it was 'simulated by rolling a cannon-ball down a wooden trough, lightning by cut-out zig-zags that were briefly visible or flashing light through a shaped mask'.[131] In the context of a closed playhouse, we can imagine that a grand spectacle was less necessary: the sole production of artificial lightning must have been efficient. When performed at court, *The Tempest* probably impressed its royal audience with its numerous sound effects, all the more so as, 'owing to its solid walls and reverberant roof', the Banqueting House provided exceptional acoustics which 'modif[ied] the player's method of delivery, substituting to that loud, declamatory, blatant style of histrionics [. . .] a restrained and quieter style'.[132] The intimate atmosphere of the island, teeming with sweet songs, daily noises and loud rumblings of thunder, must have been particularly well rendered in such conditions.

First and foremost a spectacle, *The Tempest* is also a stimulating playtext. While it allows us to engage with both natural and supernatural phenomena, with illusion and reality, its constant emphasis on foul and fair weather allows several perspectives to coalesce and to form what finally appears as a vast existential pattern. This meteorological and lay morality play deals with such issues as the hidden dangers of so-called temperance, the (im)possibility of redemption, the existence of the supernatural, the limits of scientific experiment, the quest for the philosopher's stone, and the (im)materiality of pageantry. While Shakespeare makes all these questions open to debate, he fleshes out his vision of Prospero's art – and, by extension, his own – by drawing on powerful weather imagery and by exploring its discursive, symbolic and spectacular levels. Storms and tempests, he suggests, be they confined in the little world of man or represented in nature at large, teach us to know ourselves as they force us to balance our emotions with a sound understanding of environmental issues.

Significantly, in spite of the 'line-grove' which is said to 'weather-fen[d]' Prospero's cell (5.1.10), Prospero himself is far from being immune to the weather. If he appears as an anti-Lear of sorts at the beginning of the play, seemingly experimenting with the elements and dominating the weather (mainly thanks to his *grimoires* which enable him to command natural forces) from the safety of land, one gradually realises that the island's stormy climate is an analogue of Prospero's own mindscape. His 'bare island' (5.1.326) represents nothing but the bare stage of the playhouse upon which he has projected his passions and anxieties. So, even before it batters the approaching ship which carries his enemies, the tempest is primarily his own as a tormented man who harrasses Caliban and enslaves Ariel accordingly.

But we can also choose to regard the climate of the island as a mirror of the geopolitical context of the play. In *The Tempest*, legend, geography and weather all serve to conceal a daunting fiction concerned with the possible forms of sovereignty in a hostile milieu. Prospero's island is an anti-utopia where, far from being 'protected from the winds', the sea is as rough as its political climate is poisonous.[133] The omnipresence of the storm points to the necessity of purging the world – and the political sphere in particular – of its foulness. 'The finall causes and effects of the thunder, are well wayed and knowne in that the ayre first by the thunder is purged, and the euill vapours consumed, yea the pestilence and other contagiousnesse often clensed and put away', Hill reports in *A contemplation of mysteries*.[134] In 'The Storm', Herbert was to declare that 'Poets have wrong'd poore storms: such dayes are best; / They purge the aire without, within the breast' (ll. 17–18). If poets have indeed wronged poor storms in their lines, Shakespeare's *Tempest* surely stands as a superb exception to the rule. T. S. Eliot understood it better than anyone else since, in the 'What the thunder said' section of *The Waste Land*, he emphasises the redemptive power of art and reappropriates the voice of the storm. Doing so, he pays an admiring tribute to his illustrious predecessor in the poetic use of a renewed apocalyptic meteorology that fuses eastern mythologies with those of the western world in order to call for some sort of rebirth and solution to its slow, depressing decay.[135]

Notes

1. George Herbert, 'The Storm', ll. 1–6, in *The Temple* (1633).
2. Interestingly, since *The Tempest* is ranked first among the 'Comedies' in the 1623 Folio, the very first annotation of the Kodama Memorial Library copy calls attention to the climatic phenomenon of the tempest: 'feare and confusion in sea tempest' (to which the early modern commentator adds, later on, that '[c]ounsellors can not command the Weather'). See Akihiro Yamada (ed.), *The First Folio of Shakespeare*, p. 6, A1, a [TLN 1–45], and A1, b [TLN 46–87]. Online edition at <http://shakes.meisei-u.ac.jp/ALL.html#45> (last accessed 18 April 2018).
3. Gwilym Jones, 'Storms Effects in Shakespeare', p. 41. See also above, Chapter 5, p. 166.
4. In this regard, the visual experience allowed by a performance of the play at court – the only record we have is that it was produced twice at Whitehall (1 November 1611 and 20 May 1613) – may have been strongly enhanced by the display of painted cloths. As Rebecca Olson suggests, 'although we have not traditionally imagined plays like *The Tempest* as being performed against a "set" – a canvas painted to represent a seacoast, for instance – it is entirely possible that at court, it was not only Shakespeare's language that set the scene'. See Rebecca Olson, 'Painted Cloths and the Making of Whitehall's Playing Space, 1611–1612'.
5. Pseudo-Aristotle, *Cosmic Order and Divine Power*, p. 33.
6. We learn that Prospero was 'rapt in secret studies' (1.2.77) when he was Duke of Milan.
7. The Latin *tempestas* (gen. *tempestatis*) referred to 'commotion, disturbance' and was related to *tempus*, 'time, season'. Climate thus proves an obsession in *The Tempest*, a play in which most of the characters are subject to the force of the elements.
8. See Niccolò Machiavelli, *The Prince*, chap. XXIV, pp. 77–8. In this chapter, Machiavelli denounces the rulers who prefer to ignore the possibility of a storm in quiet times and who, as a result, must blame their indolence rather than Fortune (p. 78). On this, see Nathan Schlueter, 'Prospero's Second Sailing: Machiavelli, Shakespeare, and the Politics of *The Tempest*', p. 182.

9. Gwilym Jones, 'Storm Effects in Shakespeare', p. 50.
10. On plantation, deforestation and the obsession with logs in *The Tempest*, see Vin Nardizzi's chapter '"There's wood enough within": *The Tempest*'s Logs and the Resources of Shakespeare's Globe', in *Wooden Os: Shakespeare's Theatres and England's Trees*, pp. 112–35. Nardizzi, in particular, correlates the erection of Prospero's theatre to the clear-cutting of the island's woods.
11. On Shakespeare's own attitude towards enclosures, see Charlotte Scott, *Shakespeare's Nature: From Cultivation to Culture*, pp. 218–19.
12. Arthur Standish, *The commons complaint*, 'The Epistle Dedicatorie', sig. A2v.
13. Wayne Neely, *The Great Bahamas Hurricane of 1929: The Story of the Greatest Bahamian Hurricane of the Twentieth Century*, pp. 49–50.
14. See above, Chapter 5, p. 152.
15. A similar motif is found in François Rabelais's *Quart livre* (1552), in which Panurge goes through a terrible tempest and declares: 'De quatre elemens ne nous reste icy que feu et eau' ('of the four elements, what is left to us here is fire and water', my translation). See *Le quart livre*, in *Les Cinq Livres de faits et dits de Gargantua et Pantagruel*, chap. XX, p. 1004.
16. In the 1551 Latin edition, Gemma Frisius's *Charta cosmographica* is placed between sheets Ai v° and Aij r°; in that of 1584, it is situated between pages 72 and 82. For further details on Gemma Frisius's map, see Fernand Hallyn, *Gemma Frisius, arpenteur de la terre et du ciel*, pp. 43–53 and pp. 66–7.
17. Richard A. Anthes, *Tropical Cyclones: Their Evolution, Structure and Effects*, p. 10.
18. Bartolomé de las Casas, *The Spanish colonie, or Briefe chronicle of the acts and gestes of the Spaniardes in the West Indies*, trans. M.M.S., 'Of the Ile of hispaniola', sig. B: the governor's ship 'was lost upon the sea, and there were with him drowned many Spaniardes' as a result of God's wrath.
19. See Girolamo Benzoni's 1594 *Americae Pars Quarto*.
20. The engraving is reproduced in Kerry Emanuel, *Divine Wind: The History and Science of Hurricanes*, p. 31.
21. Ibid. p. 30.
22. Wayne Neely, *The Great Bahamas Hurricane of 1929*, p. 50. Neely explains that 'the Indian word was pronounced 'Furacán' or 'Furacánes' during the early years of discovery and colonization of America'.
23. See above, Chapter 5, p. 150, p. 152, and p. 169, n. 1.
24. The first use of the term would date back to 1555 (*OED*, 1.a).
25. Randle Cotgrave, *A Dictionarie of the French and English Tongues*, 'TEM', n.p.
26. See José de Acosta, *The naturall and morall historie of the East and West Indies*, chap. 16, sig. E2: 'If they [the Indians] came by sea, it was

casually, and by chance, or willingly, & of purpose. I understand by chance, being cast by force of some storme or tempest, as it happens in tempestuous times. I mean done of purpose, when they prepared fleetes to discover new lands.'

27. According to the reports written by Antonio Pigafetta, who travelled with Magellan in 1519–22, Setebos was a deity worshipped in Patagonia (*First Voyage Round the World by Magellan*, p. 53). Pigafetta's account was appended to the 1555 English translation of Pietro Martire d'Anghiera's *De orbe novo* (reprinted in *The First Three English Books on America*, ed. Edward Arber). Incidentally, Patagonia was then regarded as a land occupied by giants, monstrous creatures who owed their deformity to the particularly cold climate of the place. On this, see André Thévet's 1575 book *La Cosmographie universelle*, vol. IV, book XXI, chap. II, fol. 907v. For an analysis of 'Patagonia's Giants', see Surekha Davies, *Renaissance Ethnography and the Invention of the Human: New Worlds, Maps and Monsters*, chap. 5, pp. 148–82, especially p. 173.
28. This island is situated both in the New World and in the Mediterranean, even if the Bermuda triangle, with its moist climate and dangers, seems here to remain a fairly distant reference. As a result, given the proximity of Italy and the many allusions to Virgil's *Aeneid* in the play, the Mediterranean world remains by far the most important.
29. William Strachy [*sic*], 'A true reportory of the wracke, and redemption of Sir THOMAS GATES Knight; upon, and from the Ilands of the Bermudas [. . .]', 1610, in Samuel Purchas's 1625 edition of *Purchas his pilgrimes*, lib. IX, chap. VI, p. 1738: 'These Ilands are often afflicted and rent with tempests, great strokes of thunder, lightning and raine in the extreamity of violence.'
30. Once again, as in *Measure for Measure* and *King Lear*, the tempest supplies a vivid image of hell while it is also eroticised as a metaphor of women's vaginas, a place where early modern devils were apparently very active.
31. See Robert C. Fulton, '*The Tempest* and the Bermuda Pamphlets: Source and Thematic Intention'.
32. Steve Mentz, 'Airy Spirits: Winds, Bodies, and Ecological Force in Early Modern England', p. 24.
33. On Strachey's description of the hurricane, see Steve Mentz, 'Hurricanes, Tempests, and the Meteorological Globe', especially pp. 262–4 ('The Sea-Venture in 1609'). As Mentz puts it, 'Strachey's description of the storm's arrival contains detail to recognize the typical counterclockwise rotating northeast winds on the left-hand side of the rotating storm.'
34. William Strachy, 'A true reportory', lib. IX, chap. VI, p. 1734.
35. Ibid. p. 1735.
36. Ibid.
37. Ibid. p. 1737.

38. The life of Saint Elmo (also known as Saint Erasmus of Formia), the patron saint of sailors, is told in Jacopo de Voragine's *Golden Legend*. In the early versions of the story, the emphasis was laid on his sermon at sea and on his immunity to lightning. In the *Golden Legend*, the tempest does not arise when the saint is at sea, but when he is tortured. The storm strikes his opponents while he remains protected from the stroke of lightning, an obvious manifestation of divine justice. See John R. Decker, 'Introduction: Spectacular Unmaking: Creative Destruction, Destructive Creativity', p. 6.
39. Thomas Hill, *A contemplation of mysteries*, fol. 21v. 'Castor and Pollux' was the name given by early modern sailors to St Elmo's fire.
40. Ibid. fol. 22r.
41. See Alden T. Vaughan and Virginia Mason Vaughan, *Shakespeare's Caliban: A Cultural History*, p. 40.
42. Gabriel Egan, *Green Shakespeare: From Ecopolitics to Ecocriticism*, p. 153.
43. B. J. Sokol, *A Brave New World of Knowledge: Shakespeare's The Tempest and Early Modern Epistemology*, p. 113.
44. Gabriel Egan, *Green Shakespeare*, p. 153.
45. J. M. Nosworthy, 'The Narrative Sources of *The Tempest*', p. 286.
46. See Ovid, *Metamorphoses*, trans. Arthur Golding, The Eleventh Book, ll. 550–665, pp. 337–40. Drawn from the story of Ceyx and Alcyone, this passage describes a compelling storm, with 'lightnings set[ting] the waves on fire' (l. 605).
47. *Adventspostille 1522*, in Martin Luther, *Werke. Kritische Gesamtausgabe*, X, 1. Abteilung, 2. Quoted in Rienk Vermij, 'A Science of Signs: Aristotelian Meteorology in Reformation Germany', p. 650.
48. George Abbott, *An exposition vpon the prophet Ionah*, 'The III. Lecture', p. 46.
49. In 'Airy Spirits: Winds, Bodies, and Ecological Force in Early Modern England' (p. 23), Steve Mentz pleads for 'an eco-materialist understanding of Ariel', which would 'spea[k] to the interplay between invisible and material forces'.
50. Laurie Shannon, *The Accommodated Animal: Cosmopolity in Shakespearean Locales*, p. 270. On Noah's ark in connection with Shakespeare's *As You Like It*, see above, Chapter 3, pp. 93–4.
51. See for instance the 1613 version of Samuel Purchas's *Purchas his pilgrimage*, The First Booke, chap. VIII, p. 33: 'GOD made a winde to passe in commission, and, as a common umpire, to end their unnaturall strife, forcing the waters into their ancient precincts above and beneath the firmament. (*Ambrose* interpreteth this Winde of the HOLY GHOST [. . .]) Then did the Earth *remember* her first inheritance, beeing freed from the tyrannical invasion and usurpation of the waters. And what could then forget or bee forgotten, when GOD *remembred* NOAH *and all that was with him in the Arke*?'

52. On the connections between Noah's ark and the rainbow, see also Joshua Sylvester, 'Of the Rainebow', in Robert Allott (ed.), *Englands Parnassus*, pp. 353–4: '*Noah* lookes vp, and in the ayre he viewes / A semicircle of an hundred hewes.'
53. See Yves Peyré, 'Iris's "rich scarf" and "Ariachne's broken woof": Shakespeare's Mythology in the Twentieth Century'.
54. See David Scott Wilson-Okamura, 'Virgilian Models of Colonization in Shakespeare's *Tempest*', p. 723. The author notes that in Virgil's *Aeneid*, Book 4, the mention of Iris and her saffron wings is made 'in a description of Iris as she descends to Carthage in order to release the soul of Dido from her dying body'.
55. Pierre Le Loyer, *A Treatise of Specters or Straunge Sights, Visions and Apparitions*, fol. 19r.
56. In the Geneva Bible, the marginal notes to Psalm 107:23–30 focus on the story of the ark and turn the sea into a simultaneously threatening and liberating place. See Johanna Harris, '"Be plyeabell to all good counsell": Lady Brilliana Harley's Advice Letter to her Son', p. 133.
57. The shipwreck is of course a traditional romance motif, and in line with the 'experimental' nature of romances (Michael O'Connell, 'The Experiment of Romance', p. 216), the wind adumbrates the artificiality of the theatre in *The Tempest*'s opening scene. As Michael Neill argues in '"Noises, / Sounds, and Sweet Airs": The Burden of Shakespeare's *Tempest*' (p. 324), if Shakespeare followed William Strachey's description, the wind of the first scene was probably 'singing and whistling', thereby introducing, on an acoustic level, a sharp contrast with the sorry plight of the ship. The wind could therefore be interpreted as an emanation from hell as much as an instrument of providence.
58. Francis Bacon, *Historia ventorum*, in *The Works of Francis Bacon*, vol. 5, p. 154.
59. Thomas Hill, *A contemplation of mysteries*, fol. 39v.
60. Bacon explains that '[t]he west winds are more violent than the east, and do more bend and wrench trees' (*Historia ventorum*, p. 157).
61. Theophrastus, *Concerning Weather Signs*, p. 403.
62. Keith Whitlock, 'Shakespeare's *The Tempest*: Some Thought Experiments', p. 177.
63. The remaining references are in acts 2 and 5. At the beginning of the play, Leontes wants his supposedly bastard child to be abandoned 'to its own protection / And favour of the climate' (2.3.178–9). In the last act, Leontes welcomes Perdita's fiancé, Florizel, with the following words: 'The blessed gods / Purge all infection from our air whilst you / Do climate here!' (5.1.167–9; see 'Climate', *OED*, v.1 *Obs.* 'To reside in a particular region'). As Gabriel Egan notes in *Green Shakespeare*, 'Shakespeare is clearly concerned with how the weather affects biological nature: Delphos' climate makes it fertile and the exposed child will have to take its chances for life or death according to the climate of the place' (p. 125).

64. Terence Spencer, 'Shakespeare's Isle of Delphos', pp. 201–2.
65. John Argyropoulos's Latin version was particularly successful in the sixteenth century even though it dated back to 1457.
66. Aristotle, *Nicomachean Ethics*, VII.9, p. 135.
67. Nicholas Grimalde's translation of Cicero's *De officiis* was published in 1556 and it appeared in a bilingual version two years later.
68. Jennifer Richards, *Rhetoric and Courtliness in Early Modern Literature*, p. 60.
69. John Gillies, 'Shakespeare's Virginian Masque', p. 96.
70. *Airs, Waters, Places*, in *Hippocrates*, XXIV, p. 133.
71. Ibid. p. 137.
72. For an opposite (but similarly patriotic) tradition, see Virgil's encomium of Italy's temperate climate, recalled in Michael Drayton's 1612 *Poly-Olbion*, whose first Song is partly devoted to Albion's temperate climate, p. 1, 1.1–7.
73. William Vaughan, *Approved directions for health*, p. 2.
74. This is also the case of Iago in Shakespeare's *Othello*.
75. Thomas Hill, *A contemplation of mysteries*, fols. 63r–63v.
76. On alchemy, see William R. Newman's *Promethean Ambitions: Alchemy and the Quest to Perfect Nature*, which mentions *The Tempest* in chapter 3, 'The Visual Arts and Alchemy', p. 115 and p. 153. Newman, however, is more interested in the transmutations of flesh into stone than in the role played by the tempest in Shakespeare's tragicomedy. On *The Tempest* as an allegory of alchemy, see Michael Srigley, *Images of Regeneration: A Study of Shakespeare's The Tempest and its Cultural Background*. See also Peggy Muñoz Simonds, '"My charms crack not": The Alchemical Structure of *The Tempest*'. Simonds argues that the ten stages of alchemy 'help to delineate the dramatic structure of *The Tempest*' and that the first stage, that of separation or *divisio*, is repeatedly alluded to at the beginning of the play (through such words as 'split' and 'divide'), when Ariel causes the tempest and when, later on, he describes the shipwreck scene to Prospero (p. 542).
77. See Virginia Mason Vaughan and Alden T. Vaughan (eds), *The Tempest*, p. 63.
78. On purgative storms, see King John's remark in Shakespeare's eponymous play: 'So foul a sky clears not without a storm' (4.2.108).
79. Kathleen P. Long, 'Odd Bodies: Reviewing Corporeal Difference in Early Modern Alchemy', p. 77.
80. Ben Jonson, *The Alchemist and Other Plays*, p. 219: '*Subtle* [*to Face*]: I'll thunder you in pieces. I will teach you / How to beware to tempt a fury again / That carries tempest in his hand and voice.' For the view that it was Shakespeare who influenced Jonson, see David Lucking, 'Carrying Tempest in his Hand and Voice: The Figure of the Magician in Jonson and Shakespeare'.
81. Theseus's monologue in the last act of *A Midsummer Night's Dream* similarly insists on the importance and power of poetic illusion while

stressing the liberty of the poet to refashion the state of sky and earth according to his fantasy or to the vagaries of his imagination. On this, see above, Chapter 1, p. 49.

82. For the association of vents with cleansing and purgation, see 'vent', *OED*, v.2, I.1.a. The link with cleansing is made clear in the first recorded use, dating back to 1398, which is attributed to John Trevisa's *De proprietatibus rerum* xi.i. (Tollem. MS.), translated by Bartholomew de Glanville: 'And so eyer is element of bodies and spirites, for ventynge of eyer comynge to spirites is cause of clensynge and of purgacion.'
83. Quoted in Gail Kern Paster, 'Becoming the Landscape: The Ecology of the Passions in the Legend of Temperance', p. 138.
84. On the mutual contacts between body and environment, see Mary Floyd-Wilson and Garrett A. Sullivan Jr (eds), *Environment and Embodiment in Early Modern England*.
85. Aristotle, *Meteorologica*, II.8, p. 217.
86. Ibid. pp. 217–19.
87. For further details, see Craig Martin, 'Ludovico Settala's Aristotelian *Problemata* Commentary and Late-Renaissance Hippocratic Medicine', p. 34.
88. William Fulke, *A goodly gallerye with a most pleasaunt prospect*, fol. 24.
89. Thomas Hill, *A contemplation of mysteries*, fol. 34r.
90. On this, see François Laroque, 'En marge de l'idéologie: antimasque et grotesque dans le *Dr Faustus* et *La Tempête*', pp. 108–9.
91. Rebecca Totaro, '"Revolving this will teach thee how to curse": A Lesson in Sublunary Exhalation', p. 137.
92. Thomas Elyot, *The image of gouernance*, Cap. xxi, p. 39.
93. It occurs in *King John*, where Cardinal Pandolph mentions 'natural exhalation[s] in the sky' (3.4.153), and in *Henry VIII*, where Cardinal Wolsey says he 'shall fall / Like a bright exhalation in the evening' (3.2.227). Cognate words such as 'exhale' also crop up elsewhere in the canon.
94. Rebecca Totaro, '"Revolving this will teach thee how to curse"', p. 138.
95. A. Lynne Magnusson, 'Interruption in *The Tempest*', p. 54.
96. Thomas Hill, *A contemplation of mysteries*, fols. 51v–52r.
97. See Virginia Mason Vaughan and Alden T. Vaughan (eds), *The Tempest*, p. 71.
98. Leslie Thomson, 'The Meaning of Thunder and Lightning: Stage Directions and Audience Expectations', pp. 11 and 18.
99. See John Aubrey, *Brief Lives*, p. 90. Aubrey writes that Faldo, an old woman from Mortlake in Surrey, claimed that John Dee could cause a storm to abate. On the connection between black magic and weather-working witchcraft, see above, Chapter 1, p. 42, and p. 54, n. 60.

100. 'This practice was only suppressed by the Lutheran Church insisting that God alone, not human beings, was responsible for the weather, a teaching consilient with the dominant pre-Enlightenment mind', Mike Hulme writes in 'The Conquering of Climate: Discourses of Fear and their Dissolution', p. 7.
101. Quoted in C. H. L. Ewen, *Witch Hunting and Witch Trials*, p. 23.
102. Ibid. See also M. C. Bradbrook, 'The Sources of *Macbeth*', pp. 41–2.
103. Gwilym Jones, 'Storm Effects in Shakespeare', p. 48.
104. Ibid. p. 50.
105. John Dean, 'The *Odyssey* as Romance', p. 234.
106. See C. A. Maury, 'Phenomena of the *Heaven* in the Odyssey'.
107. On harpies, see also John Harrington's poem 'Of Harpies' in Robert Allott's 1600 *Englands Parnassus*, pp. 352–3. Interestingly, these lines are preceded by a short poem devoted to the wind and attributed to Joshua Sylvester (p. 351). On page 350, yet another poem attributed to the same author is entitled 'Of Deluge'. As to Sylvester's 'Of the Rainebow', it appears on page 353 of Allott's anthology. It seems that Shakespeare closely follows this weather nexus.
108. See George Chapman's 1615 translation, *Homer's Odysses*, where the word 'harpies' crops up three times (The First Booke, p. 9; The Fourteenth Booke, p. 221; The XX. Booke, p. 310).
109. Telemachus notably says that his father has been snatched away by the Harpies (*Odyssey*, 1.241).
110. On this, see Shakespeare's Sonnet 45 ('The other two, slight air and purging fire, / Are both with thee, wherever I abide; / The first my thought, the other my desire, / These present-absent with swift motion slide', ll. 1–4).
111. In *Shakespeare and Ecocritical Theory*, Gabriel Egan writes that 'the ship by which the Italians will leave the island after the end of *The Tempest* is found whole and "in all her trim" (5.1.239) because it was not split in the first place: its apparent destruction in a storm was an illusion' (p. 9). Yet, as noted by Mickaël Popelard in 'Unlimited Science: The Endless Transformation of Nature in Bacon and Shakespeare' (p. 190), 'maintaining that Prospero's storm is "insubstantial" makes it difficult to account for the [. . .] two lines spoken by Ariel: "On their sustaining garments not a blemish, / But fresher than before" (1.2.218–19)'. Moreover, Gonzalo echoes these lines in the following act (2.1.61–4).
112. Herbert's poem must be read in connection with Matthew 8:23–7, in which Christ's disciples are frightened by a storm while Christ himself is sleeping. See Gene E. Veith, *Reformation Spirituality: The Religion of George Herbert*, p. 263.
113. In his *Tetrabiblos*, Ptolemy writes that '[r]ainbows signify storms after clear weather and clear weather after storms' (quoted in Liba Taub,

Ancient Meteorology, p. 36). See also Leonard Digges, *A prognostication of right good effect fructfully augmented*, sig. B.ii: 'If in the morning the raynebow appere, it signifieth moisture [. . .] If in the euening it shewe it self, fayr weather ensueth.' If we follow Digges's logic, we may suppose that the rainbow in *The Tempest* appears in the evening, as the author writes a few lines further on that 'Thundres in the morning signifie wynd'.

114. Vincenzo Cartari, *The Fountaine of Ancient Fiction*, sigs. Liiv–Liiir.
115. See Andrew Gurr, 'Another Jonson Critic'. In Jonson's *Hymenaei*, the governing goddesses are Iris and Juno. As Gurr suggests, Shakespeare borrows the hyperbolic language which Jonson uses to describe Juno.
116. Stephen Batman, *The golden booke of the leaden goddes*, fol. 1v.
117. Ibid. fols. 1v–2.
118. Cf. Ferdinand's remark: 'So rare a wondered father and a wife / Makes this place paradise' (4.1.123–4).
119. On this, see above, Chapter 6, p. 196.
120. Lucretius, *De rerum natura*, Book IV, p. 136.
121. Jonathan Bate and Eric Rasmussen (eds), *The Tempest*, introduced by Jonathan Bate, p. 13.
122. 'Rack', *OED*, n.2, 2.a: 'A mass of cloud moving quickly, esp. above lower clouds; a mass *of* such cloud. Also occas. *fig.*' and n.2, 2.b: 'A bank of cloud, fog, or mist; a wisp of cloud or vapour. Also as a mass noun: mist, fog; sea spray.' *The Tempest* is listed in the examples illustrating the latter meaning.
123. See Stephen Orgel (ed.), *The Tempest*, p. 181, n. to 4.1.156. In 1638, Nicola Sabbattini pointed out the importance, for scenery designers, to be able to create clouds when the plot requires it (*Pratiques pour fabriquer scènes et machines de théâtre*, chap. 39, pp. 133–4).
124. On this, see Margaret Llasera, *Représentations scientifiques et images poétiques en Angleterre au XVIIe siècle. À la recherche de l'invisible*, p. 147.
125. The following development on *Hymenaei* and *The Haddington Masque* owes much to Martin Butler's '*The Tempest* and the Jonsonian Masque'. On the dissolving globe of *The Tempest*, see John Demaray, *Shakespeare and the Spectacles of Strangeness: The Tempest and the Transformation of Renaissance Theatrical Forms*, pp. 91–2. Demaray contends that the globe produced in *Hymenaei* and in *The Haddington Masque* was actually repainted and used onstage for the Whitehall performances of *The Tempest* in 1611 and 1613.
126. References are to the 2012 *Cambridge Edition of the Works of Ben Jonson*. See *Hymenaei* (vol. 2), ll. 188–9; *Haddington* (vol. 3), ll. 216–22.
127. On cloud machines on the early modern stage, see Janette Dillon, 'Chariots and Cloud Machines: Gods and Goddesses on Early English Stages'.
128. See above, Chapter 6, pp. 192–3.

129. In 'Of Mites and Motes: Shakespearean Readings of Epicurean Science', Jonathan Pollock duly notices that such passages have an Epicurean tone (p. 130).
130. John Wilders (ed.), *Antony and Cleopatra*, p. 255, n. to 4.14.10.
131. John Lennard, *William Shakespeare, King Lear*, p. 97. On the staging of storms, see above, Chapter 5, pp. 166–7. See also Gwilym Jones, *Shakespeare's Storms*, pp. 35–6.
132. Ernest Law, 'Shakespeare's *Tempest* as Originally Produced at Court', p. 156.
133. Thomas More, *Utopia*, Book Two, p. 57: 'The sea flows between these [two ends] into a huge bay protected from the winds by the encircling land, which is mostly not rough but calm, like a huge lake. Consequently almost the entire inner part of the island serves as a harbour, ships crossing it in all directions, to the general advantage of the natives.'
134. Thomas Hill, *A contemplation of mysteries*, fol. 48r.
135. On the correspondences between *The Waste Land* and *The Tempest*, see Gayle Greene, 'Shakespeare's *Tempest* and Eliot's *Waste Land*: "What the Thunder Said"'.

Conclusion: 'Under heaven's eye'

'The ayre of England is temperate, but thicke, cloudy and misty', Fynes Moryson writes in his 1617 travel diary, *An itinerary*.[1] This casual observation displays the tensions which Shakespeare's contemporaries must have felt, torn as they were between the love of their country and the acute awareness of its climatic imperfections, between what they longed for and what they actually went through. For Elizabethans and Jacobeans alike, climate could *durably* affect the nature of men whose humours were *temporarily* modified, through the skin, by good or bad weather spells. Admittedly, the debate was not then exactly formulated along those terms, but this rough recapitulation encapsulates the general ideas that circulated in many works of the period. Climate and weather-related topics were never thought of as being totally independent from a broader frame of references including cosmological and theological issues. People saw themselves as living 'under heaven's eye' (*The Comedy of Errors*, 2.1.16), as England's capricious skies were then believed to provide a number of keys to predict or simply understand the main threads of their daily existence. Even if, according to Cassius, '[m]en at sometime were masters of their fates' and '[t]he fault [. . .] is not in our stars, / But in ourselves' (*Julius Caesar*, 1.2.140–2), stars in Shakespeare are often presented as uncanny signs and harbingers of disaster: 'I believe [these prodigies] are portentous things / Unto the climate that they point upon' (1.3.31–2), Casca confides to a Cicero who prefers to sit on the fence ('men may construe things after their fashion', 1.3.34). As Benson Bobrick reminds us, the astrological predictions in Shakespeare's plays are systematically fulfilled,[2] so that, for the playwright, the heavens seem to have remained a privileged intermediary zone between the natural and the supernatural, the sublunary and the translunary worlds.

Contrary to many plays of the period, Shakespeare's drama is all-inclusive and eschews didacticism: he embraces human and natural phenomena without reducing them to conventional symbolism and he features men and women inside their usual habitat and distinctive environment without neglecting more universal issues.[3] On the Shakespearean stage as in the real world, clement skies are generally short-lived since odd celestial events and spells of bad weather often serve to work out a number of dramatic tensions, be they political, social or aesthetic. For the playwright, the 'airy region' (*Romeo and Juliet*, 2.1.63) actually 'moderate[s] the relation between the material and the metaphysical', as Jennifer Mae Hamilton puts it.[4] In *Hamlet*, where the title character is constantly engrossed in metaphysical uncertainties, the story narrated by the First Player is made particularly tense and vivid by the mention of a sudden and violent storm which suggests the hero's tortured mind as much as the upcoming denouement of the tragedy:

> But as we often see against the storm
> A silence in the heavens, the rack stand still,
> The bold winds speechless, and the orb below
> As hush as death, anon the dreadful thunder
> Doth rend the region: so, after Pyrrhus' pause,
> A rousèd vengeance sets him new a-work; (2.2.486–91)

These lines (present in both Q2 and F) epitomise Hamlet's predicament – after his overlong 'pause', he will eventually find the way to 'vengeance' – and they also look ahead to the Prince's famous dying words, '[t]he rest is silence' (5.2.310). While critics have generally correlated the end of the play to the restoration of order, Denmark's political stability, at this very moment, seems in fact once again under threat. All is calm before the storm breaks out . . . One therefore understands why the playwright repeatedly puts to the fore the complex, changing and interactive relationships between human beings and what the First Player calls 'the region'. They offer him dramatic, substantial and structural elements at the same time.

In this book, we have seen that, in *A Midsummer Night's Dream*, the weather is clearly disturbed by human behaviour, 'human' insofar as Titania and Oberon are treated as mortals. Through Theseus's speech at the beginning of the play's last act, however, the final say goes to the poet who can give to 'airy nothing' (5.1.16) 'a local habitation and a name' (5.1.17), thus foregrounding the role of liberty and artistic imagination at the expense of the traditional astrological and geo-humoralist determinism.

In *Romeo and Juliet*, a play written at about the same time as the *Dream*, Shakespeare adopts a radically different stance. Here, the titular characters remain desperately '[s]ervile to all the skyey influences' (*Measure for Measure*, 3.1.9) as stars and weather appear to dictate their moods and rule the common fate of many characters of this first love tragedy: the Prologue's allusion to the 'pair of star-crossed lovers' and the dog days which form part of the play's background do somehow foreshadow the death of several eminent representatives of the Verona youth. Even though Juliet energetically denies names any transcendental or originary power when it comes to Romeo, she ironically fails to realise that she is herself trapped by her own name, whose original rooting in the month of July defines and determines her tragic fate.

As You Like It offers a radically different spectacle, perhaps because the play was written and performed soon after the winter of 1599, when the Theatre was dismantled, its timber carried across the frozen Thames and used for the building of the Globe. Shakespeare indeed stages here a winter festival poles apart from hot and heated Verona, strongly reminiscent of the Roman Saturnalia and promoting pleasure and appetite in the midst of an otherwise hostile environment. It is difficult to decide whether the stoic Duke Senior and the melancholy Jaques project their inner turmoils on to the heavens or whether the sky dictates their moods. The forest's frost, which pervades the whole play, 'almost freezes up the heat of life' (*Romeo and Juliet*, 4.3.16) and makes Arden a fairly bleak pastoral landscape anticipating the desolate heath of *King Lear*. Furthermore, in this pseudo-green comedy, Shakespeare for the first time dwells at length on the presence and effects of the wind that eventually turns Arden's lost paradise into a poetic shelter, the only remaining sign of some form of divinity.

With *Othello*, the playwright pays further attention to the circulation of air, but he explores the links between humankind and nature under a different angle, as the shore-bound title character fails to keep in touch with the heavens, blinded as he is by Iago as well as by misguided beliefs and geo-humoral stereotypes. The plot is built on the absent scene of the battle against the Turks, a naval engagement which is aborted because of a providential tempest that provides the Venetian fleet with an easy victory. It is this unrepresented climatic event, a tragic version of the *Dream*'s 'airy nothing' (5.1.16), which implicitly brings about Desdemona's death just as it leads Othello to be the lover he cannot become instead of the warrior he should have been. The tempest which smashes the Ottoman fleet is a bolt from the blue but this divine surprise hides a private and personal disaster to come. The gusts of wind that helped the Venetians triumph over the Turks are turned into ominous forces that will secretly undo the Moor's marriage. That Othello should stifle his young wife under a pillow, thus depriving her of the capacity to breathe, changes the happy comedy of act 1 into a tragedy of *pneuma*.[5] Swollen by a surplus of air, unable to expel his inner storm, Othello ultimately breaks down before bursting into a thousand pieces.

If, in *Othello*, the Mediterranean tempest is gradually displaced inside the Moor's inner psyche, in *King Lear*, by contrast, the sovereign's despair is less internalised, thus leading the King's Men to resort, for the play's performance, to a number of special, sensational stage effects in order to materialise the violence of the storm. Yet in a private venue, we can imagine that Lear's internal rage would have been highlighted to the detriment of his physical environment. Be that as it may, in this particular tragedy, the influence of weather upon man, and of man upon weather, reveals an acute awareness of the environmental problems posed by men and women's behaviour. When Shakespeare wrote his piece, lands previously unknown had been colonised, science and technology had gone through unprecedented advances and the control of nature had become a much more attainable goal. No wonder then that, in this play, the weather is presented in an anthropomorphic way by the disgraced monarch, who wishes he could master it: 'Blow, winds, and crack your cheeks!' Lear exclaims (3.2.1). This wind is both a generative[6] and a destructive force, depending on men's abilities to adapt and evolve. Importantly, in the old king's misogynist allusion to the 'sulphurous pit' (4.5.123), which ties in with the 'sulphurous and thought-executing fires' (3.2.4) mentioned earlier on, the play also points to the reversibility of heaven and hell already hinted at in the couplet concluding Sonnet 129 ('none knows well / To shun the heaven that leads men to this hell'). Such an interaction between sexual and celestial issues is just another sign showing that thunder is a force to be reckoned with. Owing nothing to supernatural causes, bad weather becomes an existential as well as a profoundly nihilistic – and ultimately fatal – experience.

In *Anthony and Cleopatra*, the emphasis is more specifically put on the sensorial aspects of the weather, a dimension which has direct repercussions on the Egyptian slimy, sticky soil. This is not contradictory with the play's exploration of celestial issues since Shakespeare makes the sky another territory to be explored and conquered. He thus plays on the trope of the early modern traveller applied to Anthony, the Roman general who dreams of a 'new heaven' (1.1.17) as a country to be discovered and a means to extend his knowledge of the world so as to improve his 'banquet of sense'. So, if moderate, rigid Rome is clearly the political winner at the end, its values remain traditional and uninventive, while hot, wet, sensualist Egypt does pave the way for the epistemological shift from hedonism to Epicureanism, a philosophy misrepresented and misunderstood by the Stoics, that is, by Octavius Caesar and his followers, as the play implicitly suggests. With all due precautions, it is tempting to think that its Epicurean approach to clouds – poetic objects *par excellence* – as well as to weather and space foreshadows Hobbes's focus on the senses, materialist vision of nature, and attempt to liberate humanity from its superstitious beliefs.

The thin border between superstition and science is given special prominence in *The Tempest*, a late play which stages the epistemological shift from popular knowledge to natural philosophy. As in *King Lear*, the purifying, expiatory function of thunderstorms is once again resorted to while the various human reactions to adverse meteorological conditions, ranging from despair and anger to exultation and hopefulness, are being dramatised by the playwright. The despondency caused by sea gales is of course hinted at in other plays of the canon, as at the beginning of *The Merchant of Venice*, for example, when Salarino tries to guess the reasons why his friend Antonio feels 'so sad' and imagines rough seas and strong winds putting his ships in danger: 'My wind, cooling my broth, / Would blow me to an ague when I thought / What harm a wind too great might do at sea' (1.1.22–4). But if *The Tempest* effectively represents on stage '[w]hat harm a wind too great might do at sea', it further prompts us to reconsider on a much larger scale the role played by human beings in the triggering of apparently natural catastrophes, and the way storms and tempests may shape and change human conduct as well as men and women's destinies. The climatic conditions on Prospero's island influence the way the characters behave. Under the pressure of their new environment, the shipwrecked passengers engage in broken-up conversations and utter disjointed sentences, thus making their words mirror the disruptive weather of the island. But, as he presents the complex interplay between human actions and the environment, Shakespeare also touches on the issue of men and women's problematic engagement with weather patterns. In this perspective, Prospero may appear as a proto-scientist who successfully experiments with the climate and who seeks to justify his magical manipulation of the weather thanks to various political and moral purposes. Ultimately, beyond his temporary control of the weather, Prospero also works on the vagaries of the human imagination, thus appearing as a powerful master of illusion on his theatrical island.

In his intuitive anticipation of the critical climatic issues which increasingly concern us today, Shakespeare could, to a certain extent, be presented as another master of the weather. 'Men judge by the complexion of the sky / The state and inclination of the day' (*Richard II*, 3.2.190–1), a phrase which the playwright must have kept in mind while writing his plays, all the more so as the Globe's star-spangled, heaven-shaped roof provided him with an apt visual reminder of Scrope's maxim. Whenever his characters happen to blame climate for their misfortunes,[7] he suggests on the contrary that such an attitude can only work as an excuse at worst, and as a symptom at best: the real causes of disgrace and failure are to be found elsewhere and, more often than not, must be laid down to men and women's responsibility rather than to the 'ways of God to men' (Milton, *Paradise Lost*, I, 26).

While he borrowed much – perhaps even more than we sometimes imagine – from classical literature and contemporary treatises, digesting and recycling an impressive amount of meteorological knowledge, Shakespeare's stance was not that of a would-be scientist, since what he was interested in was the *representation* of climate and its various possible effects on a theatrical audience, rather than trying to define or study its possible causes. To him, the various climatic manifestations and environmental phenomena were the 'objective correlatives'[8] of his characters' actions and emotions. Clouds, thunder, lightning, earthquakes, hurricanes, eclipses and meteors – all of them weather correlates that are apprehended through the links they were supposed to have with the heavens, the planets and the elements at large – are turned by him into key energetic words endowed with special poetic force and resonance. These words, together with the images and ear-piercing sounds they were associated with (in 1615, Helkiah Crooke described the sound of thunder as 'a vehement and violent noyse'[9]), worked as accelerators and propagators of emotions and sensations among the audiences of public and private theatres alike. The technical means then available served to reinforce these verbal effects and turned them into material sensations that could instil surprise, terror or admiration in the spectators' minds. The mere evocation of the heavens sufficed to infuse the audience with a distinct awareness of nature – that is, with a sense of their early modern environment. The 'blue calico ceiling with golden stars'[10] which was to be seen in the Whitehall Cockpit theatre of 1630 is but one of the numerous examples which testify to the enduring materiality of the English skies in post-Shakespearean drama. Similarly, the numerous clouds which were increasingly used in the setting of Jacobean and Carolean plays and masques did not simply contribute to the elaboration of what Thomas M. Greene describes as 'vertical drama'.[11] They also gave the 'airscape' a substantial quality while fulfilling two main functions: that of 'incorporating the sacred into the real' and that of 'conceal[ing] the unrepresentable nature of infinity at the same time as pointing to it'.[12]

In Shakespeare's playtexts, the materiality of the sky never precludes a more literary treatment of their meteorological images. References to rain, clouds or wind remarkably exemplify the characters' dilemmas rather than simply call attention to the general physics of the local weather. In such a context, climate/*klima* is to be taken as a liminal space which reveals tensions, anxieties, expectations and oppositions.[13] It is up to the playwright – and to present-day directors – to turn this threshold zone into a dramatic locus where the trajectories, or fates, of the various dramatis personae may ultimately be grasped and mapped.

Climate is also for Shakespeare a framing device giving more coherence and density to his playtexts and providing the audience with a natural, elemental, and sometimes cosmic background. Obviously, his drama cannot and must not be seen as a window to the reality of early modern climate but as a kaleidoscope refracting the contemporary ideas which circulated in his days about our not-so-fated sky. The unsettling vagaries of sky and weather were what he sought to capture and translate into dramatic terms, essentially because they were likely to get an immediate emotional response from the spectators. He thus relied on the multiple variations and combinations of heavens, stars, wind and rain and so on, so as to reproduce, dramatise, and sometimes materialise (depending upon the staging conditions and, in the case of public playhouses, if weather conditions permitted) the shiftiness and evanescence of human life. These variations shape the plays' language, be it tempestuous, languid, dry or fluid, even though weather metaphors in Shakespeare's plays hardly ever serve the clichés of an all too familiar, frozen Petrarchan imagery. They often point to human frailties or inconsistencies as well as to dramatic turns and sudden reversals of situations. Shakespeare's multifaceted characters are all 'minded like the weather, most unquietly' (*King Lear*, 3.1.2), and each of them corresponds to, and generates, his or her own specific clime. Like us, they must repeatedly face the whims of the sky at crucial and decisive moments.

Their various, sometimes excessive reactions presuppose the advent of a major epistemological shift, if not of an epistemological crisis at the turn of the seventeenth century. In the first half of the sixteenth century, Renaissance natural philosophers, still largely influenced by Aristotelianism, remained quite sceptic as to 'the possibility of certain knowledge of meteorology', mainly because of the accidental nature of weather-related phenomena.[14] By contrast, in the second part of the century, '[t]he crack in Aristotle's authority' widened,[15] the distinction between climate and weather gradually emerged and, above all, the dramatisation of celestial events, especially in England, allowed for a more immediate access to the natural world. Shakespeare lived at a time when old beliefs were being challenged and gradually deconstructed, even though they died hard. While his is an essentially syncretic, pragmatic and non-systematic approach encompassing superstitious and erudite, religious and sceptical attitudes regarding the sky, he nonetheless chose to question ingrained and new beliefs alike while insisting on a number of facts and facets recurring from play to play: men and women's (im)possible independence from the weather, the nature and problematic existence of the void, the limits of weather experiments, and the reversibility of our physical environment in which rain, wind, heat or cold may become sources of joy as much as causes of despair. His interest in the way weather conditions actually affect human behaviour prompted him to modify, and sometimes reverse, many traditional points of view in order to show that things also work the other way round with man's ominous capacity to provoke violent climatic disorders and to generate chaos on earth.

From the determinism put forward by *Romeo and Juliet* to the materialism and relativism of *Anthony and Cleopatra*, from the archaic beliefs highlighted (and subtly dismissed) in *A Midsummer Night's Dream* to the proto-science staged in *The Tempest*, Shakespeare actually provides us with a panoramic view of the early modern 'fated sky'. Between superstition and the cold rationality prefiguring Descartes's theories, a poised stance is possible, he suggests. Influenced by Greek medical and meteorological thinking and, in particular, by an Epicurean approach to nature and the cosmos at large, his plays allow men and women to undergo a sensory and sensitive experience of the weather and of sky-related phenomena, one that could free them from dread and which could make them dream of a 'brave new world' (*The Tempest*, 5.1.183); one, above all, that enabled early modern playgoers to accept contingency and to feel unconstrained by the limits of the 'wooden O' of the playhouse and of '[t]he little O o'th'earth' (*Anthony and Cleopatra*, 5.2.81).

Far from Descartes's mind/body dualism, Locke called 'sensory experience [. . .] the only genuine source of knowledge'.[16] In June 1666, ten years after the first translation of Lucretius' *De rerum natura* into English,[17] he started writing a weather diary in which he meticulously recorded temperature, wind-direction, atmospheric pressure and the like, and whose extracts were to be published in Robert Boyle's *General History of the Air* (1692).[18] Observations on the weather allowed him to reconcile his lifelong interest in science with his passion for philosophy and to combine notes on his personal routine with observations on the quality of the air. In other words, as he wrote on the meteorological events which formed part of his daily life, he furnished his mind with a myriad sensations which, in turn, made him think. Unsurprisingly, in his *Thoughts Concerning Education* (1693), he advised parents to let their children experience the adversity of climate and to reconcile them with the cold and the rain characteristic of his country: 'Plenty of open air [. . .] not too warm and strait clothing; especially the head and feet kept cold, and the feet often used to cold water and exposed to wet.'[19] Because 'the strength of the body is chiefly in being able to endure hardships, so also does that of the mind', Locke added.[20]

As I conclude this book, I cannot help thinking that, a century before him, with his many characters bracing themselves for rain, cold and storms, and desperately trying to fashion their own selves by experiencing the vicissitudes of the angry skies, Shakespeare, in the way he represents weather, climate and environment, had already paved the way for the new scientific empiricism.

Notes

1. Fynes Moryson, *An itinerary*, Part III, Book 3, p. 146.
2. Benson Bobrick, *The Fated Sky: Astrology in History*, p. 182. For his analysis, Bobrick relies on Johnstone Parr's *Tamburlaine's Malady*.
3. This is to be contrasted with the masques and pageants of the Stuart era, as those entertainments relied on weather elements and astrological components often stripped of environmental significance in order to convey a symbolic, didactic and political message. See for instance

Thomas Middleton's civic pageant performed in 1621, *The Sun in Aries*, ed. David M. Bergeron, in Thomas Middleton, *The Collected Works*, pp. 1586–92.

4. Jennifer Mae Hamilton, *This Contentious Storm: An Ecocritical and Performance History of King Lear*, p. 20.
5. For more on air and *pneuma* in Shakespeare, see Carolyn Sale, 'Eating Air, Feeling Smells: Hamlet's Theory of Performance', p. 163. See also Leah Knight, *Reading Green in Early Modern England*, p. 54, and Carla Mazzio, 'History of Air: Hamlet and the Trouble with Instruments', pp. 169–70.
6. In Michael Maier's *Atalanta fugiens* (1617), an alchemical emblem book containing fifty emblems illustrating the alchemical process, the first emblem, accompanied by the motto 'Portavit eum ventus in ventre suo' ('The Wind carries it in his belly'), depicts the Wind (p. 1). The Wind is pregnant, which makes him a generative force. See H. M. E. de Jong (ed.), *Michael Maier's Atalanta fugiens: Sources of an Alchemical Book of Emblems*.
7. On the idea of 'Blaming Climate', see Mike Hulme, *Weathered Cultures of Climate*, chap. 6, pp. 68–80.
8. The phrase is used by T. S. Eliot in 'Hamlet and his Problems', p. 100.
9. Helkiah Crooke, *Mikrokosmographia a description of the body of man*, Book 8, chap. 16, p. 588.
10. John Astington, *English Court Theatre, 1558–1642*, p. 134.
11. See Thomas M. Greene, 'Magic and Festivity at the Renaissance Court: The 1987 Josephine Waters Bennett Lecture', p. 650.
12. Hubert Damisch, *Théorie du nuage. Pour une histoire de la peinture*, quoted in John E. Thornes, 'A Brief History of Weather in European Landscape Art', p. 364.
13. More generally, Shakespeare's apprehension of heavenly phenomena reveals the rigidly hierarchical social structure of Elizabethan society. 'When beggars die there are no comets seen; / The heavens themselves blaze forth the death of princes', Calpurnia tells Caesar in *Julius Caesar* (2.2.30–1). Calpurnia's acknowledgement that degree in the heavens reflects degree in the social order refers back to ancient times while somehow also reflecting the traditional Elizabethan view. Her lines foreshadow Ulysses' cosmic conception of the Aristotelian world in *Troilus and Cressida*: 'And therefore is the glorious planet Sol / In noble eminence enthroned and sphered / Amidst the other, whose med'cinable eye / Corrects the ill aspects of planets evil / And posts like the commandment of a king, / Sans check, to good and bad. But when the planets / In evil mixture to disorder wander, / What plagues and what portents, what mutiny? / What raging of the sea, shaking of earth?' (1.3.89–97).
14. Craig Martin, 'Conjecture, Probabilism, and Provisional Knowledge in Renaissance Meteorology', p. 282.

15. H. Howard Frisinger, *The History of Meteorology: To 1800*, p. 36. Frisinger partly ascribes the 'gradual breaking away from [Aristotle's] influence' to the Italian polymath Girolamo Cardano (1501–76).
16. Jessica Riskin, *Science in the Age of Sensibility: The Sentimental Empiricists of the French Enlightenment*, p. 10.
17. I refer to John Evelyn's 1656 translation, *An Essay on the First Book of T. Lucretius Carus De Rerum Natura.*
18. On this, see Jan Golinski, *British Weather and the Climate of Enlightenment*, p. 55.
19. John Locke, *Some Thoughts Concerning Education (Including Of the Conduct of the Understanding)*, p. 28.
20. Ibid.

Bibliography

Shakespeare editions

Bate, Jonathan and Eric Rasmussen (eds), *The Tempest*, introduced by Jonathan Bate, Basingstoke: Macmillan, The Royal Shakespeare Company, 2008.

Bevington, David (ed.), *Antony and Cleopatra*, Cambridge: Cambridge University Press, The New Cambridge Shakespeare, (1990) 2013.

Brissenden, Alan (ed.), *As You Like It*, Oxford: Oxford University Press, 1993.

Brooks, Harold (ed.), *A Midsummer Night's Dream*, London: Methuen, 1979.

Dover Wilson, John (ed.), *King Lear*, Cambridge: Cambridge University Press, (1960) 1968.

Elam, Keir (ed.), *Twelfth Night*, London: Cengage Learning, The Arden Shakespeare, 2008.

Hattaway, Michael (ed.), *As You Like It*, Cambridge: Cambridge University Press, The New Cambridge Shakespeare, (2000) 2009.

Holland, Peter (ed.), *A Midsummer Night's Dream*, Oxford: Oxford University Press, 1995.

Honigmann, E. A. J. (ed.), *Othello*, London: Thomson Learning, The Arden Shakespeare, (1997) 2006.

Levenson, Jill L. (ed.), *Romeo and Juliet*, Oxford: Oxford University Press, 2000.

Mowat, Barbara A. and Paul Werstine (eds), *As You Like It*, New York: Washington Square Press, The Folger Shakespeare Library, 1997.

Neill, Michael (ed.), *Anthony and Cleopatra*, Oxford: Oxford University Press, 1994.

Neill, Michael (ed.), *Othello*, Oxford: Oxford University Press, 2006.

Orgel, Stephen (ed.), *The Tempest*, Oxford: Oxford University Press, 1994.

Taylor, Gary, John Jowett, Terri Bourus and Gabriel Egan (eds), *William Shakespeare: The Complete Works, Modern Critical Edition*, Oxford: Oxford University Press, The New Oxford Shakespeare, 2016.

Vaughan, Virginia Mason and Alden T. Vaughan (eds), *The Tempest*, London: Thomson Learning, The Arden Shakespeare, (1999) 2001.

Weis, René (ed.), *King Lear*, A Parallel Text Edition, 2nd edn, London: Routledge, (1993) 2010.

Weis, René (ed.), *Romeo and Juliet*, London: Bloomsbury, The Arden Shakespeare, 2012.

Wells, Stanley, Gary Taylor, John Jowett and William Montgomery (eds), *William Shakespeare: The Complete Works*, 2nd edn, Oxford: Clarendon Press, 2005.

Wilders, John (ed.), *Antony and Cleopatra*, London: Routledge, The Arden Shakespeare, 1995.

Yamada, Akihiro (ed.), *The First Folio of Shakespeare, A Transcript of Contemporary Marginalia in a Copy of the Kodama Memorial Library of Meisei University*, Tokyo: Yushodo Press Co., 1998; digital version at <http://shakes.meisei-u.ac.jp/ALL.html#45> (last accessed 18 April 2018).

Primary sources

Abbott, George, *An exposition vpon the prophet Ionah*, London: Printed by Richard Field, 1600, STC (2nd edn) / 34.

Acosta, José de, *The naturall and morall historie of the East and West Indies*, trans. E. G., London: Printed by Valentine Sims for Edward Blount and William Aspley, 1604, STC (2nd edn) / 94.

Adams, Thomas, *A divine herball*, London: Printed by George Purslowe, for Iohn Budge, 1616, STC (2nd edn) / 111.

Alciato, Andrea, *Emblematum liber*, Augsburg, 1531.

Allott, Robert (ed.), *Englands Parnassus*, Imprinted at London, For N. L[ing,] C. B[urby] and T. H[ayes], 1600, STC (2nd edn) / 378.

Anghiera, Pietro Martire d', *De orbe novo (Decades)*, ed. and trans. F. A. McNutt, 2 vols., New York: Putnam, 1912.

Anghiera, Pietro Martire d', *The decades of the newe worlde or west India*, trans. Richard Eden, London: William Powell for William Seres, 1555, STC (2nd edn) / 646.

Anon., *1607. A true report of certaine wonderfull ouerflowings of waters*, Printed at London by W[illiam] I[aggard] for Edward White, 1607, STC (2nd edn) / 22915.

Anon., *Aristotles politiques*, London: Printed by Adam Islip, 1598, STC (2nd edn) / 760.

Anon., *Certayne Godly exarcises meditacions and prayers*, London, 1565, STC (2nd edn) / 10617.

Anon., *Heere beginneth the Kalender of sheephards: newly augmented and corrected*, Imprinted at London, By [J. Charlewood and G. Robinson for] John Wally, c. 1585, STC (2nd edn) / 1692:14.

Anon., *The last terrible tempestious windes and weather*, London, 1613, STC (2nd edn) / 25840.

Anon., *The true chronicle history of King Leir* (1605), in Geoffrey Bullough (ed.), *Narrative and Dramatic Sources of Shakespeare*, vol. 7, *Major Tragedies*, London: Routledge and Kegan Paul, (1973) 1978, pp. 337–402.

Aristotle, *The Generation of Animals*, trans. A. L. Peck, Cambridge, MA: Harvard University Press, Loeb Classical Library, 1943.

Aristotle, *History of Animals* (*Historia animalium*), trans. D'Arcy Wentworth Thompson, in *The Works of Aristotle*, ed. William David Ross, vol. 9, Oxford: Oxford University Press, 1952.

Aristotle, *Meteorologica*, trans. H. D. P. Lee, Cambridge, MA: Harvard University Press, Loeb Classical Library, 1978.

Aristotle, *Nicomachean Ethics*, ed. Roger Crisp, Cambridge: Cambridge University Press, 2000.

Aristotle, *Problems*, trans. W. S. Hett, vol. 1, Cambridge, MA: Harvard University Press, 1936.

Aristotle, *The problemes of Aristotle*, Edinburgh: Robert Waldgrave, 1595, sig. E3, STC (2nd edn) / 763.

Armin, Robert, *A nest of ninnies*, Printed by T. E[ast] for Iohn Deane, 1608, STC (2nd edn) / 772.7.

Aspenhall, William (ed.), *The Correspondence of Philip Sidney and Hubert Languet*, Boston: The Merrymount Press, 1912.

Aubrey, John, *Brief Lives*, ed. Oliver Lawson Dick, London: Secker and Warburg, 1960.

Bacon, Francis, *Historia ventorum*, in *The Works of Francis Bacon*, vol. 5, *Translations of the Philosophical Works 2*, ed. James Spedding, Robert Leslie Ellis and Douglas Denon Heath, Cambridge: Cambridge University Press, 2011.

Bacon, Francis, *Wisdom of the Ancients*, trans. Arthur Gorges, London, 1619.

Barough, Philip, *The methode of phisicke*, Imprinted at London by Thomas Vautroullier, 1583, STC (2nd edn) / 1508.

Batman, Stephen, *Batman vppon Bartholome his booke De proprietatibus rerum, newly corrected, enlarged and amended*, London: Thomas East, 1582, STC (2nd edn) / 1538.

Batman, Stephen, *The golden booke of the leaden goddes*, London: Thomas Marshe, 1577, STC (2nd edn) / 1583.

Bede, *The Complete Works of Venerable Bede*, ed. J. A. Giles, vol. 6, London: Whittaker and Co., 1843.

Benzoni, Girolamo, *Americae Pars Quarto*, Frankfurt: Theodor de Bry, 1594.

Bertelli, Pietro, *Diuersaru nationum habitus*, Patauij, 1594.

Blagrave, John, *The mathematical ievvel*, London: Printed by Walter Venge, 1585, STC (2nd edn) / 3119.

Blundeville, Thomas, *M. Blundevile his exercises containing sixe treatises*, London: Printed by John Windet, 1594, STC (2nd edn) / 3146.

Boaistuau, Pierre, *Certaine secrete wonders of nature*, trans. E. Fenton, Imprinted at London by Henry Bynneman, 1569, STC (2nd edn) / 3164.5.

Boccacio, Giovanni, *Famous Women* (*De mulieribus claris*, c. 1375), ed. and trans. Virginia Brown, Cambridge, MA: Harvard University Press, The I Tatti Renaissance Library, 2001.

Bodin, Jean, *The six bookes of a common-weale*, trans. Richard Knolles, London: Printed by Adam Islip, 1606, STC (2nd edn) / 3193.

Botero, Giovanni, *The worlde, or an historicall description of the most famous kingdomes and common-weales*, Imprinted at London by Edm. Bollifant, for Iohn Iaggard, 1601, STC (2nd edn) / 3399.

Brasbridge, Thomas, *The poore mans ieuuel, that is to say, A treatise of the pestilence*, London: For George Byshop, 1578, STC (2nd edn) / 3549.

Brooke, Arthur, *The Tragicall Historye of Romeus and Juliet*, in Geoffrey Bullough (ed.), *Narrative and Dramatic Sources of Shakespeare*, vol. 1, *Early Comedies, Poems*, London: Routledge and Kegan Paul, 1961, pp. 284–363.

Bullein, William, *The gouernment of health*, London: Printed by Valentine Sims, 1595, STC (2nd edn) / 4042.

Bullough, Geoffrey (ed.), *Narrative and Dramatic Sources of Shakespeare*, vol. 7, *Major Tragedies*, London: Routledge and Kegan Paul, (1973) 1978.

Burton, Robert, *The Anatomy of Melancholy*, Oxford: Benediction Classics, 2016.

Calvin, John, *Sermons of Master Iohn Caluin, vpon the booke of Iob*, trans. Arthur Golding, [London:] Printed by [Henry Bynneman for] Lucas Harison and George Byshop, [1574], STC (2nd edn) / 4445.

Camerarius, Joachim, *The history of strange wonders*, London: Printed by Roulande Hall, 1561, STC (2nd edn) / 18507.

Carr, Ralph, *The Mahumetane or Turkish historie containing three bookes*, London: Printed by Thomas Este, 1600.

Carroll, D. Allen (ed.), *Greene's Groatsworth of Wit Bought with a Million of Repentance*, Binghamton, NY: Medieval and Renaissance Texts and Studies, 1994.

Cartari, Vincenzo, *The Fountaine of Ancient Fiction*, trans. Richard Linch, London, 1599.

Casas, Bartolomé de las, *Apologética historia sumaria*, ed. V. Avril Castello, 3 vols. [Obras completes: Bartolomé de las Casas 6–8], Madrid: Alianza Editorial, 1992.

Casas, Bartolomé de las, *The Spanish colonie, or Briefe chronicle of the acts and gestes of the Spaniardes in the West Indies*, trans. M.M.S., London: Printed by Thomas Dawson, 1583, STC (2nd edn) / 4739.

Cawley, A. C. (ed.), *The Wakefield Pageants in the Towneley Cycle*, Manchester: Manchester University Press, 1958.
Chapman, George, *Bussy d'Ambois*, ed. N. S. Brooke, Manchester: Manchester University Press, 1999.
Chapman, George (trans.), *Homer's Odysses*, London: Printed by Richard Field [and W. Jaggard], 1615, STC (2nd edn) / 13637.
Charron, Pierre, *Of wisdome*, trans. Samson Lennard, London, 1608, STC (2nd edn) / 5051.
Cicero, *On Divination, Book 1*, trans. with commentary by David Wardle, Oxford: Clarendon Press, 2006.
Clever, William, *The flower of phisicke*, London, 1590, STC (2nd edn) / 5412.
Combe, Thomas, *Theater of Fine Devices* (1593), introduction by Mary V. Silcox, Aldershot: Scolar Press, 1990.
Cotgrave, Randle, *A Dictionarie of the French and English Tongues*, London: Adam Islip, 1611.
Covell, William, *Polimenteia*, Cambridge and London: Printed by Iohn Legate, printer to the Vniversitie of Cambridge [and J. Orwin in London], 1595, STC (2nd edn) / 5883.
Crooke, Helkiah, *Mikrokosmographia a description of the body of man*, London: Printed by William Jaggard, 1615, STC (2nd edn) / 6062.
Cummings, Brian (ed.), *The Book of Common Prayer: The Texts of 1549, 1559, and 1662*, Oxford: Oxford University Press, (2011) 2013.
Cuningham, William, *The cosmographical glasse*, London, 1559, STC (2nd edn) 6119.
Da Porto, Luigi, *Istoria novellamente ritrovata di due nobili amanti* (1530), in Adolph Caso (ed.), *Romeo and Juliet: Original Text of Masuccio Salernitano, Luigi Da Porto, Matteo Bandello, William Shakespeare*, trans. Maurice Jonas, Boston: Dante University of America Foundation, 1992, pp. 23–52.
Das, Nandini (ed.), *Robert Greene's Planetomachia (1585)*, Aldershot: Ashgate, 2007.
Dekker, Thomas, *The Shoemaker's Holiday*, ed. R. L. Smallwood and Stanley Wells, Manchester: Manchester University Press, 1979.
Dekker, Thomas and John Webster, *Westward Ho*, in Fredson Bowers (ed.), *The Dramatic Works of Thomas Dekker*, vol. 2, Cambridge: Cambridge University Press, 1955.
Digges, Leonard, *A prognostication of right good effect fructfully augmented*, London: Printed by Thomas Gemini, 1555, STC (2nd edn) / 435.35.
Drayton, Michael, *Poly-Olbion*, London: Printed by Humphrey Lownes, 1612, STC (2nd edn) / 7226.
Duplessis-Mornay, Philippe, *A woorke concerning the trewnesse of the Christian religion*, trans. Philip Sidney and Arthur Golding, London: Printed by John Charlewood and George Robinson, 1587, STC (2nd edn) 18149.

Eden, Richard (trans.), *The decades of the newe worlde* (London, 1555), in Edward Arber (ed.), *The First Three English Books on America*, Birmingham, 1885.

Elyot, Thomas, *The image of gouernance*, London: In officina Thomae Berthcleti [*sic*] typis impress, Anno. M.D.XLI. [1541], STC (2nd edn) / 7664.

Epicure, *Letter to Pythocles*, in *Diogenes Laertius: Lives of Eminent Philosophers*, trans. R. D. Hicks, vol. 2, London and Cambridge, MA: Harvard University Press, Loeb Classical Library, 1958.

Erasmus, Desiderius, *The Praise of Folie*, [trans.] Sir Thomas Chaloner, ed. Clarence H. Miller, Early English Text Society, vol. 257, London: Oxford University Press, 1965.

Evelyn, John (trans.), *An Essay on the First Book of T. Lucretius Carus De Rerum Natura*, London: Gabriel Bedle & Thomas Collins, 1656.

Ficino, Marsilio, *Platonic Theology*, trans. Josephine L. Burroughs, *Journal of the History of Ideas* 5 (April 1944).

Florio, John, *Perpetuall and naturall prognostications of the change of weather*, London: Printed by Iohn Wolfe, 1591, STC (2nd edn) / 403.3.

Florio, John, *A Worlde of Wordes*, Printed at London by Arnold Hatfield for Edward Blount, 1598, STC (2nd edn) / 11098.

Fulke, William, *A goodly gallerye with a most pleasaunt prospect*, London: Printed by William Griffith, 1563, STC (2nd edn) / 11435.

Gascoigne, George, *A delicate diet, for daintiemouthde droonkardes*, London: By [John Charlewood for] Richard Jones, 1576, STC (2nd edn) / 11640.

Gascoigne, George, *A Devise of a maske for the right honourable Viscount Mountacute A Hundreth Sundrie Flowres*, ed. George W. Pigman, Oxford: Clarendon Press, 2000.

Giraldi Cinthio, Giovanni Battista, *Gli Hecatommithi Third Decade, Seventh Novella* (Venice, 1565), trans. Bruno Ferraro, in Michael Neill (ed.), *Othello*, Oxford: Oxford University Press, 2006, Appendix C, pp. 434–44.

Googe, Barnabe (trans.), *The Zodiacke of Life*, London: Printed by Henry Middleton, 1576, STC (2nd edn) / 19151.

Gresham, Edward, *A new almanacke and prognostication for the yere of our Lord God*, London: Printed for the Company of the Stationers, 1607, STC (2nd edn) / 452.7.

Herbert, George, *The Temple* (1633), in *The English Poems of George Herbert*, ed. Helen Wilcox, Cambridge: Cambridge University Press, 2007.

Herodotus, *The famous hystory of Herodotus*, trans. B. R. [Barnabe Riche?], London: Printed by Thomas Marshe, 1584, STC (2nd edn) / 13224.

Heywood, John, *A dialogue conteinyng the nomber in effect of all the prouerbes in the englishe tongue*, London: Printed by Thomas Berthelet, 1546, STC (2nd edn) / 1329.

Heywood, John, *The playe of the weather*, London: Printed by Ihon Awdeley, c. 1573, STC (2nd edn) / 13307.

Heywood, Thomas, *Philocothonista, or the Drunkard, Opened, Dissected, and Anatomized*, London: Robert Raworth, 1635.

Heywood, Thomas, *Troia Britanica*, London: Printed by William Jaggard, 1609, STC (2nd edn) / 13366.

Hill, Thomas, *A contemplation of mysteries*, London: Printed by Henry Denham, 1574, STC (2nd edn) / 13484.

Hill, Thomas, *The profitable arte of gardening nowe the thirde time set forth*, Imprinted at London by Henrie Bynneman, 1574, STC (2nd edn) / 13493.

Hippocrates, *Airs, Waters, Places*, in *Hippocrates*, trans. W. H. S. Jones, vol. 1, Cambridge, MA: Harvard University Press, Loeb Classical Library, 1923.

Hippocrates, *On Ancient Medicine*, ed. and trans. Mark J. Schiefsky, Leiden: Brill, 2005.

Holinshed, Raphael, *The firste [laste] volume of the chronicles of England, Scotlande, and Irelande*, London: Printed for John Hunne, 1577, STC (2nd edn) / 13568b.

Holland, Philemon (trans.), *The historie of the vvorld*, London: Printed by Adam Islip, 1601, STC / 20029.

Hooker, Richard, *Of the lavves of ecclesiasticall politie Eyght bookes*, Printed at London by John Windet, 1593, STC (2nd edn) / 13712.

Hume, Martin A. S. (ed.), *Calendar of State Papers, Spain (Simancas)*, vol. 4, *1587–1603*, London: Her Majesty's Stationery Office, 1899; available at British History Online, <http://www.british-history.ac.uk/cal-state-papers/simancas/vol4/pp411-425> (last accessed 30 April 2018).

Hyginus, Caius Julius, *Fabularum liber*, Basel: Ioan Hervagium, 1535.

Jacobus Rex [James I], *His Maiesties Lepanto*, London, 1603.

Johnson, Thomas, *Cornucopiae*, London: Printed by J. Danter for William Barley, 1595, STC (2nd edn) / 14707.

Jonson, Ben, *The Alchemist and Other Plays*, ed. Gordon Campbell, Oxford: Oxford University Press, 1995.

Jonson, Ben, *The Cambridge Edition of the Works of Ben Jonson*, ed. D. Bevington, M. Butler and I. Donaldson, 7 vols., Cambridge: Cambridge University Press, 2012.

Jonson, Ben, *The Masque of Blackness*, in David Lindley (ed.), *Court Masques: Jacobean and Caroline Entertainments 1605–1640*, Oxford: Oxford University Press, 1995.

Knolles, Richard, *The generall historie of the Turkes*, London: Printed by Adam Islip, 1603, STC (2nd edn) / 15051.

Lactantius, *The Divine Institutes*, in Rev. Alexander Roberts, Sir James Donaldson and Arthur Cleveland Coxe (eds), *The Ante-Nicene Fathers: The Writings of the Fathers Down to A.D. 325*, vol. 7, New York: Cosimo Classics, 2007.

Langley, Thomas, *An abridgement of the notable woorke of Polidore Vergile*, London: Printed by Richard Grafton, 1546.

Le Loyer, Pierre, *A Treatise of Specters or Straunge Sights, Visions and Apparitions*, London, 1603.

Lemnius, Levinus, *The Secret Miracles of Nature*, London, 1658.

Locke, John, *Some Thoughts Concerning Education (Including Of the Conduct of the Understanding)*, ed. John William Adamson, Mineola, NY: Dover Publications, Philosophical Classics, Cambridge University Press, (1922) 2007.

Lodge, Thomas, *Rosalynde. Euphues golden legacies* (1590), in Geoffrey Bullough (ed.), *Narrative and Dramatic Sources of Shakespeare*, vol. 2, *Comedies (1597–1603)*, London: Routledge and Kegan Paul, 1958.

Lucretius, *Titi Lucretii Cari De rerum natura libri sex*, ed. Denis Lambin [Dionysius Lambinus], Paris and Lyon: Rovillium [Guillaume Rouillé and Philippc Rouillé], 1563.

Lucretius, *De rerum natura*, trans. W. H. D. Rouse, revised by Martin Ferguson Smith, Cambridge, MA: Harvard University Press, Loeb Classical Library, 1992.

Lucretius, *De rerum natura*, trans. R. C. Trevelyan, Cambridge: Cambridge University Press, 1937.

Luther, Martin, 'Vorlesungen über 1. Mose', in *Kritische Gesamtausgabe* 42–4, Weimar: Böhlau, 1911–1913, vol. 42.

Luther, Martin, *Werke. Kritische Gesamtausgabe* 10, 1. Abteilung, 2. Hälfte, Weimar: Böhlau, 1925.

Machiavelli, Niccolò, *The Prince*, trans. with notes by Georges Bull, introduction by Anthony Grafton, London: Penguin, (1961) 1999.

Maier, Michael, *Atalanta fugiens*, Oppenheim, 1617.

Mallarmé, Stéphane, 'The Swan' (1885), trans. Barbara Johnson, in Chaviva Hošek and Patricia Parker (eds), *Lyric Poetry: Beyond New Criticism*, Ithaca: Cornell University Press, p. 371.

Manilius, *Astronomica*, ed. and trans. G. P. Goold, Cambridge, MA: Harvard University Press, Loeb Classical Library, (1977) 1997.

Marlowe, Christopher, *Doctor Faustus. A- and B-Texts* (1604, 1616), ed. David Bevington and Eric Rasmussen, Manchester: Manchester University Press, The Revels Plays, 1993.

Marlowe, Christopher, *Tamburlaine the Great*, ed. J. S. Cunningham and Eithne Henson, Manchester: Manchester University Press, 1998.

Middleton, Thomas, *The Sun in Aries* (1621), ed. David M. Bergeron, introduced by Michael Berlin, in *The Collected Works*, ed. Gary Taylor and John Lavagnino, Oxford: Clarendon Press, 2007, pp. 1586–92.

Middleton, Thomas, *The Triumphs of Truth* (1613), ed. David M. Bergeron, in *The Collected Works*, ed. Gary Taylor and John Lavagnino, Oxford: Clarendon Press, 2007, pp. 963–76.

Milton, John, *Paradise Lost*, ed. John Leonard, London: Penguin Classics, 2000.

Montaigne, Michel de, *Montaigne's Essays*, 3 vols., London: The Nonesuch Press, 1928.

More, Thomas, *Utopia*, trans., ed. and introduced by Dominic Baker-Smith, London: Penguin, 2012.

Morton of Berwick, Thomas, *A treatise of the nature of God*, London: Printed by Thomas Creed for Robert Dexter, 1599, STC (2nd edn) / 18198.

Moryson, Fynes, *An itinerary*, London: Printed by John Beale, 1617, STC (2nd edn) / 18205.

Nashe, Thomas, *Pierce Penniless his Supplication to the Devil* (1592), in *The Unfortunate Traveller and Other Works*, ed. J. B. Steane, Harmondsworth: Penguin, (1972) 1978, pp. 49–145.

Nashe, Thomas, *Summer's Last Will and Testament*, in *Thomas Nashe: Selected Works*, ed. Stanley Wells, Abingdon: Routledge, Routledge Revivals, (1964) 2015.

Olaus Magnus, *Historia de gentibus septentrionalibus*, Rome: Johannes Maria de Viottis, 1555.

Ovid, *Metamorphoses*, trans. Arthur Golding, ed. Madeleine Forey, London: Penguin, 2002.

Pigafetta, Antonio, *First Voyage Round the World by Magellan*, accompanied by original documents, with notes and an introduction by Lord Stanley of Alderley, Cambridge: Cambridge University Press, (1874) 2010.

Plato, *Symposium*, trans. Alexander Nehamas and Paul Woodruff, in *Complete Works*, ed. John M. Cooper, Indianapolis: Hackett, 1997.

Platter, Thomas, *Thomas Platter's Travels in England*, trans. Clare Williams, London: Jonathan Cape, 1937.

Plutarch, 'The Life of Octavius Caesar Augustus', in *The lives of the noble Grecians and Romanes*, trans. Thomas North, 1603, in Geoffrey Bullough (ed.), *Narrative and Dramatic Sources of Shakespeare*, vol. 8, *The Roman Plays*, Routledge and Kegan Paul, 1977, pp. 321–3.

Plutarch, *The lives of the noble Grecians and Romanes*, trans. Thomas North, London: Printed by Thomas Vautroullier and John Wight, 1579, STC (2nd edn) / 20066.

Pory, John (trans.), *A geographical historie of Africa*, London: Printed by Eliot's Court Press, 1600, STC (2nd edn) / 15481.

Pseudo-Aristotle, *Cosmic Order and Divine Power: Pseudo-Aristotle, On the Cosmos*, introduction, text, translation and interpretative essays by Johan C. Thom, Renate Burry, Clive Chandler, Hans Daiber, Jill Craye, Andrew Smith, Hidemi Takahashi and Anna Tzvetkova-Glaser, ed. Johan C. Thom, Tübingen: Mohr Siebeck, 2014.

Purchas, Samuel, *Purchas his pilgrimage*, London: Printed by William Stansby for Henrie Fetherstone, 1613, STC (2nd edn) / 20505.

Rabelais, François, *Les Cinq Livres de faits et dits de Gargantua et Pantagruel*, ed. Marie-Madeleine Fragonard, Paris: Gallimard, Quarto, 2017.
Rainoldes, John, *Th'Overthrow of Stage-Playes* (1599), New York: Johnson Reprint, 1972.
Rastell, John, *Interlude of the Four Elements*, in A. W. Pollard (ed.), *English Miracle Plays, Moralities and Interludes*, 8th edn, Oxford: Clarendon, (1927) 1950, pp. 97–105.
Sabbattini, Nicola, *Pratiques pour fabriquer scènes et machines de théâtre* (1638), introduction by Louis Jouvet, Clermont-Ferrand: Ides et Calendes, 2015.
Scot, Reginald, *The discouerie of witchcraft*, London: Printed by [Henry Denham for] William Brome, 1584, STC (2nd edn) / 21864.
Seneca, *Natural Questions*, books 4–7, ed. Jeffrey Henderson, trans. Thomas H. Corcoran, Cambridge, MA: Harvard University Press, Loeb Classical Library, 1972.
Spenser, Edmund, *The Faerie Queene*, ed. A. C. Hamilton, London: Longman, (1977) 1998.
Spenser, Edmund, *The Shepheardes Calender* (1579), in *The Poetical Works of Edmund Spenser*, ed. J. C. Smith and E. De Selincourt, London: Oxford University Press, 1912.
Standish, Arthur, *The commons complaint*, London: Printed by William Stansby, 1611, STC (2nd edn) / 23200.
Stow, John, *The annales of England* [. . .] *continued*, London: Printed by [Eliot's Court Press and F. Kingston for] Ralfe Newbery, 1600, STC (2nd edn) / 23335.
Strachy [*sic*], William, 'A true reportory of the wracke, and redemption of Sir THOMAS GATES Knight; upon, and from the Ilands of the Bermudas [. . .]', 1610, in Samuel Purchas, *Purchas his pilgrimes*, London: Printed by William Stansby, 1625, STC (2nd edn) / 20509.
Tacitus, Cornelius, *Agricola and Germany*, trans. Anthony R. Birley, Oxford: Oxford University Press, 1999.
Theophrastus, *Concerning Weather Signs* (*De signis tempestates*), in *Enquiry into Plants*, vol. 2, books 6–9, trans. Arthur F. Hort, London: Heinemann; Cambridge, MA: Harvard University Press, Loeb Classical Library, (1916) 1977.
Thévet, André, *La Cosmographie universelle*, Paris, 1575.
Vaughan, William, *Approved directions for health*, London: Printed by T. S[nodham] for Roger Iackson, 1612, STC (2nd edn) / 24615.
Vaughan, William, *Naturall and artificial directions for health*, London: Printed by Richard Bradocke, 1600, STC (2nd edn) / 24612.
Vegetius, *The foure bookes of Flauius Vegetius Renatus*, trans. John Sadler, London: Printed by Thomas Marshe, 1572, STC 24631.
Virgil, *Eclogues, Georgics, Aeneid 1–6*, trans. H. Rushton Fairclough, revised by G. P. Goold, Cambridge, MA: Harvard University Press, Loeb Classical Library, (1916) 1999.

Vitruvius, *The Ten Books on Architecture*, trans. Ingrid D. Rowland and Michael J. Dewar, Cambridge: Cambridge University Press, 1999.

Walkington, Thomas, *The Opticke Glasse of Humors*, London, 1631.

Watson, Thomas, *The hekatompathia or Passionate centurie of loue*, London, 1582.

Webster, John, *The Works of John Webster*, ed. David Gunby, David Carnegie and Antony Hammond, 3 vols., Cambridge: Cambridge University Press, 1995.

Weis, René (ed.), *The Duchess of Malfi*, in *The Duchess of Malfi and Other Plays*, Oxford: Oxford University Press, 1996.

Whithorne, Peter, *Certaine vvayes for the ordering of souldiours in battelray*, Imprinted at London, by W. Williamson for Ihon Wight, 1573, UMI reel number 312: 5b.

Whitney, Geoffrey, *A Choice of Emblemes*, London, 1586.

Whittingham, William, *The Bible and Holy Scriptures conteyned in the Olde and Newe Testament*, Printed at Geneva, 1561, STC (2nd edn) / 2095.

Whythorne, Thomas, *The Autobiography of Thomas Whythorne* [c. 1569–76], ed. James Marshall Osborn, London: Oxford University Press, 1962.

Wright, Thomas, *The Passions of the Mind in General*, ed. William Webster Newbold, New York: Garland, 1986.

Secondary sources

Abraham, Lyndall, 'Alchemical Reference in *Antony and Cleopatra*', *Sydney Studies in English* 8 (1982), pp. 100–4.

Adelman, Janet, *Suffocating Mothers: Fantasies of Maternal Origin in Shakespeare's Plays, Hamlet to The Tempest*, New York: Routledge, 1992.

Ailles, Jennifer L., 'Ecocritical Heywood and *The Play of the Weather*', *Early Theatre* 16.2 (2013), pp. 185–96.

Alexander, Gavin, 'Dickenson, John (c. 1570–1635/6), Author and Government Official', in *Oxford Dictionary of National Biography*, Oxford: Oxford University Press, 2004; <http://www.oxforddnb.com/view/10.1093/ref:odnb/9780198614128.001.0001/odnb-9780198614128-e-7601> (last accessed 26 April 2018).

Anderson, Katharine, 'Looking at the Sky: The Visual Context of Victorian Meteorology', *The British Journal for the History of Science* 36.3 (September 2003), pp. 301–32.

Anthes, Richard A., *Tropical Cyclones: Their Evolution, Structure and Effects*, Boston: American Meteorological Society, 1882.

Archer, John Michael, *Old Worlds: Egypt, Southwest Asia, India, and Russia in Early Modern English Writing*, Stanford: Stanford University Press, 2001.

Armstrong, Philip, 'Preposterous Nature in Shakespeare's Tragedies', in Michael Neill and David Schalkwyk (eds), *The Oxford Handbook of Shakespearean Tragedy*, Oxford: Oxford University Press, 2016, pp. 104–19.

Astington, John, *English Court Theatre, 1558–1642*, Cambridge: Cambridge University Press, 1999.

Austin, John Langshaw, *How to Do Things with Words*, Oxford: Clarendon Press, 1962.

Bakhtin, Mikhail, *Rabelais and his World*, trans. Hélène Iswolsky, Bloomington: Indiana University Press, (1968) 1984.

Bakker, Fredericus Antonius, *Epicurean Meteorology: Sources, Method, Scope and Organization*, Leiden: Brill, 2016.

Baldwin, T. W., *William Shakspere's Small Latine and Lesse Greeke*, 2 vols., Urbana: University of Illinois Press, 1944.

Bates, Catherine, 'Love and Courtship', in Alexander Leggatt (ed.), *The Cambridge Companion to Shakespearean Comedy*, Cambridge: Cambridge University Press, (2002) 2003, pp. 102–22.

Bedford, Ronald, Lloyd Davis and Philippa Kelly (eds), *Early Modern Autobiography: Theories, Genres, Practices*, Ann Arbor: University of Michigan Press, 2006.

Bednarz, James P., 'Imitations of Spenser in *A Midsummer Night's Dream*', *Renaissance Drama*, New Series, 14 (1983): *Relations and Influences*, pp. 79–102.

Behringer, Wolfgang, *A Cultural History of Climate*, London: Polity Press, 2009.

Benson, Sean, *Shakespeare, Othello and Domestic Tragedy*, London: Continuum, 2012.

Berry, Philippa, 'Between Idolatry and Astrology: Modes of Temporal Repetition in *Romeo and Juliet*', in Dympna Callaghan (ed.), *A Feminist Companion to Shakespeare*, 2nd edn, Malden: Wiley Blackwell, (2000) 2016, pp. 378–92.

Berry, Philippa, *Shakespeare's Feminine Endings: Disfiguring Death in the Tragedies*, London: Routledge, 1999.

Blanchot, Maurice, *The Writing of the Disaster* (1980), trans. Ann Smock, Lincoln: University of Nebraska Press, 1995.

Bobrick, Benson, *The Fated Sky: Astrology in History*, New York: Simon & Schuster, (2005) 2006.

Boehrer, Bruce, *Environmental Degradation in Jacobean Drama*, Cambridge: Cambridge University Press, 2013.

Boia, Lucian, *The Weather in the Imagination*, trans. Roger Leverdier, London: Reaktion Books, 2005.

Bono, Barbara J., *Literary Transvaluation: From Vergilian Epic to Shakespearean Tragicomedy*, Berkeley: University of California Press, 1984.

Borlik, Todd A., *Ecocriticism and Early Modern English Literature: Green Pastures*, New York: Routledge, 2011.

Bradbrook, M. C., 'The Sources of *Macbeth*', in *Shakespeare Survey*, vol. 4, *Interpretation*, ed. Allardyce Nicoll, Cambridge: Cambridge University Press, 1951, pp. 35–48.

Brayton, Dan, 'Shakespeare and Slime: Notes and the Anthropocene', in Jennifer Munroe, Edward J. Geisweidt and Lynne Bruckner (eds), *Ecological Approaches to Early Modern English Texts: A Field Guide to Reading and Teaching*, Farnham: Ashgate, 2015, pp. 81–90.

Brayton, Dan, *Shakespeare's Ocean: An Ecocritical Exploration*, Charlottesville: University of Virginia Press, 2012.

Brogan, Stephen, *The Royal Touch in Early Modern England: Politics, Medicine and Sin*, Woodbridge: The Boydel Press, The Royal Historical Society, 2015.

Brown, Pamela Allen, '*Othello* Italicized: Xenophobia and the Erosion of Tragedy', in Michele Marrapodi (ed.), *Shakespeare, Italy, and Intertextuality*, Manchester: Manchester University Press, 2004, pp. 145–57.

Brownlow, F. W., *Shakespeare, Harsnett, and the Devils of Denham*, Newark: University of Delaware Press, 1993.

Bruckner, Lynne and Dan Brayton (eds), *Ecocritical Shakespeare*, New York: Routledge, 2016.

Burnett, Mark Thornton, *Constructing Monsters in Shakespeare's Drama and Early Modern Culture*, Basingstoke: Palgrave Macmillan, 2002.

Burt, Stephen, *The Weather Observer's Handbook*, Cambridge: Cambridge University Press, 2012.

Burton, Jonathan, *Traffic and Turning: Islam and English Drama, 1579–1624*, Newark: University of Delaware Press, 2005.

Butler, Chris, '"The howling of Irish wolves": *As You Like It* and the Celtic Essex Circle', in Willy Maley and Rory Loughnane (eds), *Celtic Shakespeare: The Bard and the Borderers*, London: Routledge, (2013) 2016, pp. 89–102.

Butler, Martin, '*The Tempest* and the Jonsonian Masque', in Sophie Chiari and John Mucciolo (eds), *Performances at Court in Shakespeare's Era*, forthcoming.

Butler, Sara M., *The Language of Abuse: Marital Violence in Later Medieval England*, Leiden: Brill, 2007.

Caldwell, Melissa M., *Skepticism and Belief in Early Modern England: The Reformation of Moral Value*, Abingdon: Routledge, 2017.

Cameron, Euan, *Enchanted Europe: Superstition, Reason, and Religion 1250–1750*, Oxford: Oxford University Press, 2010.

Capp, Bernard, *Astrology and the Popular Press: English Almanacs 1500–1800*, London: Faber and Faber, 1979.

Cavert, William M., *The Smoke of London: Energy and Environment in the Early Modern City*, Cambridge: Cambridge University Press, 2016.

Chen-Morris, Raz, '"The Quality of Nothing": Shakespearean Mirrors and Kepler's Visual Economy of Science', in Ofer Gal and Raz Chen-Morris (eds), *Science in the Age of Baroque*, Dordrecht: Springer, pp. 99–118.

Chiari, Sophie, 'Le temps et ses objets dans l'oeuvre de Shakespeare', in Line Cottegnies, Anne-Valérie Dulac and Anne-Marie Miller-Blaise (eds), *Transformer le corps masculin: armes, armures et objets personnels gravés*, Paris: CNRS Éditions, forthcoming.

Chiari, Sophie, '"Vat is de clock, Jack?": Shakespeare and the Technology of Time', in Sophie Chiari and Mickaël Popelard (eds), *Spectacular Science, Technology and Superstition in the Age of Shakespeare*, Edinburgh: Edinburgh University Press, 2017, pp. 197–220.

Clark, Stuart, *Thinking with Demons: The Idea of Witchcraft in Early Modern Europe*, Oxford: Oxford University Press, 1997.

Clericuzio, Antonio, *Elements, Principles and Corpuscles: A Study of Atomism and Chemistry in the Seventeenth Century*, Dordrecht: Springer, 2000.

Colie, Rosalie L., *Shakespeare's Living Art*, Princeton: Princeton University Press, 1974.

Connell, Sophie M., *Aristotle on Female Animals: A Study of the Generation of Animals*, Cambridge: Cambridge University Press, 2016.

Considine, John, 'Hill, Thomas (*c.*1528–*c.*1574)', in *Oxford Dictionary of National Biography*, Oxford: Oxford University Press, 2004; <http://www.oxforddnb.com/view/article/13303> (last accessed 26 April 2018).

Cook, Ann Jennalie, *The Privileged Playgoers of Shakespeare's London, 1575–1642*, Princeton: Princeton University Press, 1981.

Cooper, Alix, *Inventing the Indigenous: Local Knowledge and Natural History in Early Modern Europe*, Cambridge: Cambridge University Press, 2007.

Coursen, Herbert R., *Watching Shakespeare on Television*, Rutherford: Fairleigh Dickinson Press; London: Associated University Presses, 1993.

Craig, John, 'Bodies at Prayer in Early Modern England', in Natalie Mears and Alec Ryrie (eds), *Worship and the Parish Church in Early Modern Britain*, London: Routledge, (2013) 2016, pp. 173–96.

Crane, Mary Thomas, *Losing Touch with Nature: Literature and the New Science in Sixteenth-Century England*, Baltimore: Johns Hopkins University Press, 2014.

Crane, Mary Thomas, 'Optics', in Henry S. Turner (ed.), *Early Modern Theatricality*, Oxford: Oxford University Press, 2013, pp. 250–69.

Crane, Mary Thomas, 'Roman World, Egyptian Earth: Cognitive Difference and Empire in Shakespeare's *Antony and Cleopatra*', *Comparative Drama* 43.1 (Spring 2009), pp. 1–17.

Crowley, Roger, *Empires of the Sea: The Siege of Malta, the Battle of Lepanto, and the Contest for the Center of the World*, London: Faber, 2009.

Cunin, Muriel, '*King Lear*: Fabric of the Human Body and Anatomy of the World', in François Laroque, Pierre Iselin and Sophie Alatorre (eds),

'And that's true too': New Essays on King Lear, Newcastle upon Tyne: Cambridge Scholars Publishing, 2009, pp. 90–103.

Curtis, Mary Ann, 'The Joining of Male and Female: An Alchemical of Transmutation in *Antony and Cleopatra*', *The Upstart Crow* 12 (1992), pp. 116–26.

Damisch, Hubert, *A Theory of Cloud: Toward a History of Painting*, trans. Janet Lloyd, Stanford: Stanford University Press, (1972) 2002.

Darriulat, Jacques, *Sébastien le Renaissant*, Paris: Lagune, 1998.

Davies, Surekha, *Renaissance Ethnography and the Invention of the Human: New Worlds, Maps and Monsters*, Cambridge: Cambridge University Press, 2016.

Davis, Nick, *Early Modern Writing and the Privatization of Experience*, London: Bloomsbury, 2013.

Dean, John, 'The *Odyssey* as Romance', *College Literature* 3.3 (Fall 1976): *The Homeric Epic*, pp. 228–36.

Decker, John R., 'Introduction: Spectacular Unmaking: Creative Destruction, Destructive Creativity', in John R. Decker and Mitzi Kirkland-Ives (eds), *Death, Torture and the Broken Body in European Art, 1300–1650*, Farnham: Ashgate, 2015, pp. 1–18.

Degenhardt, Jane Hwang, *Islamic Conversion and Christian Resistance on the Early Modern Stage*, Edinburgh: Edinburgh University Press, 2010.

Degroot, Dagomar, 'Exploring the North in a Changing Climate: The Little Ice Age and the Journals of Henry Hudson, 1607–1611', *Journal of Northern Studies* 9.1 (2015), pp. 69–91.

Demaray, John, *Shakespeare and the Spectacles of Strangeness: The Tempest and the Transformation of Renaissance Theatrical Forms*, Pittsburgh: Duquesne University Press, 1998.

Desai, R. W., '"What means Sicilia? He something seems unsettled": Sicily, Russia, and Bohemia in *The Winter's Tale*', *Comparative Drama* 30.3 (Fall 1996), pp. 311–24.

Dessen, Alan, 'Mist and Fog on the Elizabethan and Jacobean Stage', in Virginia Mason Vaughan, Fernando Cioni and Jacquelyn Bessel (eds), *Speaking Pictures: The Visual/Verbal Nexus of Dramatic Performance*, Madison, NJ: Fairleigh Dickinson University Press, 2010, pp. 106–18.

DeWall, Nichole, '"A Plague 'o Both Your Houses": Shakespeare and Early Modern Plague', unpublished dissertation, Northeastern University, Boston, April 2008.

Dick, Steven J., *Discovery and Classification in Astronomy: Controversy and Consensus*, Cambridge: Cambridge University Press, 2013.

Didi-Huberman, Georges, *Ninfa profunda. Essai sur le drapé-tourmenté*, Paris: Gallimard, 2017.

DiGangi, Mario, *Sexual Types: Embodiment, Agency, and Dramatic Character from Shakespeare to Shirley*, Philadelphia: University of Pennsylvania Press, 2011.

Dillon, Janette, 'Chariots and Cloud Machines: Gods and Goddesses on Early English Stages', in Lloyd Edward Kermode, Jason Scott-Warren and Martine van Elk (eds), *Tudor Drama before Shakespeare, 1485–1590: New Directions for Research, Criticism, and Pedagogy*, New York: Palgrave Macmillan, pp. 111–29.

Dobson, Mary J., *Contours of Death and Disease in Early Modern England*, Cambridge: Cambridge University Press, 1997.

Dodson, Sarah, 'Notes on the Earthquake in *Romeo and Juliet*', *Modern Language Notes* 65.2 (February 1950), p. 144.

Dooley, Brendan (ed.), *A Companion to Astrology in the Renaissance*, Leiden: Brill, 2014, pp. 429–34.

Dusinberre, Juliet, 'Pancakes and a Date for *As You Like It*', *Shakespeare Quarterly* 54 (2003), pp. 371–405.

Dutton, Richard, *Shakespeare, Court Dramatist*, Oxford: Oxford University Press, 2016.

Egan, Gabriel, *Green Shakespeare: From Ecopolitics to Ecocriticism*, London: Routledge, 2006.

Egan, Gabriel, *Shakespeare and Ecocritical Theory*, London: Bloomsbury, 2015.

Eliot, T. S., 'Hamlet and his Problems', in *The Sacred Wood: Essays on Poetry and Criticism*, New York: Barnes & Noble, 1960, pp. 95–103.

Ellis, Antony, *Old Age, Masculinity and Early Modern Drama: Comic Elders on the Italian and Shakespearean Stage*, Farnham: Ashgate, 2009.

Elton, William R., *King Lear and the Gods*, Lexington: University Press of Kentucky, (1966) 1988.

Emanuel, Kerry, *Divine Wind: The History and Science of Hurricanes*, Oxford: Oxford University Press, 2005.

Engler, Balz, 'Othello's Great Heart', *English Studies* 68 (1987), pp. 129–36.

Estok, Simon C., *Ecocriticism and Shakespeare: Reading Ecophobia*, New York: Palgrave Macmillan, 2011.

Evans, Gwynne Blakemore, *Elizabethan Jacobean Drama: The Theatre in its Time*, New York: New Asterdam Books, (1988) 1990.

Everett, Barbara, '*Romeo and Juliet*: The Nurse's Story', *Critical Quarterly* (Summer 1972), pp. 129–39.

Ewen, C. H. L., *Witch Hunting and Witch Trials*, London, 1929.

Fagan, Brian, *The Little Ice Age: How Climate Made History, 1300–1850*, New York: Basic, 2001.

Feerick, Jean E., 'A "Nation . . . Now Degenerate": Shakespeare's *Cymbeline*, Nova Britannia, and the Role of Diet and Climate in Reproducing Races', *Early American Studies* 1.2 (Fall 2003), pp. 30–71.

Feerick, Jean E., *Strangers in Blood: Relocating Race in the Renaissance*, Toronto: University of Toronto Press, 2010.

Findlay, Alison, *Women in Shakespeare: A Dictionary*, London: Continuum, 2010.

Fleming, James Dougal, *The Mirror of Information in Early Modern England: John Wilkins and the Universal Character*, New York: Palgrave Macmillan, 2017.

Floyd-Wilson, Mary, *English Ethnicity and Race in Early Modern Drama*, Cambridge: Cambridge University Press, 2003.

Floyd-Wilson, Mary, 'English Mettle', in Gail Kern Paster, Katherine Rowe and Mary Floyd-Wilson (eds), *Reading the Early Modern Passions: Essays in the Cultural History of Emotions*, Philadelphia: University of Pennsylvania Press, 2004, pp. 130–46.

Floyd-Wilson, Mary, 'Transmigrations: Crossing Regional and Gender Boundaries in *Antony and Cleopatra*', in Viviana Comensoli and Anne Russel (eds), *Enacting Gender on the English Renaissance Stage*, Urbana: University of Illinois Press, 1999, pp. 73–96.

Floyd-Wilson, Mary and Garrett A. Sullivan Jr (eds), *Environment and Embodiment in Early Modern England*, Basingstoke: Palgrave Macmillan, 2007.

Francis, Toni, 'New Directions: Imperialism as Devilry: A Postcolonial Reading of *Doctor Faustus*', in Sarah Munson Deats (ed.), *Doctor Faustus: A Critical Guide*, London: Continuum, (2010) 2012, pp. 111–23.

Franklin, Colleen, '"An Habitation of Devils, a Domicill for Unclean Spirits, and a Den of Goblings": The Marvellous North in Early Modern English Literature', in Helen Ostovich, Mary V. Silcox and Graham Roebuck (eds), *The Mysterious and the Foreign in Early Modern England*, Cranbury: Associated University Presses, pp. 27–38.

Frisinger, H. Howard, *The History of Meteorology: To 1800*, New York: Science History Publications, 1977.

Frye, Northrop, *Anatomy of Criticism: Four Essays*, with a new foreword by Harold Bloom, Princeton: Princeton University Press, (1957) 2000.

Fuller, Mary C., *Remembering the Early Modern Voyage: English Narratives in the Age of European Expansion*, New York: Palgrave Macmillan, 2008.

Fulton, Robert C., '*The Tempest* and the Bermuda Pamphlets: Source and Thematic Intention', *Interpretations* 10.1 (1978), pp. 1–10.

Gaard, Greta, Simon C. Estok and Serpil Oppermann, *International Perspectives in Feminist Ecocriticism*, New York: Routledge, 2013.

Gaignebet, Claude, *À plus hault sens*, 2 vols., Paris: Maisonneuve & Larose, 1986.

Garber, Marjorie, *Shakespeare After All*, New York: Anchor Books, 2004.

Gatti, Hilary, *Giordano Bruno and Renaissance Science*, Ithaca: Cornell University Press, 1999.

Gill, Mary Louise, *Aristotle on Substance: The Paradox of Unity*, Princeton: Princeton University Press, 1989.

Gillespie, Raymond, *Devoted People: Belief and Religion in Early Modern England*, Manchester: Manchester University Press, 1997.

Gillies, John, *Shakespeare and the Geography of Difference*, Cambridge: Cambridge University Press, 1994.
Gillies, John, 'Shakespeare's Virginian Masque', in Stephen Orgel and Sean Keilen (eds), *Shakespeare and History*, New York: Garland Publishing, 1999, pp. 83–118.
Glete, Jan, 'Warfare at Sea 1450–1815', in Jeremy Black (ed.), *War in the Early Modern World 1450–1815*, London: Routledge, pp. 25–52.
Golinski, Jan, *British Weather and the Climate of Enlightenment*, Chicago: University of Chicago Press, 2007.
Gombrich, Ernst, *Art and Illusion: A Study in the Psychology of Pictorial Representation*, London: Phaidon, 1960.
Gómez, Nicolás Wey, 'The Politics of Light: Al-Kindi's Geometrical Optics and the Vindication of the American Tropics in Bartolomé de las Casas's *Apologética historia sumaria* (1527–1561)', in Walter S. Melion and Lee Palmer Wandel (eds), *Early Modern Eyes*, Leiden: Brill, 2010, pp. 11–54.
Gómez, Nicolás Wey, *The Tropics of Empire: Why Columbus Sailed South to the Indies*, Cambridge, MA, and London: MIT Press, 2008.
Goodland, Katharine, 'Inverting the Pietà in Shakespeare's *King Lear*', in Regina Buccola and Lisa Hopkins (eds), *Marian Moments in Early British Drama*, London: Routledge, 2007, pp. 65–74.
Grant, Mark, *Galen on Food and Diet*, London: Routledge, 2000.
Graves, R. B., *Lighting the Shakespearean Stage, 1567–1642*, Carbondale: Southern Illinois University Press, 1999.
Green, Ian, *Humanism and Protestantism in Early Modern English Education*, Farnham: Ashgate, 2009.
Greenblatt, Stephen, *The Rise and Fall of Adam and Eve*, New York: W. W. Norton & Company, 2017.
Greenblatt, Stephen, *The Swerve: How the Renaissance Began*, London: The Bodley Head, 2011.
Greene, Gayle, 'Shakespeare's *Tempest* and Eliot's *Waste Land*: "What the Thunder Said"', *Orbis Litterarum* 34.4 (December 1979), pp. 287–300.
Greene, Thomas M., 'Magic and Festivity at the Renaissance Court: The 1987 Josephine Waters Bennett Lecture', *Renaissance Quarterly* 40.4 (Winter 1987), pp. 636–59.
Grigsby, Bryon Lee, *Pestilence in Medieval and Early Modern English Literature*, New York and London: Routledge, 2004.
Groves, Beatrice, *Texts and Traditions: Religion in Shakespeare 1592–1604*, Oxford: Clarendon Press, 2007.
Gurr, Andrew, 'Another Jonson Critic', *The Ben Jonson Journal* 18 (2011), pp. 27–44.
Hackett, Helen, '*A Midsummer Night's Dream*', in Richard Dutton and Jean Howard (eds), *A Companion to Shakespeare's Works, Volume III: The Comedies*, Malden: Blackwell Publishing, 2003, pp. 338–57.

Halio, Jay L., *Shakespeare in Performance: A Midsummer Night's Dream*, 2nd edn, Manchester: Manchester University Press, (1995) 2003.
Hall, Jonathan, *Anxious Pleasures: Shakespearean Comedy and the Nation-State*, Madison and Teaneck: Fairleigh Dickinson University Press; London: Associated University Presses, 1995.
Hallyn, Fernand, *Gemma Frisius, arpenteur de la terre et du ciel*, Paris: Champion, 2008.
Hamilton, Jennifer Mae, *This Contentious Storm: An Ecocritical and Performance History of King Lear*, London: Bloomsbury, 2017.
Harris, Alexandra, *Weatherland: Writers and Artists under English Skies*, London: Thames & Hudson, 2016.
Harris, Henry, *Things Come to Life: Spontaneous Generation Revisited*, Oxford: Oxford University Press, 2002.
Harris, Johanna, '"Be plyeabell to all good counsell": Lady Brilliana Harley's Advice Letter to her Son', in James Daybell and Andrew Gordon (eds), *Women and Epistolary Agency in Early Modern Culture, 1450–1690*, London: Routledge, 2016, pp. 128–48.
Harris, Jonathan Gil, 'Shakespeare and Race', in Margreta de Grazia and Stanley Wells (eds), *The New Cambridge Companion to Shakespeare*, Cambridge: Cambridge University Press, (2001) 2010, pp. 201–16.
Harris, Jonathan Gil, *Untimely Matter in the Time of Shakespeare*, Philadelphia: University of Pennsylvania Press, 2008.
Hassel, R. Chris, Jr, *Renaissance Drama and the English Church Year*, Lincoln: University of Nebraska Press, 1979.
Healy, Margaret, *Fictions of Disease in Early Modern England: Bodies, Plagues and Politics*, Basingstoke: Palgrave Macmillan, 2001.
Healy, Margaret, *Shakespeare, Alchemy and the Creative Imagination: The Sonnets and A Lover's Complaint*, Cambridge: Cambridge University Press, 2011.
Heilman, Robert B., *Magic in the Web: Action and Language in Othello*, Lexington: University Press of Kentucky, 1956.
Heninger, S. K., Jr, 'Sequences, Systems, Models: Sidney and the Secularization of Sonnets', in Neil Fraistat (ed.), *Poems in their Place: Intertextuality and Order of Poetic Collections*, Chapel Hill: University of North Carolina Press, 1986, pp. 66–94.
Hiltner, Ken, *What Else Is Pastoral? Renaissance Literature and the Environment*, Ithaca: Cornell University Press, 2011.
Hoenselaars, A. J., *Images of Englishmen and Foreigners in the Drama of Shakespeare and his Contemporaries: A Study of Stage Characters and National Identity in English Renaissance Drama, 1558–1642*, London: Associated University Presses, 1992.
Holderness, Graham, *Shakespeare and Venice*, Farnham: Ashgate, 2010.
Hulme, Mike, 'Climate', in Bruce R. Smith (ed.), *The Cambridge Guide to the Worlds of Shakespeare*, vol. 1, *Shakespeare's World, 1500–1660*, Cambridge: Cambridge University Press, 2016, pp. 29–34.

Hulme, Mike, 'The Conquering of Climate: Discourses of Fear and their Dissolution', *The Geographical Journal* 174.1 (March 2008), pp. 5–16.

Hulme, Mike, *Weathered Cultures of Climate*, Los Angeles: Sage, 2017.

Hunter, Robert G., 'Cleopatra and the "Oestre Junonicque"', in *Shakespeare Studies* 5 (1969), pp. 236–9.

Iselin, Pierre, '"Strange things toward": The Dismembered Perspective of *King Lear*', in François Laroque, Pierre Iselin and Sophie Alatorre (eds), *'And that's true too': New Essays on King Lear*, Newcastle upon Tyne: Cambridge Scholars Publishing, 2009, pp. 104–21.

Iyengar, Sujata, 'Shakespeare's Embodied Ontology of Gender, Air and Health', in Sujata Iyengar (ed.), *Disability, Health, and Happiness in the Shakespearean Body*, New York: Routledge, 2015, pp. 176–92.

Jankovic, Vladimir, *Reading the Skies*, Chicago: University of Chicago Press, 2001.

Johnson, Francis R., *Astronomical Thought in Renaissance England: A Study of English Scientific Writings from 1500 to 1645*, New York: Octagon Books, 1968.

Jolly, Margrethe, *The First Two Quartos of Hamlet: A New View of the Origins and Relationship of the Texts*, Jefferson, NC: McFarland & Company, Inc., 2014.

Jones, Gwilym, *Shakespeare's Storms*, Manchester: Manchester University Press, 2015.

Jones, Gwilym, 'Storm Effects in Shakespeare', in Farah Karim Cooper and Tiffany Stern (eds), *Shakespeare's Theatre and the Effects of Performance*, London: Bloomsbury, The Arden Shakespeare, pp. 33–50.

Jong, H. M. E. de (ed.), *Michael Maier's Atalanta fugiens: Sources of an Alchemical Book of Emblems*, Leiden: Brill, 1969.

Jouanna, Jacques, *Hippocrates*, trans. M. B. DeBevoise, Baltimore: Johns Hopkins University Press, 1999.

Kahn, Coppélia, 'The Absent Mother in *King Lear*', in Margaret W. Ferguson, Maureen Quilligan and Nancy J. Vickers (eds), *Rewriting the Renaissance: Discourse of Sexual Differences in Early Modern Europe*, Chicago: University of Chicago Press, 1985, pp. 33–49.

Kelly, Morgan and Cormac Ó Gráda, 'The Waning of the Little Ice Age: Climate Change in Early Modern Europe', *Journal of Interdisciplinary History* 44.3 (Winter 2014), pp. 301–25.

Kempe, Michael, 'Noah's Flood: The Genesis Story and Natural Disasters in Early Modern Times', *Environment and History* 9.2 (May 2003): *Coping with the Unexpected – Natural Disasters and their Perception*, pp. 151–71.

Kennedy, Rebecca F., C. Sydnor Roy and Max L. Goldman (eds), *Race and Ethnicity in the Classical World: An Anthology of Primary Sources in Translation*, Indianapolis and Cambridge: Hackett Publishing Company, 2013.

Kernan, Alvin B., *Shakespeare, the King's Playwright: Theater in the Stuart Court, 1603–1613*, New Haven: Yale University Press, 1995.

Klein, Bernhard, 'Antony and Cleopatra', in Michael Neill and David Schalkwyck (eds), *The Oxford Handbook of Shakespearean Tragedy*, Oxford: Oxford University Press, 2016, pp. 452–67.

Knight, Leah, *Reading Green in Early Modern England*, New York: Routledge, (2014) 2016.

Kogel, Renée, *Pierre Charron*, Geneva: Droz, 1972.

Kott, Jan, *Shakespeare Our Contemporary*, New York: W. W. Norton & Company, (1961) 1974.

Lamb, H. H., *Climate, History and the Modern World*, 2nd edn, London: Routledge, (1982) 1995.

Laoutaris, Chris, *Shakespearean Maternities: Crises of Conception in Early Modern England*, Edinburgh: Edinburgh University Press, 2008.

Laroque, François, 'En marge de l'idéologie: antimasque et grotesque dans le *Dr Faustus* et *La Tempête*', *Actes des congrès de la Société française Shakespeare* 3 (1981), pp. 99–114.

Laroque, François, '"Motley's the only wear": *As You Like It* ou la bigarrure', in Jean-Paul Debax and Yves Peyré (eds), *As You Like It: essais critiques*, Toulouse: Presses Universitaires du Mirail, 1998, pp. 77–92.

Laroque, François, 'Roméo et Juliette, entre violence et jouissance', *Revue Silène* (2007), pp. 1–14.

Laroque, François, 'The Science of Astrology in Shakespeare's Sonnets, *Romeo and Juliet* and *King Lear*', in Sophie Chiari and Mickaël Popelard (eds), *Spectacular Science, Technology and Superstition in the Age of Shakespeare*, Edinburgh: Edinburgh University Press, 2017, pp. 29–42.

Laroque, François, *Shakespeare's Festive World: Elizabethan Seasonal Entertainment and the Professional Stage*, trans. Janet Lloyd, Cambridge: Cambridge University Press, (1991) 1993.

Law, Ernest, 'Shakespeare's *Tempest* as Originally Produced at Court', in Patrick M. Murphy (ed.), *The Tempest: Critical Essays*, New York: Routledge, 2010, pp. 150–72.

Le Goff, Jacques, *Birth of Purgatory*, trans. Arthur Goldhammer, Chicago: University of Chicago Press, 1981.

Le Roy Ladurie, Emmanuel, *Times of Feast, Times of Famine: A History of Climate since the Year 1000*, trans. Barbaray Bray, Garden City: Doubleday, 1971.

Lennard, John, *William Shakespeare, King Lear*, Penrith: Humanities-Ebooks LLP, 2010.

Levine, Laura, *Men in Women's Clothing: Anti-theatricality and Effeminization, 1579–1642*, Cambridge: Cambridge University Press, 1994.

Lewis, Rhodri, *Hamlet and the Vision of Darkness*, Princeton: Princeton University Press, 2017.

Lindheim, Nancy, '*King Lear* as Pastoral Tragedy', in Rosalie Colie and F. T. Flahiff (eds), *Some Facets of King Lear: Essays in Prismatic Criticism*, London: Heinemann, 1974, pp. 169–84.

Llasera, Margaret, *Représentations scientifiques et images poétiques en Angleterre au XVIIe siècle. À la recherche de l'invisible*, Paris: CNRS Éditions, 1999.

Lock, Julian, 'Brasbridge, Thomas (1536/7–1593)', in *Oxford Dictionary of National Biography*, Oxford: Oxford University Press, 2004; online edn, January 2015, <http://www.oxforddnb.com/view/article/3285> (last accessed 26 April 2018).

Long, Kathleen P., 'Odd Bodies: Reviewing Corporeal Difference in Early Modern Alchemy', in Kathleen P. Long (ed.), *Gender and Scientific Discourse in Early Modern Culture*, Farnham: Ashgate, 2010, pp. 63–86.

Lovejoy, Arthur O., *The Great Chain of Being: A Study of the History of an Idea*, with a new introduction by Peter J. Stanlis, New Brunswick, NJ: Transaction Publishers, (1936) 2009.

Lucking, David, 'Carrying Tempest in his Hand and Voice: The Figure of the Magician in Jonson and Shakespeare', *English Studies: A Journal of English Language and Literature* 85 (2004), pp. 297–310.

Lurbe, Pierre, 'Theologico-political Issues in Richard Hooker's *Laws of Ecclesiastical Polity* and Thomas Hobbes's *Leviathan*', *Revue LISA/LISA e-journal* 9.1 (2011); <http://lisa.revues.org/4050> (last accessed 2 May 2018).

Macauley, David, *Elemental Philosophy: Earth, Air, Fire, and Water as Environmental Ideas*, New York: SUNY Press, 2010.

McCombie, Frank, 'Medium and Message in *As You Like It* and *King Lear*', in *Shakespeare Survey*, vol. 33, *King Lear*, ed. Kenneth Muir, Cambridge: Cambridge University Press, (1980) 2002, pp. 67–80.

McGough, Laura J., *Gender, Sexuality, and Syphilis in Early Modern Venice: The Disease that Came to Stay*, London: Palgrave Macmillan, 2010.

MacMullan, Katherine Vance, 'Death Imagery in *Antony and Cleopatra*', *Shakespeare Quarterly* 14.4 (Autumn 1963), pp. 399–410.

Magnusson, A. Lynne, 'Interruption in *The Tempest*', *Shakespeare Quarterly* 37.1 (Spring 1986), pp. 52–65.

Maley, Willy, 'Critical Review: "Great thing of us forgot"? New British Angles on *King Lear*', in Andrew Hiscock and Lisa Hopkins (eds), *King Lear: A Critical Guide*, New York: Continuum Renaissance Drama, 2011, pp. 156–79.

Marienstras, Richard, *New Perspectives on the Shakespearean World*, Cambridge: Cambridge University Press, 1985.

Markley, Robert, 'Summer's Lease: Shakespeare in the Little Ice Age', in Ivo Kamps, Thomas Hallocks and Karen Raber (eds), *Early Modern Ecostudies: From the Florentine Codex to Shakespeare*, New York: Palgrave Macmillan, 2008, pp. 131–42.

Martin, Craig, 'Conjecture, Probabilism, and Provisional Knowledge in Renaissance Meteorology', *Early Science and Medicine* 14.1/3 (2009): *Evidence and Interpretation: Studies on Early Science and Medicine in Honour of John E. Murdoch*, pp. 265–89.

Martin, Craig, 'Ludovico Settala's Aristotelian *Problemata* Commentary and Late-Renaissance Hippocratic Medicine', in Peter Distelzweig, Benjamin Goldberg and Evan R. Ragland (eds), *Early Modern Medicine and Natural Philosophy*, Heidelberg: Springer, pp. 19–42.

Martin, Craig, *Renaissance Meteorology: Pomponazzi to Descartes*, Baltimore: Johns Hopkins University Press, 2011.

Martindale, Charles and Michelle, *Shakespeare and the Uses of Antiquity: An Introductory Essay*, London: Routledge, (1990) 1994.

Matthes, François E., 'Report of Committee on Glaciers', *Transactions, American Geophysical Union* 20 (1939), pp. 518–23.

Maury, C. A., 'Phenomena of the *Heaven* in the Odyssey', *The Classical Weekly* 11.14 (28 January 1918), pp. 105–6.

Mazzio, Carla, 'Coda: Scepticism and the Spectacular: On Shakespeare in an Age of Science', in Sophie Chiari and Mickaël Popelard (eds), *Spectacular Science, Technology and Superstition in the Age of Shakespeare*, Edinburgh: Edinburgh University Press, 2017, pp. 237–44.

Mazzio, Carla, 'History of Air: Hamlet and the Trouble with Instruments', *South Central Review* 26.1, pp. 153–96.

Mentz, Steve, 'Airy Spirits: Winds, Bodies, and Ecological Force in Early Modern England', in *The Shakespearean International Yearbook*, vol. 15, *Shakespeare and the Human*, general ed. Tom Bishop and Alexa Huang, guest ed. Tiffany Jo Werth, Farnham: Ashgate, 2015, pp. 21–38.

Mentz, Steve, *At the Bottom of Shakespeare's Ocean*, Harrisburg, PA: Continuum, 2009.

Mentz, Steve, 'Hurricanes, Tempests, and the Meteorological Globe', in Howard Marchitello and Evelyn Tribble (eds), *The Palgrave Handbook of Early Modern Literature and Science*, London: Palgrave Macmillan, 2017, pp. 257–76.

Mentz, Steve, 'Strange Weather in *King Lear*', *Shakespeare* 6.2 (2010), pp. 139–52.

Miller-Blaise, Anne-Marie, '"No temple but the wood[en]": Stagecraft and the Staging of the Theatre's Materiality in Shakespeare's *As You Like It*', in Sophie Chiari, Sophie Lemercier-Goddard and Michèle Vignaux (eds), *New Perspectives on Shakespeare's As You Like It*, Clermont-Ferrand: Presses Universitaires Blaise Pascal, 2017, pp. 197–215.

Milward, Peter, 'Wright, Thomas (*c.*1561–1623)', in *Oxford Dictionary of National Biography*, Oxford: Oxford University Press, 2004; <http://www.oxforddnb.com/view/article/30059> (last accessed 26 April 2018).

Morton, Timothy, *Ecology Without Nature: Rethinking Environmental Aesthetics*, Cambridge: Cambridge University Press, 2009.

Mullaney, Steven, *The Reformation of Emotions in the Age of Shakespeare*, Chicago: University of Chicago Press, 2015.

Mulryne, J. R., 'Cleopatra's Barge and Antony's Body: Italian Sources and English Theatre', in Michele Marrapodi (ed.), *Shakespeare, Italy and Intertextuality*, Manchester: Manchester University Press, 2004, pp. 197–215.

Munroe, Jennifer, 'Shakespeare and Ecocriticism Reconsidered', *Literature Compass* 12.9 (2015), pp. 461–70.

Munroe, Jennifer, Edward J. Geisweidt and Lynne Bruckner, *Ecological Approaches to Early Modern English Texts: A Field Guide to Reading and Teaching*, New York: Routledge, 2016.

Nadal, Robert and Arnaud Zucker, 'Les éclipses', in Arnaud Zucker (ed.), *L'encyclopédie du ciel. Mythologie, Astronomie, Astrologie*, Paris: Robert Laffont, coll. 'Bouquins', pp. 474–81.

Nagel, Alexander, 'Structural Indeterminacy in Early-Sixteenth-Century Italian Painting', in Alexander Nagel and Lorenzo Pericolo (eds), *Subject as Aporia in Early Modern Art*, London: Routledge, (2010) 2016, pp. 17–42.

Nardizzi, Vin, *Wooden Os: Shakespeare's Theatres and England's Trees*, Toronto: University of Toronto Press, 2013.

Neelands, W. David, 'Predestination', in Torrance Kirby (ed.), *A Companion to Richard Hooker*, Leiden and Boston: Brill, 2008, pp. 185–220.

Neely, Wayne, *The Great Bahamas Hurricane of 1929: The Story of the Greatest Bahamian Hurricane of the Twentieth Century*, Bloomington: iUniverse LLC, 2013.

Neill, Michael, '"Noises, / Sounds, and Sweet Airs": The Burden of Shakespeare's *Tempest*', in *The Tempest*, ed. Tiffany Grace, Boston: Wadsworth, Cengage Learning, 2012, pp. 309–34.

Nero, Ruth, 'Tragic Form in *Romeo and Juliet*', *Studies in English Literature, 1500–1900* 9.2 (Spring 1969): *Elizabethan and Jacobean Drama*, pp. 241–58.

Newman, Karen, '"And wash the Ethiop white": Femininity and the Monstrous in *Othello*', in *Fashioning Femininity and English Renaissance Drama*, foreword by Catharine R. Stimpson, Chicago: University of Chicago Press, 1991, pp. 71–94.

Newman, William R., *Promethean Ambitions: Alchemy and the Quest to Perfect Nature*, Chicago: University of Chicago Press, (2004) 2005.

Nosworthy, J. M., 'The Narrative Sources of *The Tempest*', *The Review of English Studies* 24.96 (October 1948), pp. 281–94.

Nutall, Anthony David, *A New Mimesis: Shakespeare and the Representation of Reality*, with a new preface by its author, New Haven: Yale University Press, (1983) 2007.

O'Callaghan, Michelle, 'Pastoral', in Michael Hattaway (ed.), *A Companion to English Renaissance and Culture*, Malden: Blackwell Publishing, 2002, pp. 307–16.

O'Connell, Michael, 'The Experiment of Romance', in Alexander Leggatt (ed.), *The Cambridge Companion to Shakespearean Comedy*, Cambridge: Cambridge University Press, 2002, pp. 215–29.

Olson, Rebecca, 'Painted Cloths and the Making of Whitehall's Playing Space, 1611–1612', in Sophie Chiari and John Mucciolo (eds), *Performances at Court in Shakespeare's Era*, forthcoming.

Orlin, Lena Cowen, 'Desdemona's Disposition', in Shirley Nelson Garner and Madelon Sprengnether (eds), *Shakespearean Tragedy and Gender*, Bloomington and Indianapolis: Indiana University Press, 1996, pp. 171–92.

Oster, Emily, 'Witchcraft, Weather and Economic Growth in Renaissance Europe', *Journal of Economic Perspectives* 18.2 (Winter 2004), pp. 215–28.

Parker, Patricia, 'Barbers, Infidels, and Renegades: *Antony and Cleopatra*', in Lena Cowen Orlin (ed.), *Center or Margin: Revisions of the English Renaissance in Honor of Leeds Barroll*, Selinsgrove: Susquehanna University Press, 2006, pp. 54–90.

Parker, Patricia, 'The Novelty of Different Tongues: Polyglot Punning in Shakespeare (and Others)', in François Laroque and Franck Lessay (eds), *Esthétique de la nouveauté à la Renaissance*, Paris: Presses de la Sorbonne Nouvelle, 2001, pp. 41–58.

Parr, Johnstone, *Tamburlaine's Malady*, Kingsport: University of Alabama Press, 1953.

Passannante, Gerard, *The Lucretian Renaissance: Philology and the Afterlife of Tradition*, Chicago: University of Chicago Press, 2011.

Paster, Gail Kern, 'Becoming the Landscape: The Ecology of the Passions in the Legend of Temperance', in Mary Floyd-Wilson and Garrett A. Sullivan Jr (eds), *Environment and Embodiment in Early Modern England*, Basingstoke: Palgrave Macmillan, 2007, pp. 137–52.

Paster, Gail Kern, *The Body Embarrassed: Drama and the Disciplines of Shame in Early Modern England*, Ithaca: Cornell University Press, 1993.

Paster, Gail Kern, 'Leaky Vessels: The Incontinent Women of City Comedy', *Renaissance Drama* 18 (1987), pp. 43–65.

Paster, Gail Kern, '"Minded Like the Weather": The Tragic Body and its Passions', in Michael Neill and David Schalkwyk (eds), *The Oxford Handbook of Shakespearean Tragedy*, Oxford: Oxford University Press, 2016, pp. 202–17.

Paster, Gail Kern, 'The Tragic Subject and its Passions', in Claire McEachern (ed.), *The Cambridge Companion to Shakespearean Tragedy*, Cambridge: Cambridge University Press, 2003, pp. 152–70.

Perkins Wilder, Lina, *Shakespeare's Memory Theatre: Recollection, Properties, and Character*, Cambridge: Cambridge University Press, 2010.

Peyré, Yves, 'Iris's "rich scarf" and "Ariachne's broken woof": Shakespeare's Mythology in the Twentieth Century', in Jill Levenson, Jonathan Bate and Dieter Mehl (eds), *Shakespeare in the Twentieth Century: The Selected*

Proceedings of the International Shakespeare Association World Shakespeare Congress, Los Angeles, 1996, Newark: University of Delaware Press, 1998, pp. 280–93.
Peyré, Yves, '"Travels in the clouds": Metamorphosis, Doubt, and Reason in the Renaissance', in Jean-Marie Maguin and Michèle Willems (eds), *French Essays on Shakespeare and his Contemporaries*, London: Associated University Presses, 1995, pp. 11–38.
Pillai, Tripthi, 'Constructing Experiences and Charting Narratives: The Future in/of *A Midsummer Night's Dream*', in Regina Buccola (ed.), *A Midsummer Night's Dream: A Critical Guide*, London: Continuum Renaissance Drama, 2010, pp. 150–71.
Pollock, Jonathan, 'Of Mites and Motes: Shakespearean Readings of Epicurean Science', in Sophie Chiari and Mickaël Popelard (eds), *Spectacular Science, Technology and Superstition in the Age of Shakespeare*, Edinburgh: Edinburgh University Press, 2017, pp. 119–32.
Popelard, Mickaël, 'Unlimited Science: The Endless Transformation of Nature in Bacon and Shakespeare', in Sophie Chiari and Mickaël Popelard (eds), *Spectacular Science, Technology and Superstition in the Age of Shakespeare*, Edinburgh: Edinburgh University Press, 2017, pp. 171–96.
Raven, James, 'Printing and Printedness', in Hamish Scott (ed.), *The Oxford Handbook of Early Modern European History, 1350–1750*, vol. 1, *Peoples and Place*, Oxford: Oxford University Press, 2015, pp. 214–43.
Reynolds, Philip Lyndon, *Food and the Body: Some Peculiar Questions in High Theology*, Leiden: Brill, 1999.
Rhodes, Neil, 'Time', in Andrew Hadfield, Matthew Dimmock and Abigail Shinn (eds), *The Ashgate Research Companion to Popular Culture in Early Modern England*, London: Routledge, (2014) 2016, pp. 283–94.
Richards, Jennifer, *Rhetoric and Courtliness in Early Modern Literature*, Cambridge: Cambridge University Press, 2003.
Ringler, William A. and Steven W. May, 'An Epilogue Possibly by Shakespeare', *Modern Philology* 70 (1972), pp. 138–9.
Riskin, Jessica, *Science in the Age of Sensibility: The Sentimental Empiricists of the French Enlightenment*, Chicago: University of Chicago Press, 2002.
Roberts, Jeanne Addison, *The Shakespearean Wild: Geography, Genus and Gender*, Lincoln: University of Nebraska Press, 1991.
Ronk, Martha, 'Locating the Visual in *As You Like It*', in Harold Bloom (ed.), *William Shakespeare's As You Like It*, Philadelphia: Chelsea House Publishers, 2004, pp. 165–88.
Roud, Steve, *The English Year*, London: Penguin, 2008.
Rubenstein, Mary-Jane, *Worlds without End: The Many Lives of the Multiverse*, New York: Columbia University Press, 2014.
Rubinstein, Frankie, *A Dictionary of Shakespeare Sexual Puns and their Significance*, Basingstoke: Macmillan, (1984) 1989.

Rusche, Harry, 'Edmund's Conception and Nativity in *King Lear*', *Shakespeare Quarterly* 20.2 (Spring 1969), pp. 161–4.

Ryrie, Alec, *Being Protestant in Reformation Britain*, Oxford: Oxford University Press, 2013.

Sale, Carolyn, 'Eating Air, Feeling Smells: Hamlet's Theory of Performance', *Renaissance Drama* 35 (2006), pp. 145–68.

Schäfer, Daniel, *Old Age and Disease in Early Modern Medicine*, trans. Patrick Baker, London: Routledge, (2011) 2016.

Schlueter, Nathan, 'Prospero's Second Sailing: Machiavelli, Shakespeare, and the Politics of *The Tempest*', in Stephen W. Smith and Travis Curtright (eds), *Shakespeare's Last Plays: Essays in Literature and Politics*, Lanham: Lexington Books, 2002, pp. 178–95.

Schoenbaum, Samuel, *William Shakespeare: A Documentary Life*, New York: Oxford University Press in association with The Scolar Press, 1975.

Schoenfeldt, Michael C., *Bodies and Selves in Early Modern England: Physiology and Inwardness in Spenser, Shakespeare, Herbert, and Milton*, Cambridge: Cambridge University Press, 1999.

Scott, Charlotte, *Shakespeare's Nature: From Cultivation to Culture*, Oxford: Oxford University Press, 2014.

Secor, Philip B., *Richard Hooker: Prophet of Anglicanism*, Toronto: Burns & Oates, (1999) 2000.

Segal, Charles, 'Lucretius, Epilepsy, and the Hippocratic *On Breaths*', *Classical Philology* 65.3 (July 1970), pp. 180–2.

Shannon, Laurie, *The Accommodated Animal: Cosmopolity in Shakespearean Locales*, Chicago: University of Chicago Press, 2013.

Shannon, Laurie, '"Poore wretch . . . laid all naked on the bare earth": Human Negative Exceptionalism among the Humanists', in *The Shakespearean International Yearbook*, vol. 15, *Shakespeare and the Human*, general ed. Tom Bishop and Alexa Huang, guest ed. Tiffany Jo Werth, Farnham: Ashgate, 2015, pp. 205–10.

Shinn, Abigail, '"Extraordinary discourse of vnnecessarie matter": Spenser's *Shepheardes Calender* and the Almanac Tradition', in Matthew Dimmock and Andrew Hadfield (eds), *Popular Culture in Early Modern England*, Farnham: Ashgate, 2009, pp. 137–50.

Shoaf, Richard Allen, *Lucretius and Shakespeare on the Nature of Things*, Newcastle upon Tyne: Cambridge Scholars Publishing, 2014.

Simonds, Peggy Muñoz, '"My charms crack not": The Alchemical Structure of *The Tempest*', *Comparative Drama* 31.4 (Winter 1997–8), pp. 538–70.

Smith, Bruce R., *The Acoustic World of Early Modern England: Attending to the O-Factor*, Chicago: University of Chicago Press, 1999.

Smith, Philip, 'Narrating Global Warming', in Jeffrey C. Alexander, Ronald N. Jacobs and Philip Smith (eds), *The Oxford Handbook of Cultural Sociology*, Oxford: Oxford University Press, 2012, pp. 745–62.

Sohmer, Steve, '12 June 1599: Opening Day at Shakespeare's Globe', *Early Modern Literary Studies* 3.1 (1997), <http://purl.oclc.org/emls/03-1/sohmjuli.html> (last accessed 9 May 2018).

Sohmer, Steve, 'The "Double Time" Crux in *Othello* Solved', *English Literary Renaissance* 32.2 (Spring 2002): *Studies in Drama and Culture*, pp. 214–38.

Sokol, B. J., *A Brave New World of Knowledge: Shakespeare's The Tempest and Early Modern Epistemology*, London: Associated University Presses, 2003.

Sousa, Geraldo U. de, *At Home in Shakespeare's Tragedies*, London: Routledge, 2010.

Spencer, Terence, 'Shakespeare's Isle of Delphos', *Modern Language Review* 47 (1952), pp. 199–202.

Srigley, Michael, *Images of Regeneration: A Study of Shakespeare's The Tempest and its Cultural Background*, Stockholm: Uppsala, 1985.

Staub, Susan C., '"Bred Now of Your Mud": Land, Generation, and Maternity in *Antony and Cleopatra*', in Andrew Shifflett and Edward Gieskes (eds), *Renaissance Papers 2012*, Rochester: Camden House, 2013, pp. 67–82.

Staub, Susan C., 'While She Was Sleeping: Spenser's "goodly storie" of Chrysogone', in Karen Bamford and Naomi J. Miller (eds), *Maternity and Romance Narratives in Early Modern England*, London: Routledge, (2015) 2016, pp. 13–32.

Stern, Tiffany, 'Before the Beginning; After the End: When Did Plays Start and Stop?', in Margaret Jane Kidnie and Sonia Massai (eds), *Shakespeare and Textual Studies*, Cambridge: Cambridge University Press, pp. 358–74.

Storm, Melvin, 'Uxor and Alison: Noah's Wife in the Flood Plays and Chaucer's Wife of Bath', *Modern Language Quarterly* 48 (1987), pp. 303–19.

Taub, Liba, *Ancient Meteorology*, London: Routledge, 2004.

Taylor, Gary and Rory Loughnane, 'The Canon and Chronology of Shakespeare's Works', in Gary Taylor and Gabriel Egan (eds), *The New Oxford Shakespeare: Authorship Companion*, Oxford: Oxford University Press, 2017, pp. 417–602.

Thomas, Keith, *Religion and the Decline of Magic: Studies in Popular Beliefs in Sixteenth and Seventeenth Century England*, London: Weidenfeld & Nicolson, 1971.

Thomas, Sidney, 'The Earthquake in *Romeo and Juliet*', *Modern Language Notes* 64.6 (June 1949), pp. 417–19.

Thomson, Leslie, 'The Meaning of Thunder and Lightning: Stage Directions and Audience Expectations', *Early Theatre* 2 (1999), pp. 11–24.

Thornes, John E., 'A Brief History of Weather in European Landscape Art', *Weather* 55.10 (2000), pp. 363–75.

Tiffany, Grace, *Love's Pilgrimage: The Holy Journey in English Renaissance Literature*, Newark: Delaware University Press, 2006.

Tilley, Maurice Palmer, *A Dictionary of Proverbs in England in the Sixteenth and Seventeenth Centuries*, Ann Arbor: University of Michigan Press, 1950.

Tooley, M. J., 'Bodin and the Medieval Theory of Climate', *Speculum* 28 (1953), pp. 64–84.

Torrance Kirby, W. J., 'Richard Hooker's Theory of Natural Law in the Context of Reformation Theology', *The Sixteenth Century Journal* 30.3 (Autumn 1999), pp. 681–703.

Totaro, Rebecca, '"Revolving this will teach thee how to curse": A Lesson in Sublunary Exhalation', in Jennifer C. Vaught (ed.), *Rhetorics of Bodily Disease and Health in Medieval and Early Modern England*, Farnham: Ashgate, 2010, pp. 135–54.

Traub, Valerie, *Desire and Anxiety: Circulations of Sexuality in Shakespearean Drama*, New York: Routledge, 1992.

Traub, Valerie, *The Renaissance of Lesbianism in Early Modern England*, Cambridge: Cambridge University Press, 2002.

Turner, Henry S., *Shakespeare's Double Helix*, London: Continuum, 2007.

Van Es, Bart, 'Reviving the Legacy of Indoor Performance', in Andrew Gurr and Farah Karim-Cooper (eds), *Moving Shakespeare Indoors: Performance and Repertoire in the Jacobean Playhouse*, Cambridge: Cambridge University Press, 2014, pp. 237–51.

Vaughan, Alden T. and Virginia Mason Vaughan, *Shakespeare's Caliban: A Cultural History*, Cambridge: Cambridge University Press, (1991) 1996.

Vaughan, Virginia Mason, *Othello: A Contextual History*, Cambridge: Cambridge University Press, 1994.

Veith, Gene E., *Reformation Spirituality: The Religion of George Herbert*, Eugene: Wipf and Stock Publishers, (1985) 2013.

Vermij, Rienk, 'A Science of Signs: Aristotelian Meteorology in Reformation Germany', *Early Science and Medicine* 15.6 (2010), pp. 648–74.

Vesperini, Pierre, *Lucrèce. Archéologie d'un classique européen*, Paris: Fayard, 2017.

Waters, David, *The Iberian Bases of the English Art of Navigation in the Sixteenth Century*, Coimbra: Reprint of the Journal of the University of Coimbra, 1970.

Watson, Robert N., *Back to Nature: The Green and the Real in the Late Renaissance*, Philadelphia: University of Pennsylvania Press, (2006) 2008.

Watt, Tessa, *Cheap Print and Popular Piety, 1550–1640*, Cambridge: Cambridge University Press, 1991.

Weis, René, *Shakespeare Unbound: Decoding a Hidden Life*, New York: Henry Holt and Company, 2007.

Weis, René, 'Was There a Real Shakespeare?', in Andrew Hadfield (ed.), *Were Early Modern Lives Different? Writing the Self in the Renaissance*, London: Routledge, 2014, pp. 35–48.

White, Martin, 'London Professional Playhouses and Performances', in Jane Milling and Peter Thomson (eds), *The Cambridge History of British Theatre*, vol. 1, *Origins to 1660*, Cambridge: Cambridge University Press, (2004) 2015, pp. 298–340.

White, Martin, *The Shakespeare Handbooks: A Midsummer Night's Dream*, New York: Palgrave Macmillan, 2009.

Whitlock, Keith, 'Shakespeare's *The Tempest*: Some Thought Experiments', *Sederi* 10 (1999), pp. 167–84.

Wiggins, Martin, in association with Catherine Richardson, *British Drama 1533–1642: A Catalogue*, vol. 4, *1598–1602*, Oxford: Oxford University Press, 2014.

Wiles, David, *Shakespeare's Almanac: A Midsummer Night's Dream, Marriage and the Elizabethan Calendar*, Cambridge: D. S. Brewer, 1993.

Williams, Gordon, *A Glossary of Shakespeare's Sexual Language*, London: Athlone, 1997.

Wilson, F. P., 'Illustrations of Social Life IV: The Plague', in *Shakespeare Survey*, vol. 15, *The Poems and Music*, ed. Allardyce Nicoll, Cambridge: Cambridge University Press, 1962, pp. 125–9.

Wilson, Richard, 'A Brute Part: *Julius Caesar* and the Rites of Violence', *Cahiers Élisabéthains* 50 (1996), pp. 19–32.

Wilson, Richard, *Shakespeare in French Theory: King of Shadows*, London: Routledge, 2007.

Wilson, Rob et al., 'Last Millennium Northern Hemisphere Summer Temperatures from Tree Rings, Part I: The Long Term Context', *Quaternary Science Reviews* 134 (2018), pp. 1–18.

Wilson Bowers, Kristy, *Plague and Public Health in Early Modern Seville*, Rochester: University of Rochester Press, 2013.

Wilson-Okamura, David Scott, 'Virgilian Models of Colonization in Shakespeare's *Tempest*', *English Literary History* 70.3 (Fall 2003), pp. 709–37.

Wolpert, Lewis, *The Unnatural Nature of Science*, Cambridge, MA: Harvard University Press, 1993.

Wood, David Houston, '"Very Now": Time and the Intersubjective in *Othello*', in *Time, Narrative and Emotion in Early Modern England*, London: Routledge, (2009) 2016, pp. 77–102.

Wright, Louis B., 'Elizabethan Sea Drama and its Staging', *Anglia – Zeitschrift für Englische Philologie* 51 (1927), pp. 104–18.

Wright, Louis B., *Middle-Class Culture in Elizabethan England*, Ithaca: Cornell University Press, 1935.

Wrightson, Keith, *Earthly Necessities: Economic Lives in Early Modern Britain*, New Haven: Yale University Press, 2000.

Yates, Frances A., *The Occult Philosophy in the Elizabethan Age*, London: Routledge, (1979) 2001.

Zachman, Randall C., 'Contemplating the Living Image of God in Creation', in Ulrich Möller, Michael Weinrich and Michael Welker (eds), *Calvin Today: Reformed Theology and the Future of the Church*, London: T&T Clark International, 2011, pp. 33–47.

Zirkle, Conway, 'Animal Impregnated by the Wind', *Isis* 25.1 (May 1936), pp. 95–130.

Index

Abbott, George, 189, 226
abstinence, 68, 73, 77n49
Acosta, José de, 178, 223, 247n26
Adams, Thomas, 142n40
Adelman, Janet, 140n23, 197
Aeolus, 45, 86
Africa, 112, 120–1
air / airy, 3, 8, 15, 18, 20, 23, 33, 40, 45–7, 49, 69–70, 83, 86, 100, 117, 130–9, 158, 168, 180–1, 183, 192–3, 198, 202, 206, 225, 227–9, 232, 235–6, 238, 241–3, 245, 257–9, 264
alchemy, 232–3
Allott, Robert, 122, 139n6, 250n52, 253n107
almanac(s), 11–12, 34, 67, 80, 183
amimetobion, 188, 190
anatomy, 17, 36, 152, 155, 164, 168
anger, 10, 50, 66, 84, 87, 123–4, 151, 153, 180, 182, 229, 242, 260
anti-pastoral / anti-pastoralism, 46, 91–2, 101
antipodes, 44, 54n67
antiquity, 11, 63, 84, 130, 162, 193, 197, 205
aphrodisiac / anti-aphrodisiac, 136
apocalypse / apocalyptic imagery, 1, 44, 65, 152, 155, 157, 161, 164–5, 168, 183, 198–9, 200–2, 207, 218–19, 246
apotheosis / anti-apotheosis, 139, 198, 201–2, 207
Argos, 182
Aristotle, 13–14, 16, 28n63, 29n75, 33, 35, 50n9, 51n17, 75n18, 84, 124, 130, 178, 181, 209n27, 212nn70–1, 226, 230, 234–6, 263
Armin, Robert, 189
Armstrong, Philip, 4
astrology, 12, 57, 71, 157–8, 186
ataraxy, 16, 91, 160, 206
atom(s) / atomism, 17, 21, 46, 88–9, 92, 160, 168, 176, 191, 205, 239
autumn / autumnal, 37, 40, 112, 115, 132, 204

Bacon, Francis, 44, 46, 55n75, 228, 250n60
Baldwin, T. W., 16
Banqueting House, 162, 165, 168
Batman, Stephen, 58, 68, 71, 104n32, 241
battle(s), 7, 111, 126–9, 177, 185, 201, 205, 207, 258
bear(s) (animal), 132–3, 192–3
bear (constellation), 133
Bede, 96
Behringer, Wolgang, 11
Benzoni, Girolamo, 222
Bermuda, 223–5
Berry, Philippa, 32
Bible, 1, 85, 94, 227
Blackfriars, The (playhouse), 217, 219, 227, 244
Blanchot, Maurice, 73
Blundeville, Thomas, 8
Boaistuau, Pierre, 176
Bobrick, Benson, 74n3, 256
Boccacio, Giovanni, 33
Bodin, Jean, 14–15, 75n15, 84, 121, 143n61, 146n94, 177, 179–80, 182–6
Botero, Giovanni, 14, 179
Brasbridge, Thomas, 69–70
Brayton, Dan, 199, 211n62, 214n108
breath, 4, 20, 65, 68–9, 99, 117, 132–3, 135–6, 158, 193, 243
Brooke, Arthur, 59–60, 64, 78n54
Burton, Robert, 28n69, 104n26, 134, 139n2, 147n114
calendar, 11, 31, 58, 72, 99, 236
Calvin, John, 152–3
Camerarius, Joachim, 194–5
Canis Major, 71–2
cannon, 68, 135
cannonball(s), 135, 166–7, 244
canopy, 162
Capp, Bernard, 67
carnival / carnivalesque, 19, 21, 47, 99, 112, 123, 132, 207, 235
Cartari, Vincenzo, 241
catastrophe, 40, 69, 117, 125, 181, 260
celestial bodies, 16, 41, 201, 203
Ceres, 219, 229, 230, 237, 241–2
Chapman, George, 51n12, 151–2, 214n103, 253n108
Charon, 71
Charron, Pierre, 15, 84–5, 107n74
Chaucer, Geoffrey, 36, 95
Choler, 60, 66, 72–3, 123
Christendom, 129
Christmas, 31, 60, 70, 96–7, 99, 100
Church authorities, 10
Cicero, 76n26, 230
Clever, William, 72
climacteric year, 162
climatology, 14, 71, 125
clinamen, 168
Clopton bridge, 6

cloud(s), 1, 22, 39, 81, 126, 129, 190–7, 199, 203, 205–7, 217, 222, 224–5, 234–5, 242–4, 260–2
coal, 45, 73, 122–3
coldness, 33, 35, 46, 82, 91, 93, 95, 98, 100, 123–4, 150–1, 167, 176, 183, 235
colonialism, 8
Columbus, Christopher, 219–20, 222
comet(s), 236
cook(s), 77
coolness *see* coldness
corruption, 44, 69, 125, 194
cosmography, 200, 202, 207
cosmos, 17, 23, 117, 138, 157–8, 192, 198, 263
Cotgrave, Randle, 8, 162, 223
Coursen, Herbert R., 114
court performance(s), 37, 98, 138, 162, 165–7, 243–4
Covell, William, 63
Crane, Mary Thomas, 5–6, 25n20, 204–6
Crane, Ralph, 217
Crooke, Helkiah, 5, 17, 125, 261, 92, 122
crop(s) *see* harvest(s)
Cummings, Brian, 43
Cupid, 35, 46, 115, 122, 242
Curtain, The (playhouse), 38

Da Vinci, Leonardo, 193–4
dampness *see* wetness
Danaë, 19, 182
danger(s), 69–70, 86, 92, 122, 125, 128, 137, 162, 245, 260
Darriulat, Jacques, 194
Davis, Nick, 117
dearth, 41–2, 173
death, 20, 58, 61, 64, 67, 70–2, 91–2, 95–6, 111, 123, 126, 130, 135–6, 153, 176–7, 187, 191, 197–200, 202–3, 205, 207, 217–18, 220, 224, 227, 236, 244, 257–8
Dee, John, 237
Dekker, Thomas, 149n130, 169n3, 190
deluge *see* flood
Descartes, René, 205, 263–4
despair, 36, 66, 73, 201, 242, 259–60, 263
destruction, 40, 45, 111, 122, 182, 202–3
determinism, 49, 57, 84, 123, 154, 197, 207, 258, 263
dew, 43, 45, 125, 224
Digges, Leonard, 12–13, 41, 153, 254n113
Digges, Thomas, 6
discord / discordance, 46, 229
discordia concors, 65
disease(s), 16, 40, 58, 66, 68–9, 91, 114, 131, 134, 136, 164, 235–6
dissolution, 21, 192, 196–7, 200, 206
Dobson, Mary J., 58
dog days, 20, 66–8, 71, 158, 258
Drebbel, Cornelis, 237
drought / dryness, 58, 82, 124, 182, 187

Dusinberre, Juliet, 98–9, 109n97
Dutton, Richard, 37

earthquake(s), 62–3, 69, 133, 234, 236
eclipse(s), 114, 116–17, 138, 157, 197–8
ecocriticism, 18
Eden, 44, 85
Egan, Gabriel, 187–8, 226, 250n63
Eliot, T. S., 245–6
Elizabeth I, 4, 35, 50n8, 109n91, 215n117
Elmo (Saint), 225
Elton, William R., 161
Elyot, Thomas, 235–6
emblem(s), 65, 138
emotions, 18, 38, 111–13, 151, 176, 245, 261
enclosure(s), 219
Epicure / Epicureanism, 16–17, 88–9, 91–2, 161, 168, 178, 190, 196, 205–6
epilepsy, 133–4
Erasmus, Desiderius, 106n55, 193
eroticism, 136, 181
Estok, Simon C., 164
etymology, 8, 121, 160, 181
exhalation(s), 104, 152, 234–5, 236

Fagan, Brian, 162–3
Famagusta, 129
fantasia, 193
fate, 20, 65–6, 71, 113–14, 118, 138, 240, 256, 258, 262
fecundity *see* fertility
femininity, 126
fertility, 32, 35, 37, 58, 95, 101, 132, 184, 188–90, 206, 229–31
Ficino, Marsilio, 5
fire(s), 7, 57, 60–1, 65–7, 70, 74, 83–4, 122, 137, 152, 156, 161, 163, 178, 181, 189, 194, 198–9, 201, 217, 220, 225, 227, 232, 240, 259
fireworks, 165–6
First Folio, 37, 40, 126–7, 164–5, 167
flood / flooding, 6, 40, 42, 46, 93–4, 97, 181–3, 202, 227–8
Florio, John, 7–8, 11, 55n77, 89, 113
Floyd-Wilson, Mary, 3, 13, 28n61, 143n63
flux, 21, 87, 126, 190, 193, 205
fog(s), 3, 31, 40, 42, 45
folly *see* madness
forecast(s), 80
frigidity, 151
frost(s) / frosty / frostiness, 20, 40, 45, 61, 82, 86–7, 90, 98, 228, 258
Frye, Northrop, 36
Fulke, William, 215n112, 234–5
Fuller, Mary C., 8–10

gale(s) *see* wind(s)
Galen / Galenic medicine, 16, 29n76, 86, 135, 137, 147n114

Gascoigne, George, 118, 146n92
gender(s) / gendered, 36, 48–9, 100–1, 115, 125, 182, 184, 187, 204
Genesis, 39, 44, 93, 227
geography, 7, 32, 48, 121, 190, 245
geo-humoralism, 3, 16, 21, 84, 119, 122, 177, 186, 231, 258
Gillies, John, 198, 230
Giraldi Cinthio, Giovanni Battista, 112–13, 124, 143n60
Glete, Jan, 127
globe (earth), 23, 44, 48, 66, 116, 242–3
Globe, The (playhouse), 38, 93, 95, 97–8, 117, 138, 162–3, 166–8, 258, 261
gold, 138, 182, 192, 232–3
golden world, 81, 87, 101
Googe, Barnabe, 89–91, 98
Goodland, Katharine, 161
Greene, Robert, 17, 29n83, 64, 114, 140n16, 191
Greene, Thomas M., 262
grotesque, 21, 62, 115, 118, 123, 131, 203, 206, 235
Gurr, Andrew, 254n115

hail, 21, 42, 47, 182–3, 238, 241
Hall, Jonathan, 124
Hamilton, Jennifer Mae, 257
Harris, Jonathan Gil, 121–2, 174nn76,79
harvest(s), 1, 4, 10, 18, 32, 35, 40–1, 46, 64, 67, 164, 190, 241
Hassel Jr., Chris R., 165
hate / hatred, 20, 60, 67, 69–70, 73–4, 127–8
heat, 14, 20, 47, 49, 58–61, 65–7, 70, 72–3, 82, 87, 91, 100, 119, 121–4, 131, 151, 155–6, 176, 178, 180–4, 186, 190, 209n27, 230, 242, 258, 263
hell, 36, 61, 71, 123, 130, 133, 153, 159, 166, 190, 201, 207, 218, 220, 224–5, 259
Hellespont, 92
hemisphere, 4
Herbert, George, 218, 240–1, 245, 253n112
Herodotus, 185–6, 205
Heywood, John, 12, 45, 144n80
Heywood, Thomas, 106n45, 119
Hill, Thomas, 12, 75n18, 77n49, 225–6, 228, 232, 235–6, 245
Hippocrates, 13–14, 16, 28n62, 78n66, 84, 86, 117, 143n63, 231
Holinshed, Raphael, 103n19
Holland, Philemon, 50n10, 75n22, 103–4n22, 155, 169n7, 194
Homer, Homeric, 22, 239, 253n108
Hooker, Richard, 21, 154, 170n23

horse(s), 33, 65, 115, 120, 192–6
hubris, 1, 201
Hudson, Henry, 4
Hulme, Mike, 4, 7, 108n85, 253n100
humidity *see* wetness
humours, 13, 15–16, 18–19, 21, 44, 73, 111, 117, 121–4, 183–4, 234, 243, 256
hurricane(s), 220, 222–4, 239, 261
Hyginus, Caius Julius, 72

ice, 4, 59, 80, 93, 97, 124, 182, 218
idealism, 138
imagination, 38, 49, 101, 111, 168, 193, 203, 227, 258, 261
Indians, 222
influence(s), 2, 13–18, 21, 37–9, 46, 57–9, 62–3, 74, 83, 88, 101, 112–13, 117, 121, 124, 132–3, 137, 154, 157, 177, 180, 188–9, 258–9
intemperance, 118, 177, 184
Invincible (or Spanish) Armada, 111, 128–9
Iris, 19, 219, 227, 229, 233, 239, 241
irony, 18, 90, 183, 230

James VI/I, 5, 77n42, 128, 219, 238
Jamestown, 224, 226
Johnson, Thomas, 187
Johnstoune, William, 40
Jones, Gwilym, 18, 239
Jonson, Ben, 18, 66, 77n42, 233, 241, 243–4, 251n80, 254n115
Jove *see* Jupiter
Juno, 22, 241–3
Jupiter, 52, 114–15, 156–7, 159, 180, 183, 200
justice / injustice, 20, 38, 114, 132, 136, 199, 240

Kahn, Coppélia, 165
Knolles, Richard, 14, 121, 129, 140n15, 177
knowledge / self-knowledge, 5–6, 8, 11, 48, 80, 95, 159, 186, 192, 204–5, 219, 226, 260–1, 263–4
Kott, Jan, 36

Las Casas, Bartolomé de, 143n62, 222, 247n18
laughter, 97
Le Loyer, Pierre, 227
Lemnius, Levinus, 47
Lepanto, 128–9
lightning, 1, 10, 20, 38–9, 42, 70–1, 73, 81, 134, 152, 156, 159–61, 166–7, 198, 217, 220, 229, 232, 237–9, 241, 244
liquefaction, 181
'Little Ice Age', 3, 97, 164
Locke, John, 264
Lodge, Thomas, 81–2, 91

Lucretius, 17, 19, 21, 46, 87–9, 134, 160–1, 168, 172nn49,51, 183, 190, 196–7, 207, 243, 264
Luther, Martin, 10–11, 39, 87, 105n39, 226
luxuriousness, 119, 184, 230

Machiavel / Machiavellian, 118, 120, 232
MacMullan, Katherine Vance, 202
madness, 19, 21, 99, 113, 133, 150, 164, 166, 168
magic, 18, 22, 42, 96, 98, 226–7, 232, 237, 240–1, 261
Mallarmé, Stéphane, 80, 101
Manilius, 106n45, 162
map(s), 12, 83, 203, 220
Markley, Robert, 164
Marlowe, Christopher, 8, 92, 170n9, 213n84
marriage, 20, 35–7, 57, 59, 68, 71, 81, 93, 95–6, 116, 123, 137, 176, 182, 241, 259
Mars, 123, 156, 180–1
Martin, Craig, 105n39
Martinmas, 97, 100
masculinity, 96, 193
masque, 48, 66, 229–30, 233, 241–4, 262
materialism, 205, 263
Matthes, François E., 3, 24n10
Mazzio, Carla, 5, 57
medicine, 16, 68, 130, 135, 137
Mediterranean sea / peoples / weather, 2, 127, 259
melancholy, 20, 81, 83, 95, 113, 123, 125, 132, 134, 192, 258
Mentz, Steve, 141n32, 146n96, 154, 224, 248n33, 249n49
Mercury, 180
meteor(s) / *meteoros*, 151, 201, 261
meteorology, 12, 21, 125, 152, 158, 246, 263
miasma / miasmatic theory, 68–70
Middle Ages, 10, 37
Milton, John, 104n31, 261
mindscape, 82, 164, 245
misogyny, 181, 183, 259
misrule, 18–20, 47, 66
mist(s), 32, 39, 45, 84, 128, 243, 256
Montaigne, Michel de, 15, 17, 55n77, 89, 100, 231
mood(s), 13, 83, 165, 230, 258
moon, 19, 23, 32–5, 37–41, 44, 48–9, 58, 93, 115–16, 154, 157, 197–8, 202–3
Moryson, Fynes, 256
mud, 20, 40, 46, 181, 187–91, 204
Munroe, Jennifer, 18
music / musicians, 131–2, 151, 240
mystery plays/cycles, 63, 94, 161

Nagel, Alexander, 193–4
Nashe, Thomas, 4, 12, 107n76, 157, 171n32
national stereotype, 119
nativity, 157

natural philosophy, 205, 260
nature, 5, 7, 10, 22–3, 39, 58, 91, 96, 119, 127, 152, 154–5, 159–60, 164, 190, 193, 206, 218, 226–7, 231, 237, 245, 258–60, 262
Neill, Michael, 134–5, 204, 250n57
New Jerusalem, 200
New World, 219–20, 222–3
nihilism, 21, 155, 168, 259
Noah, 19, 93–5, 227
nominalism, 64
Nosworthy, J. M., 226

onomastics, 119
Ovid, 15, 32–3, 65, 95, 107nn68,76

passion(s), 15–16, 20, 47–8, 58, 60–4, 67–8, 70
pathetic fallacy, 13, 44, 82
pestilence *see* plague
Phaeton, 19, 65–6, 71
Phebus, 45, 191
plague, 20, 68–70, 115, 153, 220, 245
planet(s), 38, 57, 62, 69, 83, 89, 114–15, 118, 132, 156–7, 180, 198
Plato, 46
Platter, Thomas, 42, 54n61
Pliny (the Elder), 15, 33, 75–6n22, 83, 103–4n22, 117, 121, 126, 155–6, 193–4
Plutarch, 179, 181, 190, 197, 201, 208n17, 211n56
Pluto, 36–7
pneuma, 21, 86, 112, 130, 134, 158, 259
Poles (North and South), 42
Pory, John, 120–1, 143n56
prayer(s), 10, 18, 42–3
private performance(s), theatre(s), 37, 167–8, 217, 261
procession(s), 10, 99
prophecies, 11
Proserpina, 36–7, 242
providence, 87, 153–4
Pseudo-Aristotle, 14, 55n78, 104n33, 217
Ptolemy, Claudius, 16, 157, 253n113)
public performance(s), theatre(s), 6, 13, 37–9, 93, 165–7, 261–2
Purchas, Samuel, 26n34, 224, 248n29, 249n51
purgation, 233

Rabelais, François, 75n21, 101, 110nn103–4, 247n15
rain(s), 3, 32, 39, 41, 43, 45, 48, 82–4, 92–3, 95, 125, 138, 150, 154, 158, 163, 166, 168, 182–3, 203, 219, 223, 228–9, 264
rainbow, 227, 241
rape, 126, 242
Rastell, John, 12
realism, 34, 164
Red Bull, The (playhouse), 13
Reformation, 1, 11, 153
revenge, 22, 31, 40, 121, 124, 151
Richmond Palace, 98–9

Ronk, Martha, 92
Rose, The (playhouse), 6, 38

Sagittarius / Sagittary, 114–15
satire, 233
Saturn, 45, 83, 97, 112, 114–15, 132, 156
Saturnalia / saturnalian, 20, 96, 97, 204, 258
Saturnism, 114
scepticism, 89
science(s), 5–6, 12, 15–18, 22, 47, 179, 186, 205, 227, 237, 241, 259–60, 263–4
Scotland, 238
sea, 33, 71, 111–13, 121, 124–30, 133–4, 160–1, 163, 177, 185, 194, 198, 201, 205, 217–20, 222, 225–7, 232–3, 237, 245, 260
Sea Venture, 224
Seneca, 63, 133
senectus mundi, 161
senses, 5, 16, 203, 243, 260
sexuality, 13, 35, 73, 152
Shakespeare, William
  1 Henry IV, 62
  1 Henry VI, 6
  3 Henry VI, 11
  All's Well That Ends Well, 2
  Anthony and Cleopatra, 52n40, 176–216, 239
  As You Like It, 19–20, 80–110, 258
  Comedy of Errors, The, 256
  Coriolanus, 98, 115, 173n66
  Cymbeline, 170–1n28, 184
  Hamlet, 19, 25n21, 110n105, 214n103, 257
  Henry V, 3, 84, 104n34, 213n89
  Henry VIII, 77n43, 252n93
  Julius Caesar, 56n84, 108n84, 166, 256, 265n13
  King John, 232, 251n78, 252n93
  King Lear, 16, 21, 52n41, 64, 76n34, 86, 91–2, 101, 118, 150–75, 172–3n59, 248n30, 197, 203, 215n118, 218, 220, 223, 233–4, 239, 248n30, 258–9
  Love's Labour's Lost, 52, 102n4
  Macbeth, 6, 20, 27n46, 42, 56n84, 123, 168, 207n2, 237
  Measure for Measure, 51n28, 218, 248n30, 258
  Merchant of Venice, The, 23, 25n17, 148n129, 212n67, 260
  Midsummer Night's Dream, A, 19–20, 31–56, 66, 84, 95, 122, 148n129, 158, 162, 168, 181, 196, 212n66, 242, 251–2n81, 257–8, 263
  Much Ado About Nothing, 107n71
  Othello, 20–1, 111–49, 172n58, 258–9
  Pericles, 226
  Richard II, 261
  Richard III, 71, 97
  Romeo and Juliet, 18, 20, 23, 32, 52n41, 57–79, 111–14, 123, 158, 172n59, 176, 257–8

Sonnets, 22, 52n40, 62, 173n59, 213n85, 253n110, 259
Taming of the Shrew, The, 95, 147n104
Tempest, The, 19, 22–3, 48, 52n41, 161, 196, 217–55, 260–1
The Winter's Tale, The, 58, 136, 226, 229, 250n63
Titus Andronicus, 52, 140n18, 223
Troilus and Cressida, 223, 265n13
Twelfth Night, 16, 29n75150
Two Gentlemen of Verona, The, 106n64
Venus and Adonis, 52
Shannon, Laurie, 155, 227
ship(s), 32–3, 112, 127–8, 218–20, 222–6, 233, 240, 245
shipwreck, 128, 163, 219, 223, 226
Shrove Tuesday, 98–100
Shrovetide *see* Shrove Tuesday
Sirius, 71–2
smoke, 45, 65
snow, 45, 48, 59–60, 61, 82, 98, 124, 180, 241
solstice, 37, 96
special / stage / sound / visual effects, 45, 165–8, 237, 244, 259
Spenser, Edmund, 10, 13, 41, 53n52, 82, 97, 189, 212n69
spontaneous generation, 125, 188
spring, 20, 36–7, 40, 59, 67, 81, 85, 88, 93, 96, 112, 132, 163, 228, 237, 242
stage direction(s), 134, 159, 165, 168, 217, 237
stain(s), 193–4
Standish, Arthur, 219
star(s), 1, 20, 23, 29n78, 35, 57, 67–8, 71, 73, 98, 114, 133, 138–9, 156–7, 161, 176, 186, 198–202, 256, 258, 261–2
sterility, 58, 80, 93, 95
storm(s), 4, 10, 19, 21–2, 46, 70, 86–7, 96, 124, 126–7, 129–30, 134, 151–5, 158–60, 163–8, 176, 183, 203, 217–20, 222–31, 233, 235–42, 244–6, 259, 259–60, 264
Stow, John, 41
Strachey, William, 224–6, 248nn29,33
suffocation, 21, 135
summer, 4, 8, 31, 37, 40–2, 47, 49, 58–61, 65–8, 70, 132, 231
sun, 11, 16, 23, 32, 39, 48, 61, 65–6, 71–3, 81, 84, 91, 100, 113, 116, 122, 124, 139, 178–9, 182, 188–9, 192, 197, 198, 202–4, 220, 231–2, 242
sunlight, 4
sunset, 45, 192, 200
superstition(s), 1, 12, 73, 114, 232, 263

Tacitus, Cornelius, 84
temper, 13, 38, 82, 124, 166, 180, 184, 232
temperament, 3, 13, 119, 123–4, 151, 183–5
temperance, 3, 179–80, 184, 229–31, 245
temperature(s), 4, 10, 18, 39–41, 66, 72, 82, 95, 100, 117, 121, 124, 178, 185, 229, 264
tempest(s), 18, 21–2, 71, 111, 128–30, 133, 137–8, 164–6, 183, 218, 220, 223–4, 227–9, 231, 233, 235, 238–41, 245, 258–60
terror, 95, 150, 152, 159, 218, 261
Theatre, The (playhouse), 31, 38, 97, 99
theologian(s), 15, 39, 152, 154, 194
theology, 15, 227
Theophrastus of Eresus, 16, 35, 107n67, 229
thermoscope, 226
Thomas, Keith, 10, 141n29
Thomson, Leslie, 237
thunder, 10, 32, 42, 150, 152–3, 156, 158–61, 163–8, 203–4, 217–18, 234–41, 244–5, 257, 259, 261
topicality / topical, 20, 44, 63, 127, 219, 226
topography, 14, 36, 84, 123
Totaro, Rebecca, 235–6
Traub, Valerie, 33, 126
travel(s) / traveller(s), 11, 42, 83, 100, 112, 120, 200, 219–20, 226, 260
travel narratives, 22, 132, 220, 223–4, 226, 256
Turkish fleet, 21, 128–9, 138

vanitas, 202
vapours, 15, 193, 245
Vaughan, Alden T. and Virginia Mason, 226
Vaughan, William, 67, 74n5, 231
Vegetius, 84
Venus, 22, 57, 112, 114–15, 123, 133, 180–1, 198, 229, 242
violence, 20–1, 49, 60–1, 73, 97, 121, 125–7, 134, 155, 160, 203, 224, 259
Virgil, 15, 22, 29n78, 76n24, 80, 95, 226–7, 229, 242, 248n28, 250n54, 251n72
Virginia, 14, 224, 231
virility *see* masculinity
Vitruvius, 83
void, 21, 168, 191, 205–6, 263
Vulcan, 122–3

water, 7, 63, 93, 111, 124, 126, 129, 133, 137, 154, 158, 176, 181–3, 187, 190–4, 198–9, 217, 220, 237, 240, 264
Watson, Thomas, 87, 105n38
Webster, John, 13, 51n11, 149n130
wedding *see* marriage
Weltanschauung, 155

wetness, 19, 32, 46–8, 82, 124–6, 182, 184, 187
Weyer, Johann, 181
White, Martin, 38
Whitehall, 162, 165, 168
Whithorne, Peter, 165
Whitney, Geoffrey, 138
Whythorne, Thomas, 151
Wiles, David, 36
Wilson, Richard, 35
Wilson Bowers, Kristy, 69
wind(s), 21, 86, 99, 127–8, 153, 218–19, 222, 224, 260
winter, 4, 13, 31, 36–7, 40–1, 47–8, 59–61, 67, 72, 80–3, 85–7, 90–3, 95–7, 99–100, 129, 132–3, 163–7, 204, 231, 237, 242, 258
wisdom, 95, 192, 237
witch(es), witchcraft, 42, 63, 114, 237–8

Zeus, 183–3
Zodiac, 38, 57, 89, 98, 162, 243